Optimal Synthesis Methods for MEMS

THE KLUWER INTERNATIONAL SERIES
IN MICROSYSTEMS
Consulting Editor: Stephen Senturia
Massachusetts Institute of Technology

Volumes published in
MICROSYSTEMS

Optimal Synthesis Methods for MEMS
G.K. Ananthasuresh
ISBN: 1-4020-7620-7

Micromachined Mirrors
Robert A. Conant
ISBN: 1-4020-7312-7

Heat Convection in Micro Ducts
Yitshak Zohar
ISBN: 1-4020-7256-2

Materials & Process Integration for MEMS
Francis E.H. Tay
ISBN 1-4020-7175-2

Microfluidics and BioMEMS Applications
Francis E.H. Tay
ISBN: 1-4020-7237-6

Optical Microscanners and Microspectrometers Using Thermal Bimorph Actuators
Gerhard Lammel, Sandra Schweizer, Philippe Renaud
ISBN 0-7923-7655-2,

Scanning Probe Lithography
Hyongsok T. Soh, Kathryn Wilder Guarini, Calvin F. Quate
ISBN 0-7923-7361-8

Microsystem Design
Stephen Senturia
ISBN: 0-7923-7246-8

Microfabrication in Tissue Engineering and Bioartificial Organs
Sangeeta Bhatia
ISBN 0-7923-8566-7

Microscale Heat Conduction in Integrated Circuits and Their Constituent Films
Y. Sungtaek Ju, Kenneth E. Goodson
ISBN 0-7923-8591-8

Micromachined Ultrasound-Based Proximity Sensors
Mark R. Hornung, Oliver Brand
ISBN 0-7923-8508-X

Bringing Scanning Probe Microscopy Up to Speed
Stephen C. Minne, Scott R. Manalis, Calvin F. Quate
ISBN 0-7923-8466-

Microcantilevers for Atomic Force Microscope Data Storage
Benjamin W. Chui
ISBN 0-7923-8358-3

Optimal Synthesis Methods for MEMS

G.K. Ananthasuresh
University of Pennsylvania

Kluwer Academic Publishers
Boston/Dordrecht/London

Distributors for North, Central and South America:
Kluwer Academic Publishers
101 Philip Drive
Assinippi Park
Norwell, Massachusetts 02061 USA
Telephone (781) 871-6600
Fax (781) 871-6528
E-Mail <kluwer@wkap.com>

Distributors for all other countries:
Kluwer Academic Publishers Group
Post Office Box 322
3300 AH Dordrecht, THE NETHERLANDS
Telephone 31 78 6576 000
Fax 31 78 6576 474
E-Mail <orderdept@wkap.nl>

Electronic Services <http://www.wkap.nl>

Library of Congress Cataloging-in-Publication
CIP info or:
Title: Optimal Synthesis Methods for MEMs
Author (s): G.K. Ananthasuresh
ISBN: 1-4020-7620-7

Printed on acid-free paper.

Printed in the United States of America

About the Cover: The cover page shows four examples of synthesized designs (top row) and the corresponding microfabricated prototypes (bottom row).

The synthesis solutions were obtained, respectively, by A. Saxena and G. K. Ananthasuresh (University of Pennsylvania), E.C.N. Silva and N. Kikuchi (University of Michigan), W. Ye and S. Mukherjee (Cornell University), and O. Sigmund (Technical University of Denmark).

The prototypes in the bottom row were microfabricated, respectively, by A. Saxena and G. K. Ananthasuresh using MCNC MUMPs, E.C.N. Silva and G. Nader (University of São Paulo, Brazil), W. Ye and S. Mukherjee (SCREAM process), and J. Jonsmann and S. Bouwstra (MIC, Technical University of Denmark).

Optimal Synthesis Methods for MEMS

Table of Contents

FOREWORD

The field of "microelectromechanical systems," or "MEMS," has gradually evolved from a "discipline" populated by a small group of researchers to an "enabling technology" supporting a variety of products in such diverse areas as mechanical and inertial sensors, optical projection displays, telecommunications equipment, and biology and medicine. Critical to the success of these products is the ability to design them, and this invariably involves detailed modeling of proposed designs. Over the past twenty years, such modeling has become increasingly sophisticated, with full suites of MEMS-oriented computer-aided-design tools now available worldwide.

But there is another equally important side to the design process – figuring out what to build in the first place. In my own book, *Microsystem Design*, I chose to emphasize the modeling aspect of design. The task of figuring out what to build was defined by a vague step called "creative thinking." I used practical product examples to illustrate the many subtle characteristics of successful designs, but I made no attempt to systematize the generation of design proposals or optimized designs. That systemization is called "synthesis," which is the subject of this book.

On a personal note, I have long recognized the need for progress on the synthesis problem. In fact, when I heard a graduate student from the University of Michigan present a paper on the constrained optimization of compliant microstructures at the 1994 Solid-State Sensors and Actuators Workshop at Hilton Head, I immediately offered that graduate student a postdoc position in my group. The name of that student was G. K. Ananthasuresh. Clearly, his interest in the synthesis problem has been maintained since that time, and, in putting together this volume of contributions from leaders in the field, he is making an important resource available to the MEMS designer. I am pleased to recommend this book to the MEMS community.

Stephen D. Senturia
Brookline, Massachusetts, U. S. A.
April 9, 2003

CONTRIBUTING AUTHORS

G. K. Ananthasuresh
Associate Professor of Mechanical Engineering and Applied Mechanics, University of Pennsylvania, Philadelphia, PA 19104-6315, U.S.A.
gksuresh@seas.upenn.edu

Edwin. T. Carlen
MEMS Development Engineer
Corning Intellisense, Wilmington, MA 01887, U. S. A.
carlenet@corning.com

Gary K. Fedder
Professor of Electrical and Computer Engineering, and Robotics
Carnegie Mellon University, Pittsburgh, PA 15213, U. S. A.
fedder@ece.cmu.edu

Carlos. H. Mastrangelo
Vice President of Engineering
Corning Intellisense, Wilmington, MA 01887, U. S. A.
mastrangch@corning.com

Subrata Mukherjee
Professor of Theoretical and Applied Mechanics and Mechanical and Aerospace Engineering, Cornell University, Ithaca, NY 14853, U.S.A.
Sm85@cornell.edu

Tamal Mukherjee
Senior Research Scientist, Electrical and Computer Engineering
Carnegie Mellon University, Pittsburgh, PA 15213, U. S. A.
tamal@ece.cmu.edu

Radha Sarma
Adjunct Associate Professor of Mechanical Engineering
University of Michigan, Ann Arbor, MI 48109, U. S. A.
radha@umich.edu

Anupam Saxena
Assistant Professor of Mechanical Engineering
Indian Institute of Technology, Kanpur, India
anupams@iitk.ac.in

Ole Sigmund
Professor, Dr. Techn., of Department of Mechanical Engineering
Technical University of Denmark, Lyngby, Denmark
Sigmund@mek.dtu.dk

Emílio Carlos Nelli Silva
Assistant Professor of Mechatronics and Mechanical Systems Engineering,
Polytechnic School from University of São Paulo, SP 05508-900, Brazil.
ecnsilva@usp.br

Wenjing Ye
Assistant Professor of Woodruff School of Mechanical Engineering
Georgia Institute of Technology, Atlanta, GA 30332-0405, U.S.A.
wenjing.ye@me.gatech.edu

PREFACE

A new technology usually begins with experimentation—the process of building prototypes to demonstrate the feasibility of a novel concept or a new technique. This is immediately followed by modeling as one wants to know how well the device works before it is built so that expensive experimentation can be reduced. These two stages have already been passed by the microsystems technology as is evident from the variety of devices and the number of modeling software companies that exist today. Modeling techniques and tools enable analysis of an existing design. The design itself is largely dependent on the experience, expertise, and the creativity of the designer. Synthesis techniques have the potential to reduce this reliance on the human designer by automatically generating designs for user-specified requirements. Synthesis methods are especially useful for a MEMS designer given the multi-disciplinary nature of an integrated MEMS device. Software tools based on synthesis techniques enable even a novice designer create new solutions. They also aid the creativity of even experienced designers. This book is a compilation of some of the synthesis efforts in the MEMS area. It is by no means an exhaustive collection and any omissions are inadvertent.

This book is a result of the group effort by many colleagues who have generously contributed their time to write the chapters. I thank all of them for their valuable part in this effort. Several others who have made significant contributions, as well as the sources of funding are acknowledged in each chapter. I am grateful to Professor Stephen D. Senturia for his advice and encouragement over the years, and for suggesting the idea of bringing out a book on the synthesis methods for MEMS.

As can be seen in all the chapters, the authors have built prototypes based on their synthesis solutions. This shows their commitment and enthusiasm to see these techniques used in real applications. Considerable details of implementation are included in all the chapters to facilitate easy application. Interested readers may contact the authors or the editor for further details or to report comments, suggestions, and errors. The methods described here can be enhanced to include many more practical constraints. These developments will be expedited as MEMS designers begin to use them in practice. Therefore, I close with the slogan—DESIGNERS WANTED.

G. K. Ananthasuresh
Philadelphia, Pennsylvania, U. S. A.

Chapter 1

INTRODUCTION

G. K. Ananthasuresh
Department of Mechanical Engineering and Applied Mechanics, University of Pennsylvania, Philadelphia, PA 19104, U. S. A.

Synopsis: Design consists of two distinct components, viz. *analysis* and *synthesis*. Extensive research on analysis of Microelectromechanical Systems (MEMS) in the last decade has led to commercially available software tools. In contrast, synthesis has received less attention. On the one hand, the integrated and the multi-disciplinary nature of MEMS creates a necessity for time-saving systematic synthesis tools, and on the other, restrictions related to microfabrication provide new opportunities to make synthesis techniques practically viable. This book is a collection of synthesis efforts that cover many, but not all, aspects of MEMS. Since synthesis has many different connotations, a simple example is presented to explain how it is viewed in this book. Optimization is used as the primary tool for synthesis in this book. The essential steps in the optimal synthesis procedure are described in this chapter. Finally, the contents of each chapter are briefly noted.

1. DESIGN OF MICROELECTROMECHANICAL SYSTEMS

Miniaturization of the macro-size engineering world is not the only fascinating aspect of Microelectromechanical Systems (MEMS) technology. It is the unprecedented integration of one or more elements that sense, actuate, compute, communicate, and even generate power that makes MEMS truly fascinating. The rapid growth seen in the last two decades might not have taken place if there were no economical advantages. To sustain this pace and to lead to more commercial applications, further developments are necessary in both design and fabrication of MEMS. Fabrication has played the pivotal role in the initial phases of this field. Design, having received significant attention in the last decade, is likely to play an equal, if not larger, role in years to come.

Anything that is ever built must be designed first. Whether this design is manual or systematic is a different matter. Experienced designers who have a thorough understanding of the device or system in question may be able to design manually and it may work successfully when it is built. The multi-disciplinary nature of MEMS presents a challenge here because expertise is often necessary in many domains. Hence, the development of systematic design methods and tools started not too long after the field emerged and is continuing at a fast pace today. Good indicators of this are the books published on the subject [1-6 and others] and the MEMS Computer Aided Design (CAD) software companies that have come up [7-14 and others] in recent years, and of course numerous papers appearing in the literature.

Design of MEMS is challenging mainly because even the simplest device combines at least two energy domains. Thus, although the equations that govern the physics at the micron scale are not much different from those of the macro length scale, the tight coupling among various equations makes the methods of solution nontrivial. Furthermore, the integration of many subsystems often manufactured and packaged on the same chip leads to additional difficulties in the design task. Based on a comprehensive view of MEMS design, Senturia [1] separated it into four categories: *system*, *device*, *physical*, and *process* levels. Design issues abound at all these levels. Noting that design consists of two distinct components, viz., *analysis* and *synthesis*, we first note to what extent each of them is automated today.

Techniques and tools that automate the analysis (i.e., simulation) at all the four levels are available, and their continued improvement is in progress. That is, starting from the bottom, from mask layouts and process flow the geometric model of a device can be built ($process \rightarrow physical$). This model is then used for simulation wherein the partial differential equations (PDEs) that govern the behavior of the device are solved. From the results of these simulations, it is possible to extract reduced order models of devices or subsystems ($physical \rightarrow device$), which are then used at the system level to predict the overall performance ($device \rightarrow system$). As noted in [15, 16], today most MEMS designers use these automated analysis capabilities to evaluate their candidate designs conceived either intuitively or based on the past experience, and refine them manually following a hit-or-miss approach. This is time consuming, and at the end there is no assurance that the final design is the best possible. A synthesis procedure can potentially alleviate this burden.

For synthesis, we need to be able to automate the above analysis steps in the opposite direction. Some of the synthesis related questions that we can ask at each level are as follows.

1. From a given set of high-level specifications, is it possible to automatically generate the best concept for system architecture with well-defined subsystems?
2. By knowing the overall system-level characteristics of a subsystem, is it possible to generate viable device configurations and their quantitative attributes?
3. From a given set of quantitative attributes of a chosen device configuration, is it possible to generate a suitable geometry?
4. And, finally, given the refined geometric model of a device, is it possible to generate a suitable microfabrication process and the mask layouts?

If the answers to these and similar *synthesis* questions were affirmative, MEMS designers would need to iterate much less than what they do today before a satisfactory, if not optimal, design is obtained. By restricting to specific classes of MEMS applications and by imposing bounds on the range of possibilities, it may be possible to address the above questions. Based on that premise, this book focuses on providing answers to the last two questions using optimization as the tool. Before discussing the specific details, the notion of synthesis is further clarified below.

2. SYNTHESIS VS. ANALYSIS

Analysis is the evaluation of something that already exists while synthesis is the creation of a new entity. Design is an amalgam of analysis and synthesis. The word "design" also implies the creation of new things, and the word "synthesis" has multiple connotations. Therefore, despite the apparent redundancy, some researchers prefer "design synthesis" to describe the synthetic component of design. The reader may refer to pages xvii-xxi of [16] for a clear definition of what synthesis means in the context of design. Here, we use simple examples to clarify how this book views synthesis and what role optimization plays in it.

Consider the trivial problem of fitting a circular arc to a cloud of points in space. The variables that define an arc include the normal to the plane of the arc, the coordinates of the center, the radius, and the starting and ending angles subtended by the two radii to the ends of the arc with a line of reference that lies in the plane of the arc and passes through the center. If one chooses to solve this problem interactively through iterative analysis, then a graphical interface can be built so that the arc can be displayed along with the cloud of points for designer-specified values for all the variables. The designer uses "prior experience or intuition" to guess the values until a satisfactory arc is obtained. This is how a MEMS designer would use a CAD software program that has only analysis capabilities.

If we were to solve this as a synthesis problem, it can be posed as an optimal synthesis problem where an objective function is defined as the sum of the squares of each of the points in the cloud from an arc defined by the aforementioned set of variables. This objective function can then be minimized using systematic optimization algorithms that determine the suitable values of the variables. This is the synthesis approach taken in this book. Clearly, the optimally synthesized arc is not likely to pass through all the points unless by coincidence in the data. If all the points do lie on a circular arc, a good globally convergent optimization algorithm will be able to find that arc. A more direct synthesis approach that does not use this notion of optimality would work for this problem if there were only three points. Then, a closed-form analytical synthesis solution can be obtained so that for any arbitrary set of three points, an arc that exactly passes through them is determined. But design problems are usually neither that simple nor an exact solution obtained for a small subset of requirements satisfies the remaining ones. To emphasize this point further, a technical design problem is considered next.

2.1 An example: mode shape synthesis of a bar

An axially deforming bar of given cross-section area along its length has its characteristic natural frequency and corresponding normal mode shapes for prescribed boundary conditions. Let us consider the problem of determining the area of cross-section in order to obtain a desired mode shape. First, the governing equation for the free axial vibration *v(x)* is given by

$$\frac{d}{dx}\left[EA(x)\frac{dv}{dx}\right]+\rho A(x)\lambda v(x)=0 \tag{1}$$

where $A(x)>0$ and $\rho>0$ are the cross-sectional area along the x-axis and the density of the bar respectively, and λ is the square of the natural frequency. For a specified normal mode shape $\overline{v}(x)$, an iterative analysis-based design approach would try to solve (i.e., simulate) Eq. (1) for different $A(x)$ functions until the resulting mode shape $v(x)$ matches $\overline{v}(x)$ closely. But with a direct synthesis approach, the tedious process of iteration can be avoided as follows.

Assume that $A(x)$ is a differentiable function in the closed interval $0 \le x \le L$, where L is the total length of the bar. Equation (1) can be re-written in as a differential equation in terms of the cross-sectional area as

$$A'(x) + \frac{\left[E\overline{v}''(x) + \rho\lambda\overline{v}(x)\right]}{E\overline{v}'(x)} A(x) = 0 \tag{2}$$

where the prime denotes the derivative with respect to x. By defining a function, $\psi(x)$ as

$$\psi(x) = \frac{E\overline{v}''(x) + \rho\lambda\overline{v}(x)}{E\overline{v}'(x)} \tag{3}$$

we can solve for the area of cross-section in Eq. (2) as shown below [17].

$$\overline{A}(x) = Ce^{-\int \psi(x)dx} \tag{4}$$

where C is a constant of integration that can be determined uniquely from the total mass of the bar, i.e.,

$$Mass = m = \rho \int_0^L \overline{A}(x) x dx \tag{5}$$

Although Eqs. (3) – (5) do solve the synthesis problem analytically and directly it may not be satisfactory from the overall goals of a design problem. For example, the synthesized $A(x)$ may be too small or too large at some points to be practical for a given manufacturing process. Hence, a constraint needs to be added. With this, an analytical solution ceases to exist. We now need to use optimization that best meets the primary goal (i.e., specified $\overline{v}(x)$) along with the new constraint on minimum and maximum areas of cross-section. Thus, optimization is inevitable in all but simple synthesis problems with no hampering constraints.

3. OPTIMIZATION AS A SYNTHESIS TOOL

An *optimal synthesis* problem for the mode shape synthesis of a bar can now be posed as follows by minimizing the integrated error between the desired mode shape $\overline{v}(x)$ and the one obtained after solving Eq. (2) for a candidate $A(x)$. Thus, here, the function $A(x)$ is the design variable. Since the design variable is a function, it becomes a problem of calculus of variations. Mathematically, the optimal synthesis problem can be written as below:

$$\underset{A(x)}{\textit{Minimize}} \quad \frac{1}{L}\int_0^L (v(x)-\bar{v}(x))^2 dx$$

Subject to

$$\frac{d}{dx}\left(EA(x)\frac{dv(x)}{dx}\right)+\rho A(x)\lambda v(x)=0 \tag{6}$$

$$A_{\min} \le A(x) \le A_{\max} \quad \text{for all } 0 \le x \le L$$

$$\int_0^L \rho x A(x)dx - m = 0$$

where the subscript $A(x)$ below *Minimize* indicates that minimization is done with respect to $A(x)$. Additionally, there will also be essential boundary condition on the bar, i.e., whether it is fixed at one or both the ends. This problem can now be solved by way of Euler-Lagrange necessary conditions for a functional minimization [18] and Karush-Kuhn-Tucker conditions for constrained optimization [19]. To be more practical, the above problem can be discretized using the finite element analysis. Then, the design variables will be areas of cross-section of N finite elements. The discretized form of Eq. (6) is as follows.

$$\underset{\mathbf{A}}{\textit{Minimize}} \quad (\mathbf{v}-\bar{\mathbf{v}})^T(\mathbf{v}-\bar{\mathbf{v}})$$

Subject to

$$\mathbf{Kv}-\lambda^2\mathbf{Mv}=0 \tag{7}$$

$$A_{\min} \le A_i \le A_{\max} \quad \text{for } i=1,\ldots,N$$

$$\sum_{i=1}^{N} \rho A_i l_i - m = 0$$

where $\mathbf{A}$ is the column vector of areas of cross-section of N bar finite elements, $\mathbf{v}$ and $\bar{\mathbf{v}}$ are obtained and desired mode shape vectors respectively, $\mathbf{K}$ is the global stiffness matrix, $\mathbf{M}$ is the global inertia matrix, l_i is the length of ith finite element (refer to [20] or any book on structural finite element analysis to see how $\mathbf{K}$ and $\mathbf{M}$ are computed). This is now a vector optimization problem that can be solved using any continuous optimization algorithm.

We now examine a problem that we may run into. An unsuspecting designer might wish that the prescribed mode should also have a prescribed natural frequency λ. With that additional requirement included in the problem statement of Eq. (6) as another constraint, it can be submitted to an optimization algorithm seeking a solution. But it will not work unless by coincidence in the data. The reason is explained below.

Returning to the analytical solution of Eq. (4), let us consider some examples shown in Table 1. It can be seen that λ gets automatically determined when the mode shape $\bar{v}(x)$ is prescribed. The reason for this becomes clear from Eqs. (3) and (4). For $\bar{A}(x)$ to be meaningful, $\psi(x)$ should be finite for all values of x. But notice that when $\bar{v}'(x)$ becomes zero, $\psi(x)$ is no longer finite. To avoid this situation, λ has to be chosen such that $\bar{v}'(x)$ is cancelled out as a factor of the numerator of $\psi(x)$. So, for an arbitrary λ, there is no valid solution. Thus, asking for more than what is possible will render an optimal synthesis problem unsolvable. A designer who tries to achieve a mode shape and a natural frequency through iterative analysis based approach might not be able to notice this subtlety. Thus, synthesis approach to solving a design problem offers more insight than the analysis-based design approach.

Table 1 Examples of solution of Eq. (4)

No.	Desired mode shape, $\bar{v}(x)$	$\psi(x)$	λ	$\bar{A}(x)$
1	$\sin\left(\frac{\pi x}{2L}\right)$	$\{E\left(-\frac{\pi^2}{4L^2}\sin\left(\frac{\pi x}{2L}\right)\right)+\rho\left(\frac{E\pi^2}{4\rho L^2}\right)\sin\left(\frac{\pi x}{2L}\right)\}/E\frac{\pi}{2L}\cos\left(\frac{\pi x}{2L}\right)$	$E\pi^2/(4\rho L^2)$	$e^{-\int 0dx}=C$
2	$-(x-1)^2+1$	$\{2-(\rho\lambda/E)x(2-x)\}/2(x-1)$	$2E/\rho$	$Ce^{x-\frac{1}{2}x^2}$
3	$x-x^2$	$\{-2+\rho\lambda/E\left(x-x^2\right)\}/(1-2x)$	$8E/\rho$	$Ce^{2(x-x^2)}$

This example of mode shape synthesis of a bar revealed a number of features of optimal synthesis problems. They can be summarized as follows.

1. Synthesis approach is direct and saves time compared to the manual, iterative analysis-based approach to design.
2. A closed-form solution is possible only in the simple cases. Once practical constraints are introduced, we benefit by introducing the notion of optimality under constraints.
3. Proper analysis should be done to ensure that a solution exists for the optimal synthesis problem that is formulated to reflect the design needs.

The general features of an optimal synthesis procedure are described next.

3.1 Components of an optimal synthesis procedure

An optimal synthesis begins with the identification of an *objective function*. Usually, the most important design requirement becomes the objective function. The next step is to identify the *design variables*. That is, what features of a design (e.g., a structure) can be freely varied in order to minimize the objective function. This is called *design parameterization*. It is a critical step because the design space in which to search to seek an optimum gets determined by the choice of design variables.

The design variables may simply be a few discrete parameters of a design such as sizes of certain things in the design. This is called *parameter optimization* or *size optimization* (e.g., Chapter 10). Some other times, the shapes may be varied. This is called *shape optimization* (e.g., Chapter 3). When the entire layout of a device is varied it is called *topology optimization* (e.g., Chapters 2, 4-7). In this, the form of the device is not assumed *a priori*. If it is a structural problem, the number of holes is not fixed nor are the forms of each hole (if any) and the structure. The amount of information fixed by the designer decreases as one moves from size to shape to topology optimization. A consequence of this is that the designer also loses some freedom in imposing practical constraints. It is best to view topology optimization as a conceptual design stage. The solutions obtained with topology optimization can be refined using shape and size optimization steps where more practical constraints can be included.

The design parameterization step is followed by the identification of constraints. Usually, the design needs that did not get to be the objective function enter as constraints. This is also an important step because the *feasible design space* (i.e., the part of the design space that satisfies all the constraints) is decided by the constraints. If proper care is not exercised, one could be left with null design space and not know that it is the case. There are techniques to avoid this situation. Interested readers may refer to [21].

After an optimization problem is formulated as described above, one must develop a procedure for *function evaluation*. That is, given a candidate set of values for the design variables, there should be a procedure to compute the objective function and constraints. This step could involve a lengthy finite element analysis or a boundary element analysis or both, or some other analysis. In MEMS design problems function evaluation could be numerically quite expensive because of the coupling among different governing equations. This is one of the reasons why the manual iterative analysis-based approach is time consuming. Although the iterative process is unavoidable in optimal synthesis procedures, it is not manual and optimization algorithms are more efficient than human designers in minimizing the number of required function evaluation steps.

A designer might choose deterministic or stochastic numerical optimization algorithms to solve an optimal synthesis problem. Either has its own advantages and disadvantages [22]. In this book, deterministic approach is adopted. Readers interested in stochastic methods may refer to [16]. When a deterministic algorithm is chosen, it is efficient to use gradient-based algorithms. The procedure for computing the gradients of the objective function and constraint expressions with respect to the design variables is called the *sensitivity analysis*, i.e., one determines how sensitive the objective and constraints are for small (infinitesimal) changes in the design variables. This can be computed either by using analytically derived formulas or numerically by using the finite difference method. The latter is of course computationally more expensive. Hence, it is common practice to try to find analytical procedures for sensitivity analysis. This can be seen in many chapters in this book.

Finally, it is the task of the optimization algorithm to update an initial guess to achieve a better design until the optimal solution is reached. It is important to note there are excellent numerical algorithms, both deterministic and stochastic, that can efficiently solve a well-posed optimal synthesis problem. An algorithm not converging to a satisfactory solution is usually an indicator of a pitfall in the problem formulation rather than the inability of an algorithm in finding the solution. Non-convexity (i.e., there being more than one minima) and non-differentiability (i.e., a discontinuity that makes continuous optimization algorithms to break down) are two problems that deserve special attention.

The discussion in this section is summarized in Fig. 1. This is essentially the common theme in Chapters 2-7 and 10. The approach in Chapters 8 and 9 is different because of the nature of the problems addressed there. An overview of the chapters follows next.

4. CONTENTS OF THE CHAPTERS

There are many synthesis problems in MEMS. They can be approached in many ways. This book focuses on a few problems and uses optimal synthesis method to solve them. The chapters in the book reflect the well-known characteristic of MEMS—operating in multiple energy domains. The actuation principle (which in reverse becomes a sensing principle in most cases) is used to separate and arrange the chapters as described next.

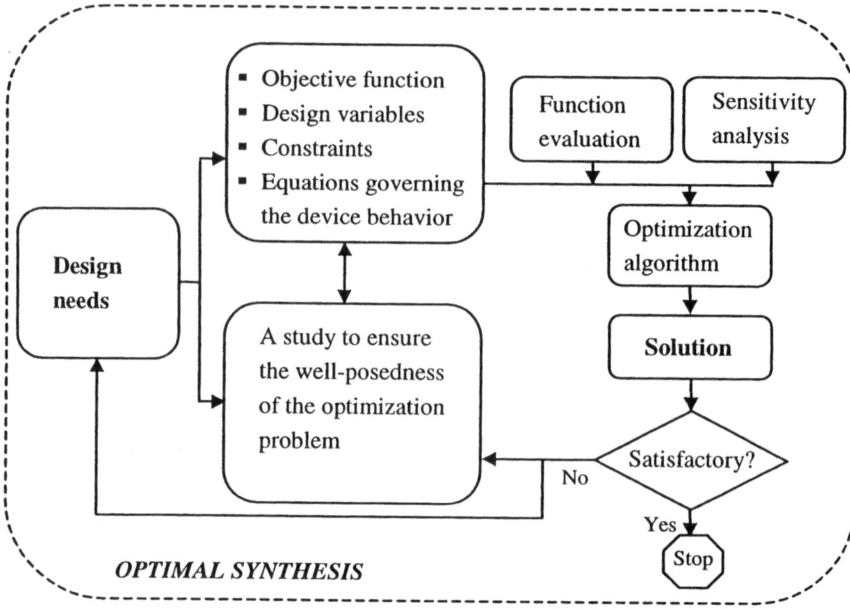

Figure 1. General steps in optimal synthesis procedure

In Chapter 2, Anupam Saxena and G. K. Ananthasuresh describe synthesis with mechanical actuation. Here, forces are assumed but their origin is not considered. Thus, purely mechanical equations govern the behavior of the device. Mechanical components of MEMS can be broadly divided into rigid-body mechanisms, structures, and elastic structures that behave like mechanisms (called *compliant mechanisms*). There are only few examples of rigid-body mechanisms in MEMS and there is a lot of literature on how to synthesize them [23]. Therefore, this discussion is not included in the chapter. Structural synthesis is also a mature field [24]. The design needs of elastic structures and compliant mechanisms include static aspects such as stiffness, flexibility, strength, motion, etc., and dynamic aspects such as natural frequency, mode shapes, and transient response. All of these are mentioned and some are described in detail.

In Chapter 3, Wenjing Ye and Subrata Mukherjee describe the synthesis of electrostatically actuated MEMS with a particular focus on comb-drives. As is necessary, the approach taken combines the elastic and electrostatic governing equations in the context of optimal synthesis.

G. K. Ananthasuresh considers the synthesis with thermal and electro-thermal actuation principles in Chapter 4. In the latter, three sets of governing equations coming from electrical, thermal, and elastic domains are used in a combined synthesis procedure.

The focus of Chapters 5 and 6, written by Emilio Carlos Nelli Silva, is on piezoelectric actuation. In Chapter 5, optimal synthesis of mechanical structures integrated with piezoelectric actuator elements, called *flextensional actuators*, is described. Chapter 6 considers piezocomposites, which are particularly suited for microfabrication principles. Here, the distinction between a mechanism and the material starts to disappear. This is further highlighted by Ole Sigmund who describes the synthesis of *periodic micro mechanisms* in Chapter 7. One can argue that they are "materials" with designed microstructure tailored to desired effective material properties. Examples described in that chapter include negative Poisson's ratio, negative and extreme thermal expansion coefficient, and desired band gaps in wave propagation.

While the chapters thus far addressed the synthesis step that goes from $device \rightarrow physical$ levels [1], the focus in Chapters 8 and 9 shifts to $physical \rightarrow process$ levels. In Chapter 8, Edwin Carlen and Carlos Mastrangelo discuss how a process can be synthesized given the geometry of a component or a subsystem. On the other hand, in Chapter 9, Radha Sarma and G. K. Ananthasuresh present a method to obtain the mask layouts for a given process and the physical geometry of a component. The focus of that chapter is on surface micromachined MEMS. Much work is done on bulk micromachined MEMS and was recently described in Chapter 5 of [16]. In order to avoid repetition, only a passing reference is made to it in Chapter 9 of this book.

Finally, in Chapter 10, Tamal Mukherjee and Garry Fedder address the important problem of system-level synthesis (i.e., going from $system \rightarrow device$). A unified circuit representation of mechanical, and electrostatic elements is combined with the optimization procedure to address some high-level synthesis requirements.

Sufficient background and implementation details of optimization are included in all the chapters. Wherever it is applicable, microfabricated prototypes are also described.

5. CLOSURE

In this chapter, the meaning of synthesis and analysis in the context of MEMS design was described. In particular, the notion of optimal synthesis as viewed in this book was clarified. An example was presented to show how a design requirement could be transformed into an optimization problem. The general steps in optimal synthesis were described and an overview of the contents of the chapters was presented.

REFERENCES

1. Senturia, S.D., *Microsystem Design*, Kluwer Academic Publishers, New York, 2001.
2. Wiegernink, R. J. and Elwenspoek, M., *Mechanical Microsensors*, Springer verlag, Berlin, 2001.
3. Gad-el-Hak, M., (ed.) *The MEMS Handbook*, CRC press, Boca Raton, FL, 2001.
4. Karniadakis, G.E. and Beskok, Al., *Micro Flows: Fundamentals and Simulation*, Springer verlag, Berlin, 2002.
5. Lyshevski, S.E., *MEMS and NEMS: Systems, Devices, and Structures*, CRC press, Boca Raton, FL, 2002.
6. Rebeiz, G. M., *RF MEMS: Theory, Design, and Technology*, John Wiley & Sons, 2003.
7. Intellisuite, Corning Intellisense, www.corningintellisense.com.
8. CoventorWare, Coventor, www.coventor.com.
9. Tanner Research, Inc., www.tanner.com.
10. MEMSCAP, www.memscap.com.
11. MEM Research, www.memresearch.com.
12. CFD Research Corp., www.cfdrc.com.
13. ANSYS/MEMS, www.ansys.com/ansys/mems/index.htm.
14. FEMLAB, Multi-physics in Matlab, www.femlab.com.
15. Antonsson, E.K., "Structured Methods for MEMS Design," a *report* submitted to NSF based on a workshop held at California Institute of Technology in November, 1995.
16. Antonsson, E.K. and Cagan J., *Formal Engineering Design Synthesis*, Cambridge University Press, Cambridge, UK.
17. Lai, E. and Ananthasuresh, G.K., "On the Design of Bars and Beams for Desired Mode Shapes," *Journal of Sound and Vibration*, Vol. 254, No. 2, 2002, pp. 393-406.
18. Weinstock, R., *Calculus of Variations with Applications to Physics and Engineering*, Dover Publications, New York, 1974.
19. Luenberger, D.G., *Introduction to Linear and Nonlinear Programming*, Addison-Wesley, Reading, MA, 1984.
20. Przemieniecki, J.S., *Theory of Matrix Structural Analysis*, Dover Publications, New York, 1968.
21. Papalambros, P.Y. and Wilde, D.J., *Principles of Optimal Design: Modeling and Computation*, Cambridge University Press, , Cambridge, UK, 2000
22. Rao, S.S., *Engineering Optimization: Theory and Practice*, John Wiley and Sons, Inc., New York.
23. Erdman, A.G., Sandor, G. N., and Kota, S., *Mechanism Design: Analysis and Synthesis*, Vol. 1, Fourth Edition, Prentice Hall, Englewood Cliffs, New Jersey.
24. Haftka, R.T. and Gurdal, Z., *Elements of Structural Optimization*, Kluwer Acedmic Publishers, New York, 1989.

Chapter 2

SYNTHESIS FOR MECHANICAL BEHAVIOR

Anupam Saxena
Department of Mechanical Engineering, Indian Institute of Technology, Kanpur, India

G. K. Ananthasuresh
Department of Mechanical Engineering and Applied Mechanics, University of Pennsylvania, Philadelphia, PA 19104, U.S.A.

Synopsis: The mechanical elements of most of the MEMS devices are based on compliant designs such as the ones found in beams, plates, and other types of elastic structures. The advantages of single-piece, assembly-free compliant designs are well known. The extruded planar geometry of MEMS devices and almost unrestricted possibilities for shapes in the plane parallel to the substrate make it possible to develop systematic synthesis methods. This chapter describes such methods for desired mechanical attributes including stiffness, flexibility, motion, strength, natural frequencies, and normal mode shapes. Topology optimization is the main focus of the chapter but some other methods are also briefly presented. Some applications are included.

1. INTRODUCTION

Moving solids and fluids, and the accompanying numerous possibilities, distinguish the fast-growing technology of Microeletromechanical Systems (MEMS) from the well-developed technology of microelectronics and Integrated Circuits (ICs). This chapter is concerned with movable solids of small size scales ranging from microns to millimeters that are fabricated using a class of methods that have come to be known as *micromachining* [1]. The controlled movement of solids is possible either through an assemblage of multiple rigid bodies connected together by kinematic joints (e.g., hinges, sliders, etc.) or by using elastic deformation. The feasibility of micron-scale elastically deformable structures was demonstrated long ago [2]. The kinematic movement using pin joints and sliders is a more recent development [3, 4]. Both types of obtaining motion of solid bodies at the micro scale have revolutionized the technology of MEMS by paving the way for a number of sensors and actuators. They have led to numerous applications some of which have been in commercial use today. They have also provided the motivation for developing improved and more versatile micromachining techniques. The novelty of miniaturized kinematic assemblages notwithstanding, the majority of MEMS devices utilize elastic deformation. The principal reason for this is that it is much easier to microfabricate single-member elastic structures.

A closer examination of the available MEMS reveals that just beams, membranes, and plates seem adequate for most applications. This is in contrast to the macro world where there is a large variety of mechanical elements such as gears, belts, pulleys, chains, cams, bearings, linkages, fasteners, etc. It is worth noting that elastic structures are still adequate to duplicate the functionality of the above macro-mechanical elements at the micro scale if they are shaped to deform appropriately. This concept has been extensively researched in the last decade under the name of *compliant mechanisms* [5-8]. An example of a compliant micromechanism is shown in Fig. 1.

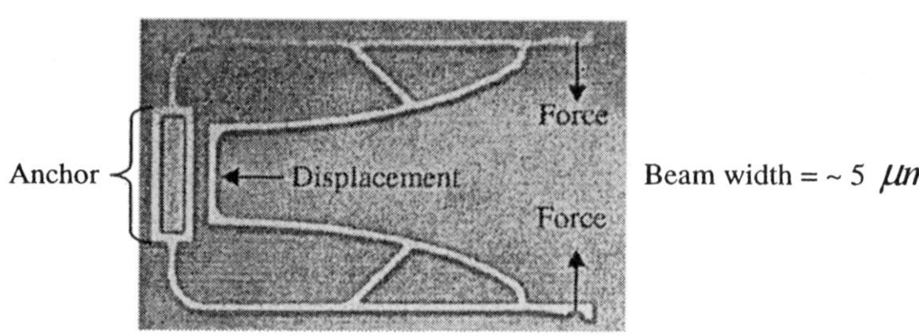

Figure 1. The SEM of an early compliant micro-mechanism made out of single crystal silicon using Michigan's boron-doped dissolved-wafer bulk micromachining [6]

There are well-developed microfabrication techniques to realize complex planar geometries of deformable structures that are as thin as a fraction of a micron and as thick as a significant fraction of a millimeter. If there are also systematic synthesis techniques to obtain the requisite geometry of a device for an application, it will considerably ease the burden on MEMS designers. In this chapter, techniques for synthesizing simple and complicated MEMS devices for various mechanical attributes are described. Section 2 contains techniques for beam-like structures. In Section 3, a technique called *topology optimization* is described for geometrically more complicated structures and mechanisms with a view to optimize for stiffness, flexibility, and strength. This is followed by Section 4, which contains a discussion of synthesis methods for dynamic characteristics including frequency, normal modes, and transient response. Concluding remarks are in Section 5.

2. SYNTHESIS OF BEAM-LIKE STRUCTURES

If the cross-section of a straight beam is uniform along its length, it is not difficult to synthesize (i.e., determining the length and cross-section's dimensions) such a beam. The same is the case with a membrane or a plate of constant thickness and regular shapes and boundary conditions. MEMS designers often rely upon easy-to-use catalog of formulas [9] for this purpose. The complexity arises when the mechanical behavior is coupled with other energy domains. This is the subject matter for Chapters 3-7. If we move from straight beams and regular plates and membranes to more complicated elastic structures, even purely mechanical synthesis becomes nontrivial. The most notable early MEMS example is the suspension design of an electrostatic comb-drive linear actuator. As shown in Fig. 2, the central shuttle mass should be able to translate in the x-direction with as little resistance as possible while being infinitely stiff in the y-direction. The requirement, as is usual in most MEMS, is that it should be a single elastic member with no rubbing parts. The elastic nature naturally provides the restoring spring force. The absence of rubbing parts eliminates friction and wear-related problems. Clearly, the use of a slider for this purpose is out of question. Several possibilities were considered [10-13] but the folded-beam design [11] is prevailing today.

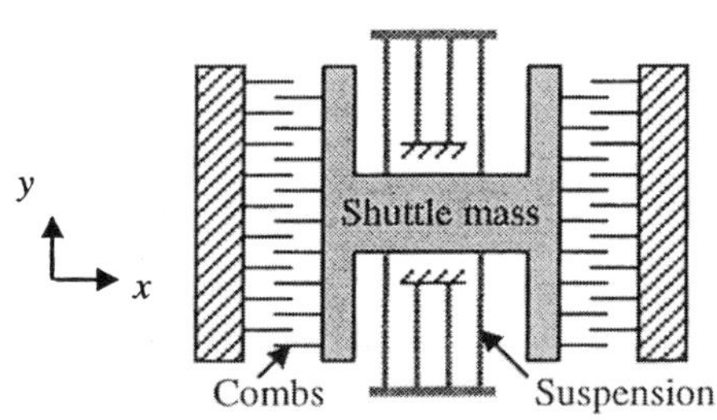

Figure 2. Comb-drive suspension, an early mechanical synthesis problem in MEMS

The folded-beam configuration was perhaps not a solution synthesized by a systematic computational algorithm, as there were no direct synthesis techniques available at that time. But today there are some guiding principles for synthesis so that desired stiffness and flexibility requirements can be met. One notable technique for structures consisting of beam-like segments is by means of a concept called *pseudo rigid-body model* [8]. The essence of this is in modeling a beam equivalently as a rigid bar with hinges consisting of some rotational (torsional) stiffness. These models have been shown to be accurate over a large range of motion. This modeling enables the MEMS designer to make use of a rich set of synthesis methods available in mechanisms and kinematics literature [14]. The reader may refer to [8] for details and some applications including orthoplanar compliant mechanisms and bistable structures.

Another technique is shape optimization of a simple or complex arrangement of beams that are either straight or curved. In [15], the shape of the centerline of a curved beam was optimized in order to synthesize gripper and other devices. In [16], a number of straight beams were arranged in a grid wherein the cross-sections of the beams as well as the grid points were moved to optimize for stiffness and flexibility. A result of this synthesis procedure was used in Sandia's microengine to amplify the displacement of the electrostatic comb-drive [17].

Next, the synthesis of more general elastic structures is described.

3. TOPOLOGY SYNTHESIS

A numerical technique called *topology optimization* is a versatile synthesis method that can automatically generate structural forms from function-level specifications. At the outset, it should be noted that the resulting solutions might not satisfy the requirements that are not included in the synthesis problem formulation. Even in those cases where a solution is not usable as it is, it will almost always aid the creative thinking of a designer. When all the crucial requirements are modeled, the technique is capable of giving practically feasible solutions that can directly be converted into mask layouts.

A mechanical synthesis problem can be posed in a general way as shown in Fig. 3. The figure shows an arbitrary region within which the device to be designed must fit. Assume that the forces and portions that are fixed are known. One may wish to design a structure or a compliant mechanism such that it is stiff or flexible or both but in different directions. One may also impose strength considerations. In other cases, it may be necessary to trace certain paths at the output points. All but the last of these are addressed in this section.

In addition to one or more of the aforementioned requirements, a resource constraint on the amount of material used is usually included. Although there is no need to minimize the material used in a micro device, it is helpful from the algorithmic viewpoint. Additionally, a small volume of material than what can fit in the available design region leads to simpler and more elegant designs. In the presence of a material resource constraint, as can be seen in Fig. 3, the structural synthesis problem is equivalent to distributing the given amount of material in a larger design region. If we discretize the region as is done in the finite element analysis, then the problem reduces to selective *retaining* or *removal* of material at different sites in the region. This is done by assigning a design variable to each site (or to an element in the discretized finite element model). This variable determines the state of the site: if it is 1, the material exists; and if it is 0, it creates a hole. In order to use continuous optimization methods, the state is smoothly varied between 0 and 1. This variable multiplies a material property such as Young's modulus at the corresponding site. Finally, an optimization algorithm determines the values of all the variables to minimize an objective function while satisfying some constraints. Similar selective retaining or removal can be done by taking a network of beam elements and defining the factors that multiply a nominal cross-section value as the design variables. If a beam's variable goes to zero, that beam will be absent in the final design. See [18, 19] for a detailed discussion of these and alternate schemes.

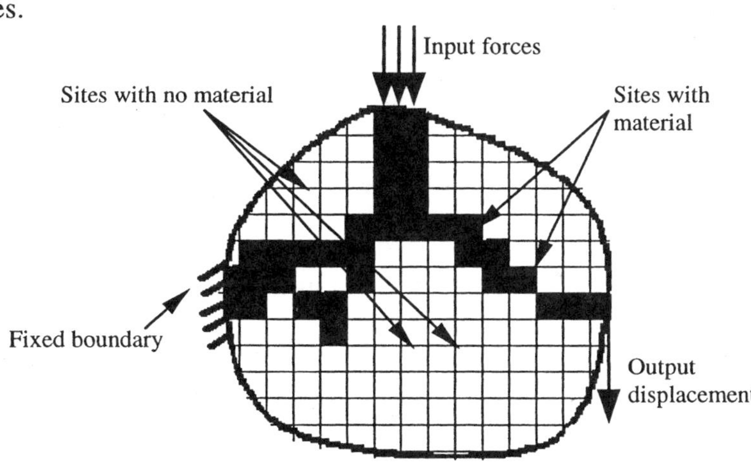

Figure 3. Schematic depicting topology synthesis problem as a material distribution problem

Functional and/or performance related requirements of an application determine the objective and constraints, which will be in terms of the design variables. In MEMS devices, the typical needs are related to stiffness, flexibility, and strength. Strain energy stored in the structure under applied loads serves as a scalar measure of stiffness. Minimizing the strain energy is equivalent to maximizing the stiffness. Likewise, maximizing the displacement at a point in the desired direction maximizes the flexibility. Strength consideration is usually dealt with constraints either imposed locally at every point or in an integrated form as one constraint for the entire structure [20-22]. Additionally, the material resource constraint may also be present. Several different formulations exist in combining one or more of these and other requirements [23-35]. See [36] for an overview. The following two formulations are described ahead: (i) flexibility-stiffness and (ii) flexibility-strength. These are adapted from [30] and [22], respectively.

3.1 Flexibility-stiffness formulation

The goal of the synthesis here is to obtain an optimal topology by maximizing the output deformation along a specified direction while simultaneously minimizing the strain energy. For input-output specifications and boundary conditions shown in Fig. 4, the output deformation Δ_{out} is computed using the unit dummy load method [37] as

$$\Delta_{out} = \mathbf{V}^T \mathbf{K} \mathbf{U} \tag{1}$$

where **V** is the displacement response due to the dummy unit load along the direction of the output deformation and **U** is the same for input loads only. Here, **K** is the global stiffness matrix for a small deformation finite element model [38]. The strain energy, *SE* is computed as

$$SE = \tfrac{1}{2} \mathbf{U}^T \mathbf{K} \mathbf{U} \tag{2}$$

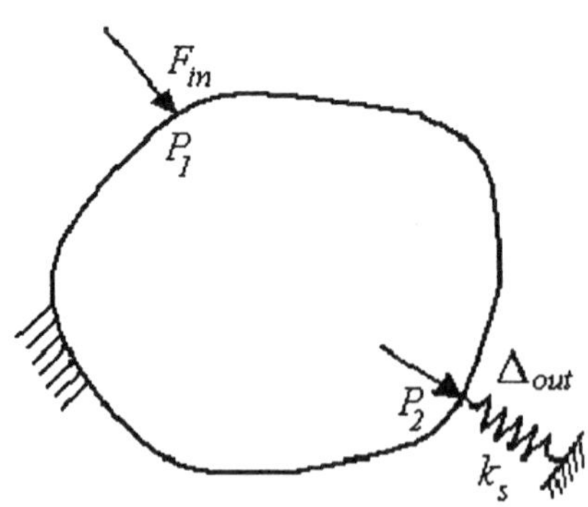

Figure 4. Flexible-stiffness formulation where the output load is represented as a spring

One way to pose the two objectives of maximizing the output deformation (flexibility requirement) and minimizing the strain energy (stiffness requirement) is to use their ratio [26]. That is,

$$\underset{\mathbf{x}}{Maximize}\ f = -\frac{\Delta_{out}}{SE}$$
$$Subject\ to\ V(\mathbf{x}) - V^* \le 0\ \ and\ \ \varepsilon \le x_i \le 1\ \forall i = 1,2,....N$$
$$\text{Equi. equations}$$

where, $V(\mathbf{x})$ is the material volume with V^* as its upper limit, x_i is the design variable representing the material state at the ith site out of N such sites, and ε is a small positive number used instead of zero for numerical convenience. In the finite element setting, the design variables are usually scaled with respect to the physical properties of the elements, namely, their cross-sections (beam element-based network approach shown in Fig. 5a) or their elastic modulus (continuum element-based approach shown in Fig. 3). It is worth noting that SE is proportional to the input displacement. Therefore, the objective function here is the ratio of the output displacement to the input displacement.

When the above problem is solved using gradient-based numerical optimization algorithms, the derivatives of the objective and constraint functions are necessary. The process of computing the derivatives is called *sensitivity analysis*. For this problem, it proceeds as follows. Differentiation of the objective function gives

$$\frac{\partial f}{\partial x_i} = -\frac{1}{SE}\frac{\partial \Delta_{out}}{\partial x_i} + \frac{\Delta_{out}}{SE^2}\frac{\partial SE}{\partial x_i} \tag{3}$$

By differentiation of Eq. (1), we get

$$\frac{\partial \Delta_{out}}{\partial x_i} = \frac{\partial \mathbf{V}^T}{\partial x_i}\mathbf{K}\,\mathbf{U} + \mathbf{V}^T\frac{\partial[\mathbf{KU}]}{\partial x_i} \tag{4}$$

By making use of the fact $\mathbf{F}$ and $\mathbf{F}_d$ (the load vectors corresponding to the input and unit dummy force respectively) in the equilibrium equations $\mathbf{F} = \mathbf{KU}$ and $\mathbf{F}_d = \mathbf{KV}$ are both independent of the design variables, Eq. (4) can be simplified to the following form.

$$\frac{\partial \Delta_{out}}{\partial x_i} = \frac{\partial \mathbf{V}^T}{\partial x_i}\mathbf{K}\,\mathbf{U} \tag{5}$$

Differentiation and rearrangement of $\mathbf{F}_d = \mathbf{KV}$ results in

$$\frac{\partial \mathbf{V}^T}{\partial x_i}\mathbf{K} = -\mathbf{V}^T \frac{\partial \mathbf{K}}{\partial x_i} \tag{6}$$

By combining (5) and (6), we get

$$\frac{\partial \Delta_{out}}{\partial x_i} = -\mathbf{V}^T \frac{\partial \mathbf{K}}{\partial x_i}\mathbf{U} \tag{7}$$

The stiffness matrix of any element $\mathbf{k}_i$ is only dependent on the corresponding design variable x_i, therefore, the expression in Eq. (7) simplifies to

$$\frac{\partial \Delta_{out}}{\partial x_i} = -\mathbf{v}_i^T \frac{\partial \mathbf{k}_i}{\partial x_i}\mathbf{u}_i \tag{8}$$

where $\mathbf{u}_i$ and $\mathbf{v}_i$ are the nodal displacement vectors for the ith element due to input and dummy loads respectively. Following similar steps, the sensitivity of strain energy can be computed as

$$\frac{\partial SE}{\partial x_i} = -\frac{1}{2}\mathbf{u}_i^T \frac{\partial \mathbf{k}_i}{\partial x_i}\mathbf{u}_i \tag{9}$$

Upon combining Eqs. (3), (8) and (9), the gradient of the flexibility-stiffness objective can be calculated. A gradient-based optimization algorithm can now be used to solve the problem. Three examples follow.

3.1.1 An accelerometer with a built-in displacement amplifier

Micro acceleration sensors are required in a variety of automotive [39], bio-mechanics [40], vibration monitoring [41], seismic [42], and other applications. An accelerometer is essentially a spring-mass system wherein the inertial forces can be sensed directly through the strain in the spring or the displacement of the inertial mass. The displacement based acceleration sensing can be achieved by measuring the change in capacitance between the moving (inertial mass) and the fixed (substrate) electrodes. For a small initial gap, the relative change in capacitance is much higher than the change in resistance in piezoresistive sensors, and therefore capacitive sensors have higher sensitivity. The accelerometer designs with capacitive sensors are less cumbersome as they offer direct means for sensing inertial accelerations. Since the electrodes can be monolithically integrated with the proof mass, these devices can be batch fabricated and, therefore, are highly economical. The sensitivity of capacitive accelerometers can be additionally improved by integrating a displacement amplifier designed using topology optimization.

Consider the symmetric half of the meshed design domain with input and output design requirements shown in Figs. 5a and 5b. Figure 5a shows the network of beam elements where the factor multiplying a nominal cross-section of each beam is a variable. On the other hand, in Fig. 5b, a discretized continuous design region consisting of four-noded plane-stress finite elements is used. Symmetry in the optimal design is desired for high transverse stiffness and thus low cross-axis sensitivity. In both cases, a force of 500 μN is applied at the input port. Polysilicon is used as the structural material for which an Young's Modulus of 169 GPa and the Poisson's ratio of 0.3. Respective widths of the beam elements are chosen as design variables on which a lower bound of 10^{-5} μm and an upper bound of 5 μm in the case of Fig. 5a. Recall that the variables reaching the lower bound are regarded as absent from the topology. In Fig. 5b, the state of each square element is a design variable with 0 denoting its non-existence, 1 denoting its existence, any other value in between its *gray* existence. This “state” variable multiplies the Young’s modulus of that element. Output displacement is desired at a point towards the top on the right edge.

By computing the gradients as described above, the flexibility-stiffness problem was solved using an optimization algorithm. The Sequential Quadratic Programming (SQP) implemented in MATLAB™'s Optimization Toolbox was used for the beam-element network case of Fig. 5a. The resultant topology of the symmetric half of the displacement amplifier is shown in Fig. 5c and the full topology is shown in Fig. 6a. At convergence, the output deformation was maximized to 6.4 μm for which the geometric advantage (the ratio of the output and input displacements) of 3.2 was obtained. Fig. 6b shows the SEM photograph of the prototype manufactured using MUMPs (Multi-User MEMS Process—a foundry process). The problem in Fig. 5b was solved using another numerical optimization technique called the *optimality criteria method* (e.g., see [19] or [36]). Its solution is shown in Fig. 5d. Although the designs in Figs. 5c and 5d are not exactly identical, much similarity can be seen. Amplification ratios 50X or more are also shown to be possible as shown in Section 3.1.3.

3.1.2 A micromechanical AND logic gate

Another example in micro-scale application is of a non-contact micro compliant AND logic gate. The design specifications include two input forces, F_1 and F_2 shown in Fig. 7a which act together to produce the intended output deformation shown. Input forces F_1 and F_2 were 500 μN each, and

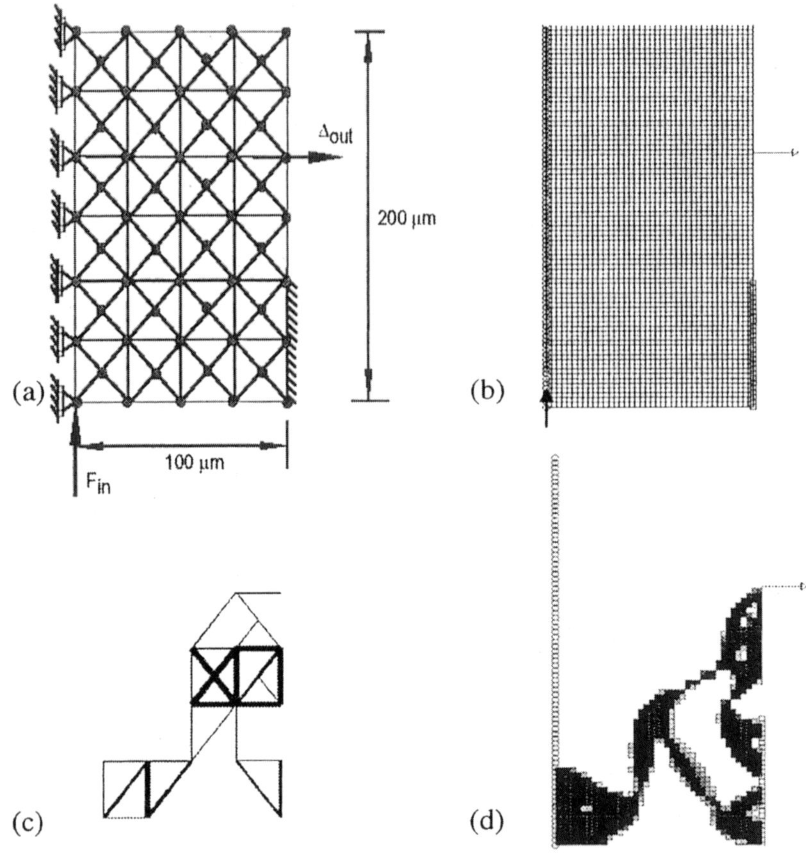

Figure 5. Accelerometer with a built-in displacement amplifier (a) specification with beam-element network (b) specifications with square plane-stress elements (c) solution of Fig. 5a (d) solution of Fig. 5b

Figure 6. (a) full image of the synthesized 3.2X displacement amplifier (b) SEM of the fabricated polysilicon prototype

elastic modulus, lower and upper bounds on design variables were chosen as the same as in the previous example. The resultant topology is shown in Fig. 7b with deformed configurations in Figs. 7c and 7d, which explain the working principle of the mechanism. With no actuation, no vertical displacement is registered at the output port (**0 + 0 = 0** case). With only one input force as actuation (F_1 in Fig. 7c), the output deformation is not significant in the upward direction. This can be considered as the **1 + 0 = 0** case. It is only with both input forces that the change in the output capacitance is detectable **1 + 1 = 1** case). The SEM of the prototype micro mechanism is shown in Fig. 8.

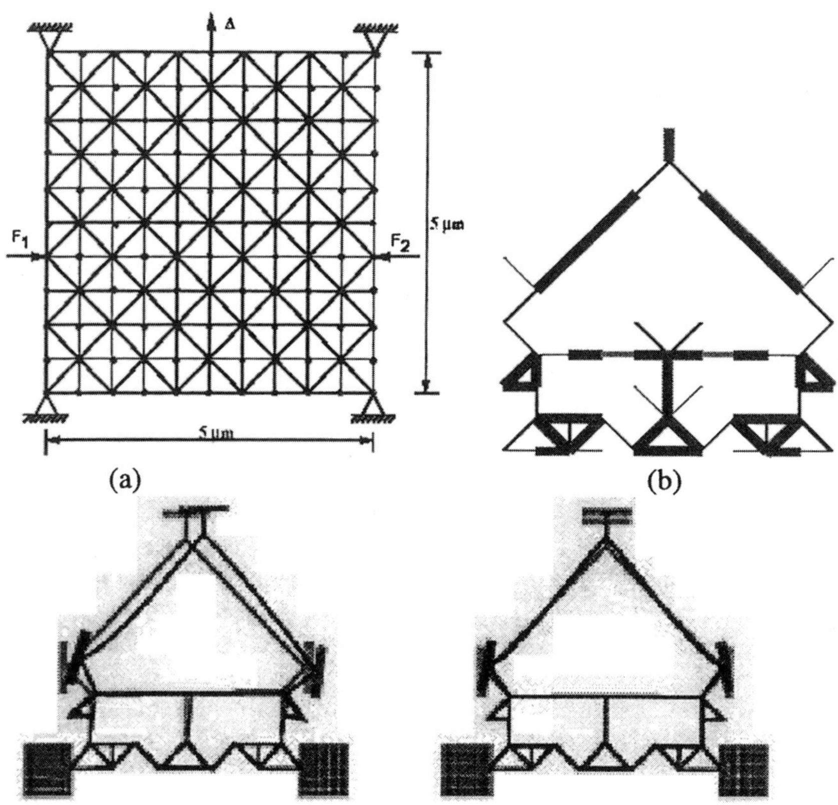

Figure 7. Synthesis example of a micro AND logic gate (a) design specifications (b) optimal topology of the AND gate (c) deformed configuration with a single load and (d) with both loads

The two examples in this section illustrate that it is possible synthesize elastic structures (or compliant mechanisms) so that the input displacement can be amplified and/or redirected as desired at the output.

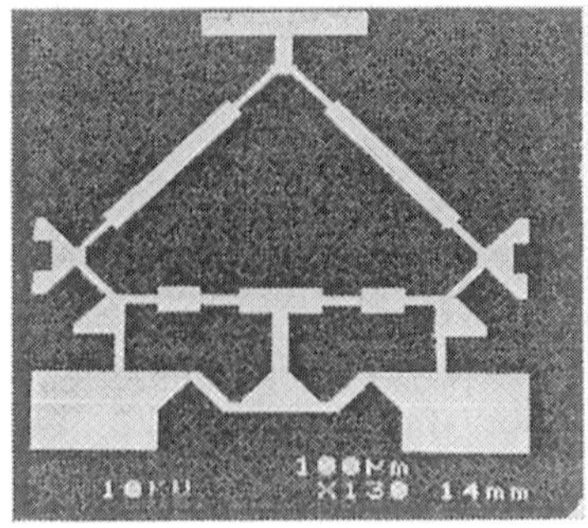

(e)

Figure 8. SEM of the micromechanical AND gate manufactured using MUMPs

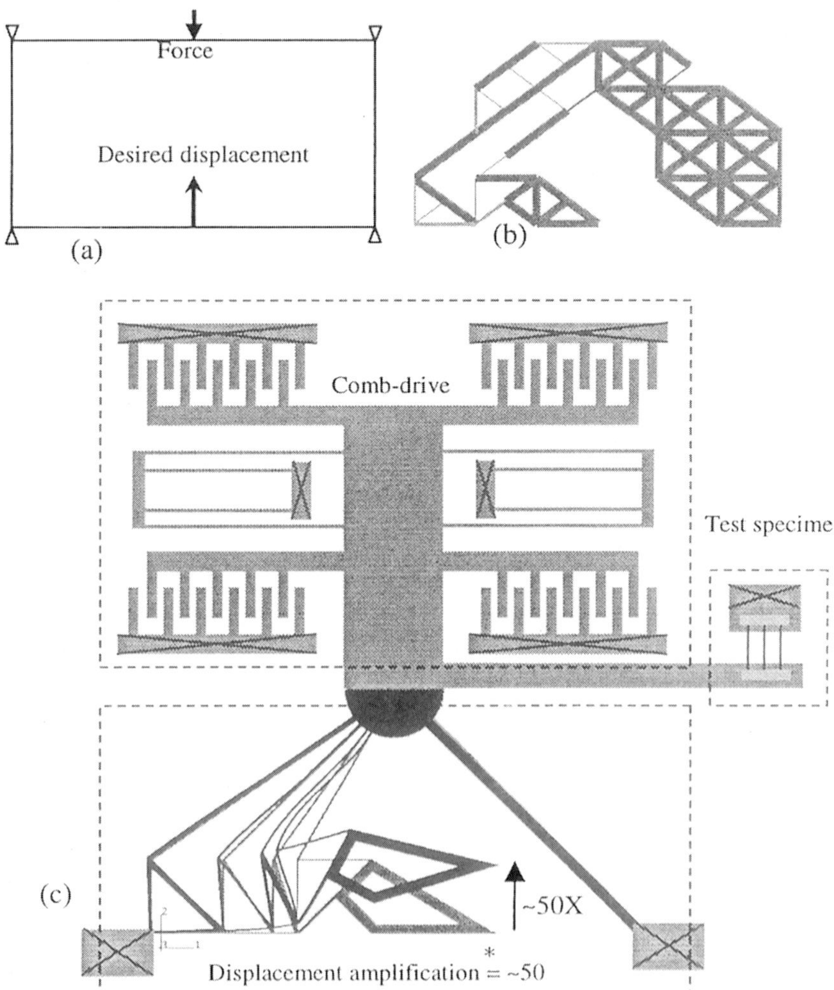

Figure 9. Compliant displacement amplifier in a micro testing machine (a) specifications for the displacement amplifier, (b) synthesized solution, (c) schematic of the test machine

3.1.3 Synthesized solutions as design aids

An example will now be presented to show that synthesis solutions aid human designer's creativity even when they are not satisfactory for a given application in their original form. Figure 9a shows the specifications for a compact displacement amplifier to be used in a sensing application to increase the sensitivity. As shown, the device has to fit within the rectangle and also operate entirely inside. Therefore, the output displacement at the bottom point is desired to move up when the top point moves down with input force. The four corners are to be anchored to the substrate. This is a non-intuitive specification. Additionally, it is intended that the output displacement be large compared to the input displacement. A solution shown in Fig. 9b was obtained by solving the flexibility-stiffness formulation presented earlier. This had an amplification ratio of about 7. In an optical force-sensing application to characterize the stiffness and strength of micro specimens mounted on a chip inside an SEM, a much larger amplification ratio is preferred. The synthesis solution in Fig. 9b inspired a human designer* to modify the solution to enhance its performance. It often comes with a thorough analysis of how the solution given by the optimization algorithm is able to work the way it does. As can be seen this modified solution, it is a 50X amplifier. Even 100X is possible by adding more of the critical "lever arms" but the fabrication dictates the minimum width of the beams. Although it is not impossible for a skilled human designer to conceptualize a design such as this, the savings in time and effort, and more importantly, the accessibility to creative concepts to all designers are useful outcomes of systematic synthesis methods.

3.2 Flexibility-strength formulation

The flexibility-stiffness formulation may not be appropriate in some cases for the design of compliant MEMS due to the following reasons. First, minimizing strain energy restrains the input deformation of the mechanism and therefore, directly interferes with the design objective of maximizing the flexibility at the output port. Thus, the resultant topology may be overly stiff. Second, minimizing strain energy restrains the stresses in the mechanism only indirectly and fails to account for local failure. The local stresses may exceed the safe working stress of the material to cause failure. A more direct approach of achieving the flexibility of compliant mechanisms and simultaneously addressing local failure issues is by maximizing the output displacement and restricting the local stresses below their safe allowable

* Luzhong Yin, post-doctoral research associate at University of Pennsylvania designed the 50X amplification mechanism shown in Fig. 9c.

limit. Of the different failure theories available, von Mises distortion energy criteria may be used [43]. The optimization problem can now be written as:

$$\underset{\mathbf{x}}{Maximize}\ \Delta_{out}$$

$$Subject\ to$$

$$\begin{aligned} & V(\mathbf{x}) - V^* \leq 0 \\ & \varepsilon \leq x_i \leq 1 \\ & \sigma_{VM_i} - \sigma_a \leq 0 \\ & \text{Equi. equations} \end{aligned}$$

where σ_{VM_i} is the von Mises stress at the ith site, and σ_a is the allowable stress. In solving the above problem, care must be taken to impose stress constraints only at the existing sites. If it is not done, a feasible design will not be obtained, as stresses can be very large at the irrelevant non-existing sites. This special handling of stress constraints is explained next.

3.2.1 Modeling stress constraints

For local stress constraints to be considered only at sites that exist, we have

$$\frac{(\sigma_{VM})_i}{\sigma_a} - 1 \leq 0 \qquad \text{for } x_i > \in \tag{10}$$

where the lower bound $\in$ suggests material presence at the site. One way to combine the two inequalities in Eq. (10) is as shown below.

$$(x_i - \in)^n \left[\frac{(\sigma_{VM})_i}{\sigma_a} - 1 \right] \leq 0 \tag{11}$$

where n (> 1) is a known positive parameter chosen *a priori*. Equation (11) implies that the local stress measure has to be less than or equal to its allowable limit if the material is present at the site ($x_i > \in$). For no material, i.e., for $x_i = \in$ which by itself makes the constraint active, the von Mises measure need not be restricted below its upper bound. Since $\in$ is very small, Eq. (11) may be modified as

$$(x_i)^n\left[\frac{(\sigma_{VM})_i}{\sigma_a}-1\right]\le 0 \tag{12}$$

The stress constraints in Eq. (12) may further be relaxed using a small relaxation parameter ε to aid the design variables approach their lower bound [20]. That is,

$$x_i^{\,n}\left[\frac{(\sigma_{VM})_i}{\sigma_a}-1\right]\le \varepsilon \tag{13}$$

or

$$(\sigma_{VM})_i \le \sigma_a\left[1+\frac{\varepsilon}{x_i^{\,n}}\right] \tag{14}$$

The effect of ε–relaxation is that for variable value higher than its lower bound, the von Mises stress in the finite element must be below or equal to the relaxed allowable stress $(\sigma_a)_{relaxed}=\sigma_a\left[1+\varepsilon/x_i^{\,n}\right]$, which is slightly larger than σ_a for a suitably chosen ε. However, in the limit the variable attains its lower bound, the upper bound for the local stress approaches (ideally) infinity. In other words, stress constraints are not imposed for non-existing elements. Eq. (14) may further be normalized and written in the following form

$$g_i=\frac{(\sigma_{VM})_i}{\sigma_a}-\frac{\varepsilon}{x_i^{\,n}}-1\le 0 \tag{15}$$

3.2.2 Sensitivity analysis for stress constraints

Implementation of local stress constraints with gradient-based optimization algorithms requires their gradients. These are calculated by differentiating Eq. (15) as follows

$$\frac{\partial g_i}{\partial x_j}=\frac{1}{\sigma_a}\frac{\partial(\sigma_{VM})_i}{\partial x_j}+\frac{n\,\varepsilon}{x_i^{\,n+1}}\frac{\partial x_i}{\partial x_j} \tag{16}$$

where $\frac{dx_i}{dx_j}$ may be determined as

$$\frac{dx_i}{dx_j} = \begin{cases} 1 \text{ for } i = j \\ 0 \text{ otherwise} \end{cases} \tag{17}$$

The von Mises stress for a 2-D plane stress problem is given by

$$(\sigma_{VM})_i^2 = \sigma_1^2 + \sigma_2^2 - \sigma_1 \sigma_2 \tag{18}$$

where σ_1 and σ_2 are the principal stresses [43]. If $[\sigma_x \ \sigma_y \ \tau_{xy}]$ represents the components of stress state at any point, then principal stresses at that point are given as

$$\sigma_{1,2} = \frac{(\sigma_x + \sigma_y)}{2} \pm \sqrt{\left[\frac{(\sigma_x - \sigma_y)}{2}\right]^2 + \tau_{xy}^2} \tag{19}$$

and so from Eq. (18)

$$(\sigma_{VM})_i^2 = \sigma_x^2 + \sigma_y^2 - \sigma_x \sigma_y + 3\tau_{xy}^2 \tag{20}$$

The expression above may be written in the matrix form as

$$(\sigma_{VM})_i^2 = \begin{bmatrix} \sigma_x & \sigma_y & \tau_{xy} \end{bmatrix} \begin{bmatrix} 1 & -1/2 & 0 \\ -1/2 & 1 & 0 \\ 0 & 0 & 3 \end{bmatrix} \begin{bmatrix} \sigma_x \\ \sigma_y \\ \tau_{xy} \end{bmatrix}$$

$$(\sigma_{VM})_i^2 = \boldsymbol{\sigma}_i^T \mathbf{W} \boldsymbol{\sigma}_i \tag{21}$$

In the finite element analysis, the stresses in i^{th} element are given by

$$\boldsymbol{\sigma}_i = \mathbf{D}_i \mathbf{B}_i \mathbf{u}_i \tag{22}$$

where $\mathbf{D}_i$ is the elastic constitutive matrix, $\mathbf{B}_i$ is the strain displacement matrix and $\mathbf{u}_i$ is the nodal displacement vector for the element. Combining Eqs. (21) and Eq. (22) yields

$$\begin{aligned}(\sigma_{VM})_i^{\,2} &= \mathbf{u}_i^T\,\mathbf{B}_i^T\,\mathbf{D}_i^T\,\mathbf{W}_i\,\mathbf{D}_i\,\mathbf{B}_i\,\mathbf{u}_i \\ &= \mathbf{u}_i^T\,\mathbf{T}_i^T\,\mathbf{W}_i\,\mathbf{T}_i\,\mathbf{u}_i \\ &= \mathbf{u}_i^T\,\mathbf{M}_i\,\mathbf{u}_i\end{aligned} \tag{23}$$

with $\mathbf{T}_i = \mathbf{D}_i\,\mathbf{B}_i$ and

$$\mathbf{M}_i = \mathbf{T}_i^T\,\mathbf{W}_i\,\mathbf{T}_i = \mathbf{B}_i^T\,\mathbf{D}_i^T\,\mathbf{W}_i\,\mathbf{D}_i\,\mathbf{B}_i \tag{24}$$

The derivative of the von Mises stress is then calculated by differentiating Eq. (23). This gives

$$2(\sigma_{VM})_i\,\frac{d(\sigma_{VM})_i}{dx_j} = \frac{d\mathbf{u}_i^T}{dx_j}\mathbf{M}_i\,\mathbf{u}_i + \mathbf{u}_i^T\,\frac{d\mathbf{M}_i}{dx_j}\mathbf{u}_i + \mathbf{u}_i^T\,\mathbf{M}_i\,\frac{d\mathbf{u}_i}{dx_j} \tag{25}$$

Since the left hand side of the above expression represents a scalar quantity, all terms in the right hand side are also scalar. Therefore,

$$\frac{d\mathbf{u}_i^T}{dx_j}\mathbf{M}_i\,\mathbf{u}_i = \left[\frac{d\mathbf{u}_i^T}{dx_j}\mathbf{M}_i\,\mathbf{u}_i\right]^T = \mathbf{u}_i^T\,\mathbf{M}_i\,\frac{d\mathbf{u}_i}{dx_j} \tag{26}$$

using which Eq. (25) can be written as

$$2(\sigma_{VM})_i\,\frac{d(\sigma_{VM})_i}{dx_j} = \mathbf{u}_i^T\,\frac{d\mathbf{M}_i}{dx_j}\mathbf{u}_i + 2\mathbf{u}_i^T\,\mathbf{M}_i\,\frac{d\mathbf{u}_i}{dx_j} \tag{27}$$

The quantity $\dfrac{d\mathbf{M}_i}{dx_j}$ is calculated by differentiating Eq. (24) as follows.

$$\frac{d\mathbf{M}_i}{dx_j} = \mathbf{B}_i^T\,\frac{d\mathbf{D}_i^T}{dx_j}\mathbf{W}_i\,\mathbf{D}_i\,\mathbf{B}_i + \mathbf{B}_i^T\,\mathbf{D}_i^T\,\mathbf{W}_i\,\frac{d\mathbf{D}_i}{dx_j}\mathbf{B}_i$$

or by using symmetry,

$$\frac{d\mathbf{M}_i}{dx_j} = 2\mathbf{B}_i^T\,\mathbf{D}_i^T\,\mathbf{W}_i\,\frac{d\mathbf{D}_i}{dx_j}\mathbf{B}_i \tag{28}$$

To compute $\frac{d\mathbf{u}_i}{dx_j}$, we can differentiate the equilibrium equation

$$\mathbf{K}\,\mathbf{U} = \mathbf{F}$$

to get

$$\frac{d\mathbf{K}}{dx_j}\mathbf{U} + \mathbf{K}\frac{d\mathbf{U}}{dx_j} = \frac{d\mathbf{F}}{dx_j} \quad \text{or} \quad \frac{d\mathbf{U}}{dx_j} = \mathbf{K}^{-1}\left[\frac{d\mathbf{F}}{dx_j} - \frac{d\mathbf{K}}{dx_j}\mathbf{U}\right] \tag{29}$$

Note that $\frac{d\mathbf{u}_i}{dx_j}$ are the components of $\frac{d\mathbf{U}}{dx_j}$ corresponding to the nodes of the ith element. That is,

$$\frac{d\mathbf{u}_i}{dx_j} = \left[\frac{d\mathbf{U}}{dx_j}\right]_i = \left[\mathbf{K}^{-1}\left[\frac{d\mathbf{F}}{dx_j} - \frac{d\mathbf{K}}{dx_j}\mathbf{U}\right]\right]_i \tag{30}$$

Expression for $\frac{d(\sigma_{VM})_i}{dx_j}$ is now obtained by substituting Eq. (28) and Eq. (30) into Eq. (27).

$$\frac{d(\sigma_{VM})_i}{dx_j} = \frac{1}{(\sigma_{VM})_i}\left[\mathbf{u}_i^T\,\mathbf{B}_i^T\,\mathbf{D}_i^T\,\mathbf{W}_i\frac{d\mathbf{D}_i}{dx_j}\mathbf{B}_i\,\mathbf{u}_i + \mathbf{u}_i^T\,\mathbf{M}_i\,\mathbf{K}^{-1}\left[\frac{d\mathbf{F}}{dx_j} - \frac{d\mathbf{K}}{dx_j}\mathbf{U}\right]_i\right] \tag{31}$$

Equation (31) represents the sensitivity of von Mises stress and when substituted along with Eq. (17) into Eq. (16), provides the sensitivities for local stress constraints. Equation (31) requires computing the inverse of the global stiffness matrix **K**, which may be avoided as follows. Let

$$\mathbf{K}^{-1}\left[\frac{d\mathbf{F}}{dx_j} - \frac{d\mathbf{K}}{dx_j}\mathbf{U}\right] = \mathbf{Q} \quad \text{or} \quad \mathbf{K}\mathbf{Q} = \left[\frac{d\mathbf{F}}{dx_j} - \frac{d\mathbf{K}}{dx_j}\mathbf{U}\right] \tag{32}$$

Noting that **K** is symmetric, positive-definite and banded, Eq. (32) may be solved efficiently for **Q** using any of the skyline, frontal or Cholesky's factorization methods, whichever is suitable.

3.2.3 An example

A rectangular region of 150 $\mu m \times 50$ μm, which is the symmetric half of the design domain, was used for a micro gripper synthesis. The region is discretized as shown in Fig. 10a with its bottom edge fixed. Input forces of 2 mN and -20 mN were applied for actuation along the horizontal and vertical directions at the top left corner of the domain. The output deformation at the bottom right node was desired to be maximized along the direction at -45° from the horizontal. Output springs of 0.01 $mN/\mu m$ and 0.1 $mN/\mu m$ were used along the vertical and horizontal directions. Due to uneven spring constants, the output port is expected to deform more along the vertically downward direction than towards right. Here, in-plane widths were chosen as design variables with Young's modulus as 2×10^3 N/mm^2 and the allowable stress σ_a as 10 N/mm^2. An upper bound of 4 mm was imposed on the design variables. The problem was solved with the relaxation parameter varying from 0.1 to 10^{-4} and the lower limit varying as ε^2 in each optimization step [22]. Figure 10b shows the optimal topology for $n = 3$ in Eq. (15) with no volume constraint.

4. SYNTHESIS FOR DYNAMIC ATTRIBUTES

The natural frequencies and normal mode shapes of a structure become important in devices that are subject to vibration and in those that operate at high speeds in the dynamic regime. Some devices, as in resonant MEMS, are intentionally operated at their natural frequencies to enhance their performance or even to serve a functional need.

4.1 Synthesis for desired natural frequencies

Optimizing the structural elements in order to have desired natural frequencies, also called *resonance frequencies*, is of interest in some applications at both the macro and micro scales. This "inverse frequency" problem is studied extensively [44], including the topology optimization problem [45-47; see 36 for a review]. In some applications, it may be desired that the natural frequency be sufficiently far away from frequencies of noise sources. This is usually the case with static-mode sensors that use elastic structures. In some others that operate in the dynamic mode, operating at resonance frequency of the structure is actually beneficial. In those cases, tuning the frequency for a particular set of conditions becomes necessary. Resonance-based sensing has been extensively used in MEMS [48, 49; see 50 for an overview].

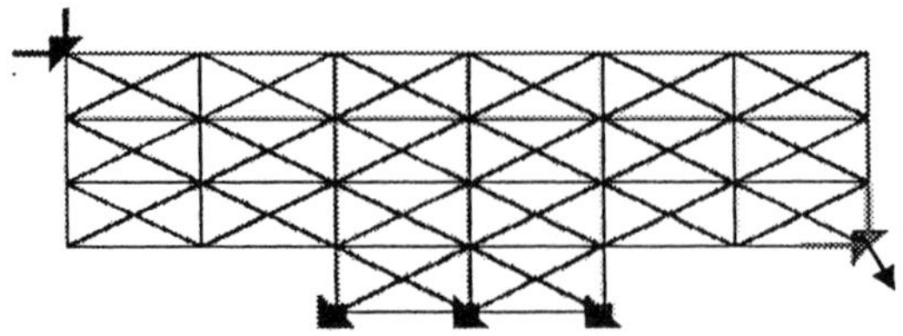
(a)

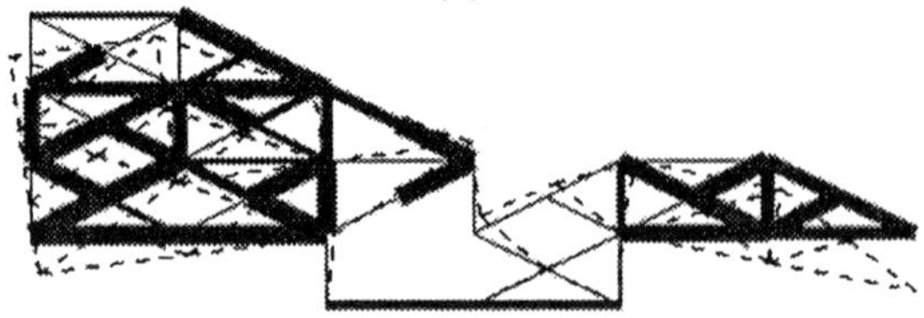
(b)

Figure 10. Synthesis of a micro gripper using the flexibility-strength formulation (a) design specifications (b) synthesized solution (dashed line indicates the deformed configuration).

Optimizing the fundamental or even higher natural frequencies is not much different from the formulations presented above for stiffness, flexibility, etc. The difference now is that the objective function is the resonance frequency and equilibrium equations to be considered are for dynamic conditions. This brings inertia of the system into the problem. The sensitivity analysis leads to some complications because when there are repeated eigenvalues (i.e., two natural frequencies being the same), there occurs a non-differentiable point that is not good for gradient-based optimization methods. There are ways to deal with this situation [51]. Hence, good methods are available if one needs to optimize a structure for a resonance frequency [52].

4.2 Synthesis for desired normal mode shapes

The "inverse mode shape" problem, on the other hand, entails the determination of the geometry of the structure such that it will have prescribed mode shapes. There are applications where a particular mode shape is critical for the performance of the device. This is especially true in the design of resonance-based micro accelerometers, gyroscopes, etc. [50] where displacement-sensing is accomplished through the measurement of capacitance (electrostatic), strain (e.g., piezoresistive), optical (laser-lever), etc. In all these cases, the sensitivity of measurement could be enhanced by adjusting the mode shapes of the structure. For instance, in the micro rate gyroscope [53], the mode shape of a ring structure can be optimized to improve the performance. Designing the shape of the cantilever probes on the atomic force microscope to attain a desired modal deflection is another example [54]. In micro locomotion systems (e.g. swimming) where repeated changes in shape propel the whole entity forward [55], energy efficiency can be achieved by designing the structure such that the normal mode shapes are the same as the required repetitive shape changes. In comparison to the frequency problem, there are far fewer methods for the synthesis of mode shapes.

An extensive line of inquiry went into the system-identification type problem. Consider a bar or a beam of inhomogeneous material construction. That is, the density along the axis of the bar/beam is non-uniform and is unknown. By conducting dynamic experiments, its natural frequencies and mode shapes can be measured. From this data, called the *spectral data*, the variable density can be recovered [56]. This approach may be applicable for some sub-micron cantilever-based sensors for molecular and bio detectors [57].

When the material of a bar or a beam (or a general elastic structure) is known and we need to tune its mode shape, it could be done by suitably synthesizing the geometry of the structure. Since beams are widely used in MEMS, this problem is discussed below in detail by adapting from [58].

4.2.1 Mode shape synthesis for beams

The problem of synthesizing the mode shape of an axially deforming bar was introduced in Section 2.1 in Chapter 1. The analytical solution presented there is applicable only in the simple cases. A numerical solution that is applicable more generally was presented in [58]. In this section, the numerical solution for the case of transversely vibrating beams is described.

It was noted in Section 2.1 of Chapter 1 that any arbitrarily specified mode shape might not be possible for the chosen boundary condition. The same is true of beams as well. The conditions for a valid mode shape depend on the boundary conditions of the beam. The discussion here is limited to the case of a cantilever beam for which the governing differential equation for free vibration is given by

$$\frac{d^2}{dx^2}\left(EI\frac{d^2u}{dx^2}\right)-\lambda\rho Au=0, \qquad \lambda=\omega^2 \tag{33}$$

where $I(x)$ and $A(x)$ are the second moment of inertia and area of cross-section along the longitudinal axis of the beam, respectively; E, L and ρ are the Young's Modulus, length, and density of the beam; $u(x)$ is the mode shape corresponding to a frequency ω. By using the finite difference derivatives, the above eigenvalue problem can be reduced to familiar linear algebra form:

$$\mathbf{Ku} = \lambda \mathbf{Mu} \tag{34}$$

where the inertia matrix **M** is a diagonal matrix with diagonal entries $m_i = \rho A_i l$, and the stiffness matrix $\mathbf{K} = \mathbf{E}\mathbf{L}^{-1}\mathbf{E}\overline{\mathbf{K}}\mathbf{E}^{\mathrm{T}}\mathbf{L}^{-1}\mathbf{E}^{\mathrm{T}}$ is pentadiagonal as shown below [59].

$$\mathbf{K} = \begin{bmatrix} a_1 & -b_1 & c_1 & 0 & 0 & \cdots & 0 & 0 \\ -b_1 & a_2 & -b_2 & c_2 & 0 & \cdots & 0 & 0 \\ c_1 & -b_2 & a_3 & -b_3 & c_3 & \cdots & 0 & 0 \\ 0 & c_2 & -b_3 & a_4 & -b_4 & \cdots & 0 & 0 \\ \vdots & \vdots & \vdots & \vdots & \vdots & \ddots & \vdots & \vdots \\ 0 & 0 & 0 & 0 & 0 & \cdots & a_{N-1} & -b_{N-1} \\ 0 & 0 & 0 & 0 & 0 & \cdots & -b_{N-1} & a_N \end{bmatrix} \tag{35}$$

with coefficients:

$$\begin{aligned} a_i &= \frac{k_i + 4k_{i+1} + k_{i+2}}{l^2} \\ b_i &= \frac{k_{i+1} + k_{i+2}}{l^2} \\ c_i &= \frac{k_{i+2}}{l^2} \end{aligned} \tag{36}$$

and $\overline{\mathbf{K}} = diag(k_i)$ with $k_i = Eh^2 A_i / l^3$, $\mathbf{L} =$ identity matrix multiplied by l, and

$$\mathbf{E} = \begin{bmatrix} 1 & -1 & 0 & \cdots & 0 \\ 0 & 1 & -1 & \cdots & 0 \\ \cdot & \cdot & \cdot & \cdot & \cdot \\ 0 & \cdots & \cdots & 1 & -1 \\ 0 & \cdots & \cdots & 0 & 1 \end{bmatrix} \tag{37}$$

The above discretization assumed finite difference grid spacing of length l where A_i is the area of cross-section across i^{th} grid spacing. A rectangular cross-section was assumed with thickness h while the width is assumed to vary along the axis. The transverse displacement in the mode shape is given by $\mathbf{u} = \{u_1, u_2, \cdots, u_N\}$. The inertia and stiffness matrices together with the mode shape vector $\mathbf{u}$ can be re-written so that the modal displacements u_i appear in the matrices and A_i as the unknown vector.

$$\mathbf{CA} = \begin{bmatrix} C_{11} & C_{12} & C_{13} & 0 & 0 & \cdots & 0 \\ 0 & C_{22} & C_{23} & C_{24} & 0 & \cdots & 0 \\ \vdots & \vdots & \ddots & \vdots & \vdots & \cdots & 0 \\ 0 & 0 & \cdots & 0 & C_{N-2,N-2} & C_{N-2,N-1} & C_{N-2,N} \\ 0 & 0 & \cdots & 0 & 0 & C_{N-1,N-1} & C_{N-1,N} \\ 0 & 0 & \cdots & 0 & 0 & 0 & C_{N,N} \end{bmatrix}_{N\times N} \begin{bmatrix} A_1 \\ A_2 \\ A_3 \\ \vdots \\ A_N \end{bmatrix}_{N\times 1} = \mathbf{0}$$

where

$$\begin{aligned} C_{1,1} &= \tfrac{Eh^2}{12l^3} u_1 - \rho\lambda l u_1 \\ C_{i,i+1=} &- \tfrac{Eh^2}{6l^3}(u_{i-1} - 2u_i + u_{i+1}) \\ C_{i,i+2=} &\tfrac{Eh^2}{12l^3}(u_i - 2u_{i+1} + u_{i+2}) \\ C_{i,i} &= \tfrac{Eh^2}{12l^3}(u_{i-2} - 2u_{i-1} - u_i) - \rho\lambda l u_i \\ C_{N,N} &= \tfrac{Eh^2}{12l^3}(u_{N-2} - 2u_{N-1} - u_N) - \rho\lambda l u_N \end{aligned} \tag{38}$$

The eigenvalue λ can be determined by solving the last row of the above equation.

$$\lambda = \frac{Eh^2}{12\rho l^4} \frac{(u_{N-2} - 2u_{N-1} + u_N)}{u_N} \tag{39}$$

Thus, as in the case of the bars, the natural frequency cannot be arbitrarily chosen when a mode shape is specified. The remaining rows of equations can be solved sequentially starting from the second row from the bottom and moving up to get:

$$A_{N-1} = -\frac{C_{N-1,N} A_N}{C_{N-1,N-1}}$$
$$A_i = -\frac{\left(C_{i,i+1} A_{i+1} + C_{i,i+2} A_{i+2}\right)}{C_{i,i}} \tag{40}$$

The solution to the area profile is not yet unique because all areas can be multiplied by a scale factor without changing the mode shape. In order to determine this scale factor, the total weight of the beam W_{total} could be used. This is an instant where the material resource constraint sets the size of the synthesized structure.

An example: Consider the following sixth order polynomial with coefficients a_i be the desired mode shape of a cantilever beam.

$$u(x) = a_o + a_1 x + a_2 x^2 + a_3 x^3 + a_4 x^4 + a_5 x^5 + a_6 x^6 \tag{41}$$

After solving for some of the unknown coefficients by applying essential and natural boundary conditions $u_o = u_o' = u_L'' = u_L''' = 0$, the polynomial can be made to depend only on three coefficients: a_4, a_5 and a_6.

$$\begin{aligned} u(x) = &\left(6a_4 L^2 + 20a_5 L^3 + 45a_6 L^4\right)x^2 \\ &- 2\left(2a_4 L + 5a_5 L^2 + 10a_6 L^3\right)x^3 \\ &+ a_4 x^4 + a_5 x^5 + a_6 x^6 \end{aligned} \tag{42}$$

To ensure that λ is positive, the following criterion must be satisfied [58] so that

$$u_L \cdot u_L^{IV} = 24L^2\left(3a_4 + 11a_5 + 26a_6 L^2\right)\left(a_4 + 5a_5 + 15a_6 L^2\right) > 0 \tag{26}$$

where u_L^{IV} is the fourth derivative of the transverse deflection evaluated at the free end of the cantilever beam. To explore the feasible choices of a_4, a_5 and a_6 for satisfying the above inequality, we let $v_1 = \frac{3a_4 + 11a_5 L}{26L^2}$ and $v_2 = \frac{a_4 + 5a_5 L}{15L^2}$ such that the inequality takes the form

$$u_L\, u_L^{IV} = 24(26)(15)L^2\left(a_6 + v_1\right)\left(a_6 + v_2\right) > 0 \tag{27}$$

Using the above condition on a_6 and by varying the values of a_4 and a_5 a family of valid cantilever mode shapes are obtained and plotted in Fig. 11. The figure shows both the fundamental (Fig. 11a) and the second (Fig. 11b) mode shapes. To further examine how the choice of a_4, a_5 and a_6 affects the mode shape and the area profile, a few fundamental mode shapes from the family of mode shapes are selected and are shown in Fig. 12. The corresponding area profiles obtained using the method described earlier are also plotted.

As mentioned earlier, changing the mode shapes of beams could enhance the sensitivity of displacement sensing. If a laser-lever as in an AFM is used, the slope at the point where the laser beam falls needs to be as large as possible to amplify the shift in the detected spot of the reflected laser [54]. If a piezoresistive strain guage is used [60], the curvature at the fixed end of the cantilever needs to be maximized to increase the strain there for a given amplitude of the resonant motion of the tip. The important thing to note is that achieving almost any shape in plane is rather easy in photolithography based microfabrication.

In addition to natural frequency and mode shapes, it is also possible to optimize for desired dynamic response as a whole for a given time-varying load [61].

5. CONCLUSIONS

Moving mechanical elements are crucial to MEMS. At the MEMS size scale, obtaining motion via elastic deformation, rather than multi-rigid-body mechanisms with joints, has many advantages including ease of fabrication and the lack of friction and wear. An additional advantage, as shown in this chapter is, the ability to systematically synthesize them even at the conceptual stage with complete quantitative details. This is not commonly possible for macro scale mechanical systems. Many attributes of elastic structures or compliant mechanisms, including stiffness, flexibility, strength, motion, natural frequencies, mode shapes, and transient response, are within the reach of synthesis methods. Some of these were described in detail in this chapter while some others are noted by citing the relevant literature. The synthesis methods such as the ones presented here and many others that have been developed and are being developed have the potential to enhance the performance of a number of MEMS devices while reducing the burden on the designers.

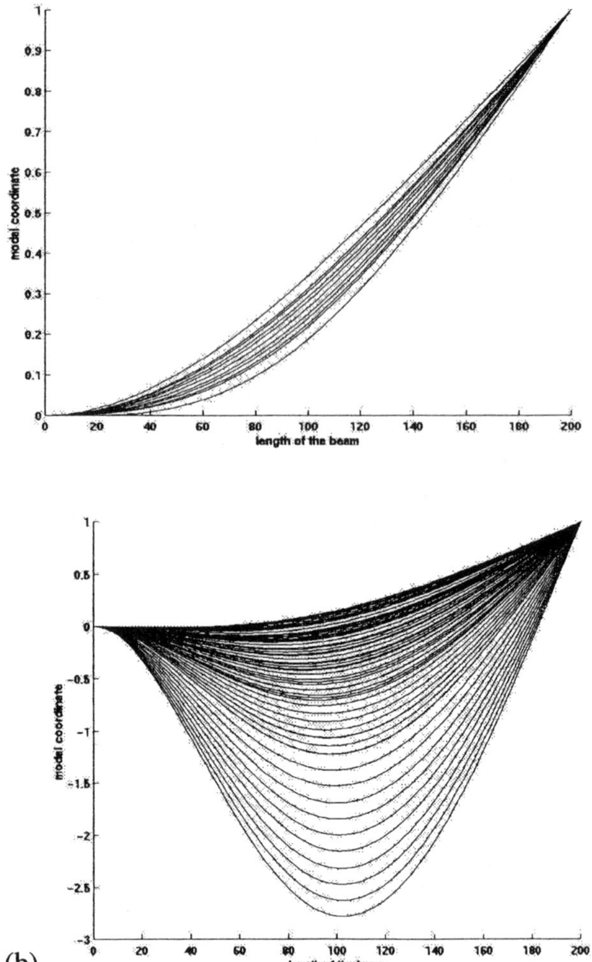

(b)

Figure 11. A family of valid primary mode shapes for a range of a_4, a_5 and a_6 for a given beam length of $L = 200$ (a) first mode (b) second mode

ACKNOWLEDGEMENTS

The grant support from National Science Foundation (DMI-9733916) and Defense Advanced Research Projects Agency (Composite CAD program, N66001-97-1-8911) is gratefully acknowledged by GKA. Dr. Luzhong Yin (University of Pennsylvania) wrote the Fortran program used to generate the solution shown in Fig. 5d. Elizabeth Lai, doctoral student at University of Pennsylvania, contributed to the mode shape synthesis work.

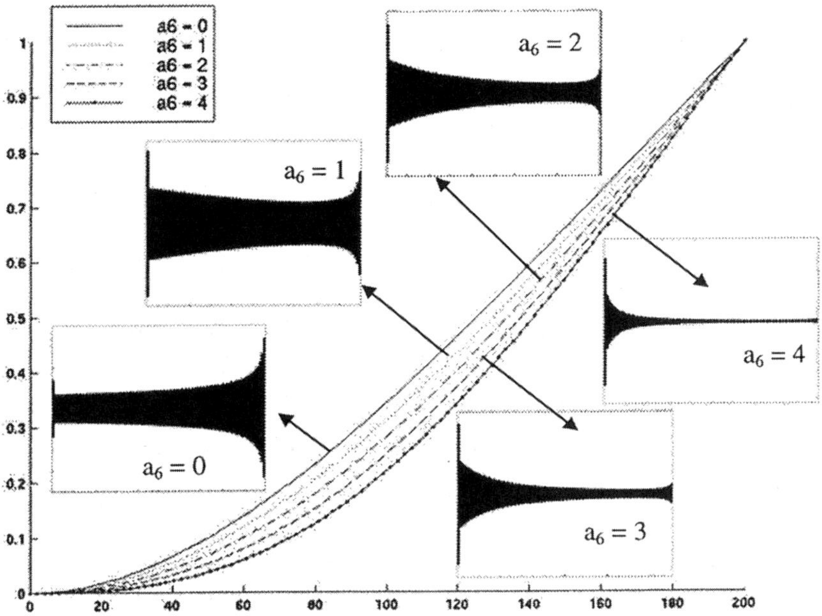

Figure 12. A family of first mode shapes for a range of a_4, a_5 and a_6 for a cantilever beam of length *L= 200* along with their synthesized area profiles

REFERENCES

1. Elwenspoek, M. and Jansen, H.V., *Silicon Micromachining*, Cambridge University Press, 1998, Cambridge, UK.
2. Petersen, K.E., "Silicon as a Mechanical Material," *Proceedings of the IEEE*, Vol. 70, No. 5, 1982, pp. 420-457.
3. Fan, L.-S., Tai, Y.C., and Muller, R.S., "Movable Micromechanical Structures for Sensors and Actuators," *IEEE Transactions on Electron Devices*, Vol. 35, 1988, pp. 724.
4. Behi, F., Mehregany, M., and Gabriel, K.G.," Microfabricated Three-Degree-of-Freedom Parallel Mechanism," *Proceedings of the IEEE workshop on MEMS*, 1990, p. 159.
5. Midha, A., "Elastic Mechanisms" in *Modern Kinematics in the Last 40 Years* (A. G. Erdman, ed.), 1992, John Wiley and Sons, Inc., New York.
6. Ananthasuresh, G.K., Kota, S. and Gianchandani, Y., "Systematic Synthesis of Microcompliant Mechanisms- Preliminary Results," Proceedings of the *National Applied Mechanisms and Robotics Conference*, Paper 82, Cincinnati, OH, Nov. 8-10, 1993, pp. 82.1-82.6
7. Ananthasuresh, G.K.; Kota, S., "Designing Compliant Mechanisms", *Mechanical Engineering* 1995, Vol. 117, No. 11, November, 1995, pp. 93-96.
8. Howell, L. L., *Compliant Mechanisms*, John Wiley and Sons, Inc., New York, 2001.

9. Roark, R.J. and Young, W.C., *Roark's Formulas for Stress and Strain*, 7th edition, McGraw-Hill, New York.
10. Cho, Y.-H. and A. P. Pisano, "Optimum Structural Design of Micromechanical Crab-Leg Flexures with Microfabrication Constraints," *Proc. ASME, DSC-v. 19, Microstructures*, Sensors and Actuators, pp. 31-50, ASME Winter Annual Meeting, Nov. 1990.
11. Tang, W.C., Nguyen, T.-C. H., Judy, M. W., and Howe, R. T. "Electrostatic Comb Drive of Lateral Polysilicon Resonators," *Sensors and Actuators A*, 21 (1990) 328-31.
12. Fedder, G.K., *Simulation of Microelectromechanical Systems, Ph.D. Thesis*, University of California at Berkeley, September 1994.
13. Saggere, L., Kota, S., and Crary, S. D., "A New Design for Suspension of Linear Microactuators," *Proc. of the 1994 International Mechanical Engineering Congress and Exposition: Dynamic Systems and Control*, Chicago.
14. Erdman, A.G. and Sandor, G.N., Kota, S., *Mechanism Design: Analysis and Synthesis*, Vol. 1, Fourth Edition, Prentice Hall, Englewood Cliffs, New Jersey.
15. Xu, D. and Ananthasuresh, G.K., "Skeletal Shape Optimization of Compliant Mechanisms", *ASME Journal of Mechanical Design*. To appear in 2003 (Vol. 125, No. 2).
16. Hetrick, J. A. and Kota, S., 1999, "An Energy Formulation for Parametric Size and Shape Optimization of Compliant Mechanisms", *ASME Journal of Mechanical Design*, Vol.121 pp 229-234.
17. Kota S., Hetrick J., Li Z., Rodgers S., Krygowski T., "Synthesizing High-Performance Compliant Stroke Amplification Systems for MEMS," *Proc. of the Thirteenth Annual International Confernece on Micro Electro Mechanical Systems*, Miyazaki, Japan, Jan. 2000.
18. Bendsøe, M.P. and Sigmund, O., "Material Interpolation Schemes in Topology Optimization," *Archives in Applied Mechanics*, Vol. 69, (9-10), 1999, pp. 635-654.
19. Yin, L. and Ananthasuresh, G.K., "Topology Optimization of Compliant Mechanisms with Multiple Materials Using a Peak Function Material Interpolation Scheme," *Structural and Multidisciplinary Optimization*, Vol. 23, No. 1, 2001, pp. 49-62.
20. Cheng, G.D., Guo, X., "ε-Relaxation Approach in Structural Optimization," *Structural Optimization*, Vol. 13, 1997, pp. 258-266.
21. Duysinx, P., and Bendsøe, M.P., "Topology optimization of continuum structures with local stress constraints," *International Journal for Numerical Methods in Engineering*, Vol. 43, 1998, pp. 1453-1478.
22. Saxena, A.; Ananthasuresh, G. K., "Topology Design of Compliant Mechanisms with Strength Considerations," *Mechanics of Structures and Mechanisms*, 2001, 29(2), pp. 199-221.
23. Ananthasuresh, G. K., Kota, S., and Kikuchi, N., "Strategies for Systematic Synthesis of Compliant MEMS," Proc. *1994 ASME Winter Annual Meeting, Symposium on MEMS*, Dynamics Systems and Control, DSC-Vol. 55-2, Chicago, IL, Nov. 1994, pp. 677-686.
24. Ananthasuresh, G. K., "Investigations on the Synthesis of Compliant Mechanisms and a New Design Paradigm for Micro-Electro-Mechanical Systems," *PhD Dissertation*, 1994, The University of Michigan, Ann Arbor, MI.
25. Sigmund, O., "On the Design of Compliant Mechanisms Using Topology Optimization," *Mechanics of Structures and Machines*, 1997, 25 (4), pp. 495-526.

26. Frecker, M.I., Ananthasuresh, G.K.,Nishiwaki, N., Kikuchi, N., Kota, S., "Topological synthesis of compliant mechanisms using multicriteria optimization," *ASME J. Mech. Design*, Vol. 119, 1997, pp.238-245.
27. Nishiwaki, S.; Frecker, M.I.; Min, S.; Kikuchi, N., "Topology Optimization of Compliant Mechanisms Using the Homogenization Method," *International Journal of Numerical Methods in Engineering*, 1998, 42, pp. 535-559.
28. Frecker, M., Kikuchi, N., and Kota, S., "Topology Optimization of Compliant Mechanisms With Multiple Outputs," *Structural and Multidisciplinary Optimization* Vol 17, 1998, pp 269-278.
29. Saggere, L. and Kota, S., "Static Shape Control of Smart Structures Using Compliant Mechanisms," *AIAA Journal*, 37(5), 1999, pp.572-578.
30. Saxena, A., Ananthasuresh, G.K., "On an optimal property of compliant topologies," *Structural and Multidisciplinary Optimization*, Vol. 19, 2000, pp. 36-49.
31. Bruns, T.E. and Tortorelli, D.A., "Topology Optimization of Nonlinear Elastic Structures and Compliant Mechanisms," *Computer Methods in Applied Mechanics and Engineering*, Vol. 190, 2001, pp. 3443-3459.
32. Saxena, A. and Ananthasuresh, G.K., "Topology Synthesis of Compliant Mechanisms for Nonlinear Force-Deflection and Curved Path Specifications," *Journal of Mechanical Design*, 2001, 123, pp. 33-42.
33. Pedersen, C.B.W.; Buhl, T.; Sigmund, O., "Topology Synthesis of Large-Displacement Compliant Mechanisms," *International Journal for Numerical Methods in Engineering*, 2001, 50, pp. 2683-2705.
34. Lau, G.K., Du, H., and Lim, M.K., "Use of Functional Specifications as Objective Functions in Topology Optimization of Compliant Mechanisms," *Computer Methods and in Applied Mechanics and Engineering*, Vol. 190, 2001, pp. 4421-4433.
35. Joo, J.-Y., Kota, S., and Kikuchi, N., "Optimal Topology Design of Compliant Mechanisms Using Frame Elements," *Mechanics and Structures of Machines*, 2002.
36. Bendsoe, M.P. and Sigmund, O., *Topology Optimization: Theory, Methods and Applications*, 2003, Springer, New York.
37. Shield, R.T., Prager, W., "Optimal structural design for given deflection," *Journal of Applied Mathematics and Physics. ZAMP*, Vol. 21, 1970, pp. 513-523.
38. Przemieniecki, J.S., *Theory of Matrix Structural Analysis*, Dover Publications, New York, 1968.
39. Mac Donald, G. A., "Review of Low Cost Accelerometers for Vehicle Dynamics," *Sensors and Actoators*, A 21- A 23, 1990.
40. Roylance, L. M. and Angell, J. B., "A Batch-Fabricated Silicon Accelerometer," *IEEE Trans. Electron Devices*, Vol ED-26, No. 12, 1979, pp. 1911-1917.
41. Samuels, H., "Single- and Dual-Axis Micromachined Acceletometers," *Analog Dialog*, Vol 30, No 4(1), 1993.
42. Lemaire, C., "Accelerometers in Earthquake Detection Systems," *Accelerometer News*, Vol 6, 1998.
43. Norton, R.L., *Machine Design—An Integrated Approach*, 1996, Prentice-Hall, Upper Saddle River, New Jersey, U.S.A.
44. Chu, M., "Inverse Eigenvalue Problems," *SIAM Review,* 40(1), 1992, pp. 1-39.
45. Diaz, A.R. and Kikuchi, N., "Solution to Shape and Topology Eigenvaue Optimization Problems Using a Homogenization Method," *International Journal for Numerical Methods in Engineering*, Vol. 35, 1992, pp. 1487-1502.

46. Ma, Z.-D., Kikuchi, N., and Cheng, H.-C., "Topological Design for Vibrating Structures," *Computer Methods in Applied Mechanics and Engineering*, Vol. 121 (1-4), 1995, pp. 259-280.
47. Tcherniak, D., "Topology Optimization of Resonating Structures Using SIMP Method," *International Journal of Numerical Methods in Engineering*, Vol. 54, No. 11, 2002, pp. 1605-1622.
48. Brand, O. and Baltes, H., "Micromachined Resonant Sensors—An Overview," *Sensors Update*, Vol. 4, H. Baltes, W. Gopel, and J. Hesse, Eds., 1996, VCH, New York, pp. 1-51.
49. Albrecht, T., Grutter, P., Horne, D., and Rugar, D., "Frequency Modulation Detection Using High-Q Cantilevers for Enhanced Force Microscope Sensitivity," *Journal of Applied Physics*, Vol. 69, No. 2, 1991, pp. 668-673.
50. Seshia, A.A., "Integrated Micromechanical resonant Sensors for Inertial Measurement Systems," *Ph.D. Thesis*, 2002, University of California, Berkeley, U.S.A.
51. Seyranian, A.P., "Sensitivity Analysis of Multiple Eigenvalues," *Mechanics of Structures and Machines*, Vol. 21, 1993, pp. 261-284.
52. Gladwell, G.M.L., *Inverse Problem in Vibration*, 1986, Dordrecht: Martinus Nijhoff.
53. Putty, M.W. and Najafi, K., "A Micromachined Vibrating Ring Gyroscope," 1994 *Technical Digest, Sensors and Actuators Workshop, Hilton Head Islands, SC,* June 13-16, 213-220.
54. Pedersen,N., "Design of Cantilever Probes for Atomic Force Microscopy (AFM)," *Engineering Optimization,* Vol. 32, No. 3, 2000, 373-392.
55. McIsaac, K.A. and Ostrowski, J.P., "Motion planning for dynamic eel-like robots," *Proceedings of the IEEE International Conference on Robotics and Automation*, Vol. 2, 2000, pp. 1695 –1700.
56. Ram, Y.M. and Gladwell, G.M.L., "Constructing a Finite Element Model of a Vibratory Rod from Eigendata," *Journal of Sound and Vibration,* Vol. 169, 1994, pp. 229-237.
57. G. Wu, H. Ji, K. Hansen, T. Thundat, R. Datar, R. Cote, M. Hagan, A. K. Chakraborty, A. Majumdar, "Origin of Nanomechanical Cantilever Motion Generated from Biomolecular Interactions," *Proc. National Acad. Science,* Vol. 98, pp. 1560-1564 (2001).
58. Lai, E. and Ananthasuresh, G.K., "On the Design of Bars and Beams for Desired Mode Shapes," *Journal of Sound and Vibration*, Vol. 254, No. 2, 2002, pp. 393-406.
59. Ram, Y.M., "Inverse Mode Problems for the Discrete Model of a Vibrating Beam," *Journal of Sound and Vibration*, Vol. 169, 1994, pp. 239-252.
60. Harley, J.A., Stowe, T.D., Kenny, T.W., "Piezoresistive Cantilevers with FemtoNewton Force Resolution," *Proceedings of the 10th International Conference on Solid-State Sensors and Actuators*, Sendai, Japan, Vol. 2, pp. 1628-1631.
61. Ou, J.S. and Kikuchi, N., "Integrated Optimal Structural and Vibration Control Design," *Structural Optimization*, Vol. 12, No. 4, 1996, pp. 209-216.

Chapter 3

SYNTHESIS OF ELECTRO-STATICALLY ACTUATED MEMS

Wenjing Ye
Woodruff School of Mechanical Engineering, Georgia Institute of Technology, Atlanta, GA 30332, U.S.A.

Subrata Mukherjee
Department of Theoretical and Applied Mechanics, Cornell University, Ithaca, NY 14853, U.S.A.

Synopsis: This chapter presents a synthesis methodology for electrostatically driven microelectromechanical systems. It addresses issues of simulation, sensitivity analysis and optimization. As an application of the general methodology, design of variable gap comb drives (shape motors) is presented. Direct simulation of the electrostatic field is carried out by the exterior, indirect boundary element method and shape sensitivities are obtained by the direct differentiation approach. The inverse problem determines the shapes of fingers of a comb drive such that the driving force is a desired function of its travel distance. Commercially available optimization codes are used to solve the synthesis problems. Numerical results are presented for shape motors that produce linear, quadratic or cubic force profiles. Based on the design, a cubic motor was fabricated using the SCREAM process. The driving forces were measured and the results were compared with the theoretical predictions.

1. INTRODUCTION

Electrostatically driven actuators are of great importance in Microelectromechanical Systems (MEMS) and have been used extensively in micromachined devices [1]. Scaled favorably in small scale, electrostatically driven actuators have advantages in low power consumption, small size, high efficiency and relatively easy fabrication compared to devices driven by other mechanisms. Applications include, but are not limited to, electrostatically driven accelerometers [2], gyroscopes[3], resonators [4,5], microswitches [6], microrelays [7], etc. Figure 1 shows Scanning Electron Microscope (SEM) pictures of a gyroscope and a laterally oscillating resonator that use electrostatic actuators (comb drives) to drive the beam into resonance.

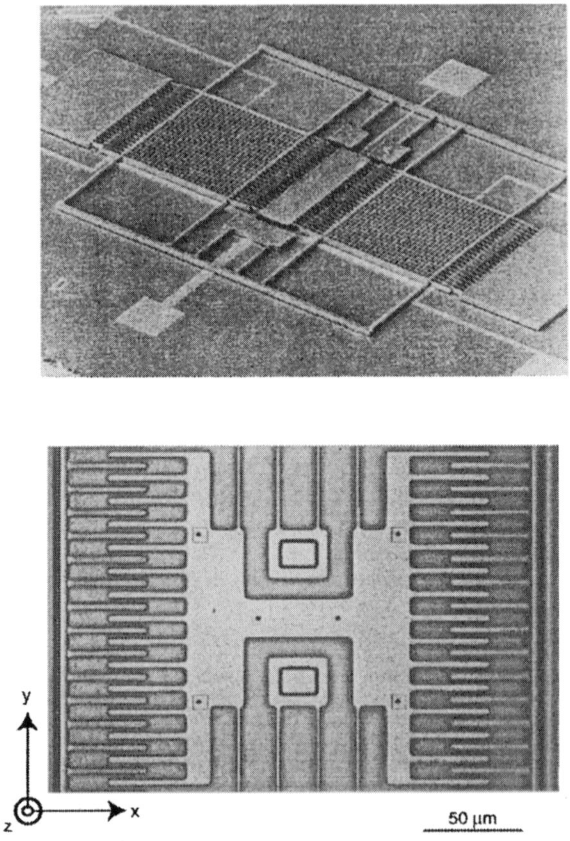

Figure 1. SEM pictures of a gyroscope (top) [ref] and a resonator (bottom) [ref].

The fundamental actuation principle behind electrostatic actuators is the attraction of two oppositely charged plates. A typical electrostatically driven actuator usually consists of two electrodes, one is fixed (fixed to the substrate) and the other is movable (released from the substrate). Figure 2 shows a schematic of a generic electrostatic actuator. When a bias voltage is applied between the two electrodes, the difference in electrostatic potentials induces opposite charges on each electrode, and therefore causes the moving electrode to move towards to the fixed electrode, resulting in mechanical motion. Figure 3 shows two electrostatic actuators commonly used in MEMS, namely a parallel-plate capacitor actuator and a comb-drive actuator. In a parallel-plate capacitor actuator, the motion is perpendicular to the plates, while in the comb drive the motion is parallel to the plates. It should be noted that both devices could also be used as sensors.

One of the most important quantities in an electrostatically driven actuator is the driving force. The magnitude of this driving force depends on the shapes and sizes of electrodes, the bias voltage, etc. In simple cases, such as parallel-plate capacitors and comb-drive actuators with uniform gap and height profiles (standard comb drives), it is possible to derive the analytical expressions for the driving force. In a parallel-plate capacitor, by neglecting the fringe field, the co-energy stored in the capacitor can be approximated by Eq. (1) [8]:

$$U = \frac{1}{2}\varepsilon\frac{V^2 A}{x} \tag{1}$$

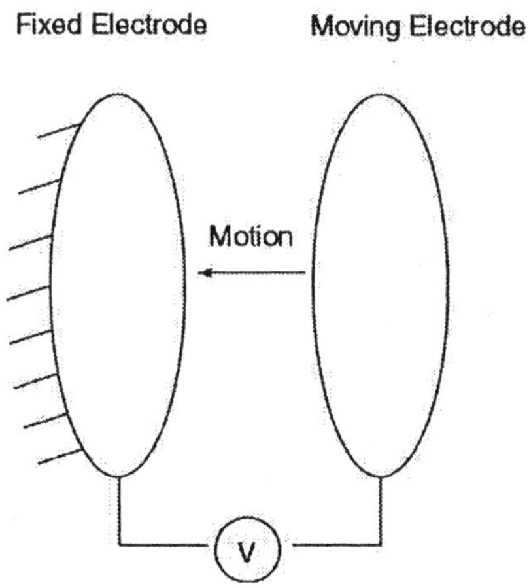

Figure 2. A schematic of a generic electrostatically driven actuator.

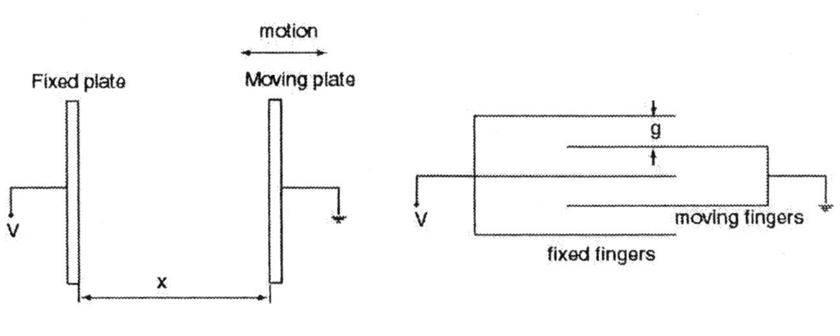

Figure 3. Schematic of a parallel-plate capacitor (left) and a comb drive (right) actuators.

Here, x is the gap between the plates, ε is the permittivity of the medium between the plates, V is the bias voltage applied between the plates and A is the overlap area of the two plates. The driving force acting on the moving plate can then be obtained by taking the derivative of Eq. (1) with respect to x as shown below.

$$|F| = \frac{1}{2}\varepsilon\frac{V^2A}{x^2} \tag{2}$$

As indicated in Eq. (2), the driving force in a parallel-plate actuator is inversely proportional to the square of the gap between the plates. Therefore, it strongly depends on the relative positions of the two plates. In contrast to parallel-plate capacitor actuators, the driving force from a comb drive is independent of the position of the moving fingers, except at the end of the travel range when two sets of fingers come into contact nullifying their positive and negative charges. In a standard comb drive with uniform gap (g) and height (h) (see Fig. 3), again when neglecting the fringe field, the stored energy in one set of fingers is

$$U = V^2\frac{\varepsilon xh}{g} \tag{3}$$

Here x is the overlap length of the two plates. The derivative of U with respect to x yields a constant driving force as shown below.

$$F = V^2\frac{\varepsilon h}{g} \tag{4}$$

Both examples clearly indicate the strong dependence of the driving force on the shapes and dimensions of electrodes. It is possible, by changing the shapes and sizes of electrodes, to obtain desired force profiles. For instance, it is of interest in some applications to have force profiles such as linear, quadratic or cubic. One example is that, in many actuator applications, large displacement motion at a low voltage is highly desirable. However, the actuator's suspension springs exhibit nonlinear response for large displacements [9]. The spring's restoring force can be modeled as $R = k_1x + k_2x^3 + \cdots$, where x is the displacement and k_i are the spring constants. A large driving force is required in order to overcome the nonlinear restoring forces. Hence, a prohibitively large voltage must be applied on conventional comb actuators in order to achieve a large range of motion. It is therefore desirable to have variable shape comb drives such that the corresponding driving force profiles have similar nonlinear terms in x as does the restoring force, for a given applied voltage.

Another example of use of a variable shape comb drive is related to tuning of MEMS. A comb drive with linear, quadratic or cubic force profile can be used for electrostatic tuning. In many MEMS applications, micromechanical resonators play an important role. In such devices, independent tuning of linear or nonlinear stiffness coefficients is an important issue [10]. Figure 4 shows a schematic picture of a tunable resonant structure using variable comb drives (C). The polynomial force (linear or cubic) produced by the variable comb drives can be used to counteract the linear or cubic restoring forces of the beams (B).

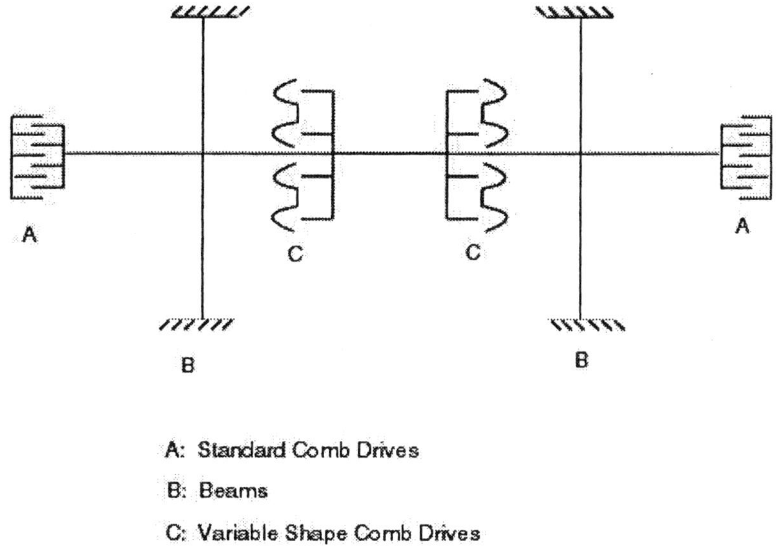

Figure 4. A schematic of a tunable resonator structure

A naïve approach to design actuators with desired force profiles is to use *ad hoc* (trial and error) methods. Such approaches are extremely time consuming. In addition, the optimal design cannot be found in most cases. A synthesis methodology is therefore highly desired to design electrostatic devices with a certain force profile. In this chapter, we describe a systematic synthesis methodology that allows us to find the optimal shapes of actuators for given desired force profiles.

This chapter is organized as follows. In the next section we lay out the governing equations for the electrostatic field. In Section 3 we discuss schemes needed for solving the synthesis problem. First, we present the proposed inverse problem formulation, then discuss the general solving scheme. Next, we describe in detail issues such as driving force simulation, sensitivity analysis, and optimization techniques. Finally in Section 4, we discuss a case study – synthesis of variable shape comb drives. We present both 2-D and 3-D designs. Preliminary fabrication results based on a 2-D design are also presented in this section.

2. GOVERNING EQUATIONS

In an ideal electrostatically driven actuator, electrodes can be modeled as conductors embedded in a uniform dielectric medium as shown in Fig. 5. Each conductor has a constant electrostatic potential. The difference in the potentials of different conductors induces charges on each conductor. The charge is distributed on the surface of each conductor and satisfies [8]

$$q_i(\mathbf{x}) = \varepsilon \frac{\partial \phi_i(\mathbf{x})}{\partial \mathbf{n}} \tag{5}$$

In Eq. (5), $q_i(\mathbf{x})$ is the surface charge density at point $\mathbf{x}$ located on the surface of ith conductor, ε is the dielectric constant of the medium, ϕ_i is the electrostatic potential of conductor i, and $\mathbf{n}$ is the inward normal to a conductor at point $\mathbf{x}$.

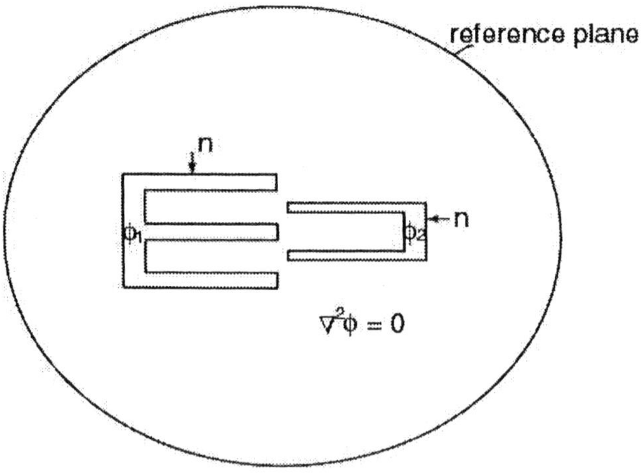

Figure 5. A system of two ideal conductors embedded in a uniform lossless dielectric medium.

The electrostatic potential ϕ in the region exterior to conductors satisfies the Laplace equation

$$\nabla^2 \phi = 0 \tag{6}$$

with the boundary conditions $\phi = \phi_i$ on conductor i, $i = 1,2,\cdots m$, where m is the number of conductors.

The electrostatic force **f** acting on the surface of a conductor is related to the charge density through Eq. (7) as below.

$$\mathbf{f} = -\frac{1}{2}\frac{q^2}{\varepsilon}\mathbf{n} \tag{7}$$

By integrating **f** along the surface of a conductor, the driving force **F** acting on this conductor can then be obtained.

$$\mathbf{F} = \int_{\Gamma} \mathbf{f}\, ds \tag{8}$$

where Γ is the surface of the conductor.

3. SHAPE SYNTHESIS OF ELECTROSTATICALLY DRIVEN ACTUATORS

The design of an electrostatically driven actuator with a desired force profile can be posed as an optimal synthesis problem. The goal of the optimization procedure is to minimize an objective function without violating the specified constraints. The optimization problem is set up as

$$\min \Psi(\mathrm{c_i}) \quad \text{subject to the constraints } a_j(c_i) \le 0$$

where the objective function Ψ is chosen as the integral of the square of the difference between the actual and desired force profiles over the range of operation of the actuator, i.e.,

$$\Psi(c_i) = \int_{l_1}^{l_2} [F(c_i,\delta) - F_e(\delta)]^2 d\delta \tag{9}$$

Here, c_i are the shape parameters of the actuator, F is the driving force, δ is the position of a moving electrode, $F_e(\delta)$ is the desired force profile that can be linear, quadratic, cubic, etc., l_1, l_2 are the initial and final positions of a moving electrode, and a_j are the constraints imposed by practical design issues, such as the minimum gap between electrodes, etc.

A general solution scheme for the proposed inverse problem is outlined in Fig. 6. Simulation and sensitivity analysis are carried out for an initial design. This information is supplied to an optimizer, which produces a better design–one that reduces the value of the objective function without violating the constraints of the problem. Iterative improvements in design continue until a preset stopping criterion is satisfied. This is the final design. There are three major steps in this scheme: simulation of the driving force, sensitivity analysis and optimization. Detailed discussion of each step is provided in the following subsections.

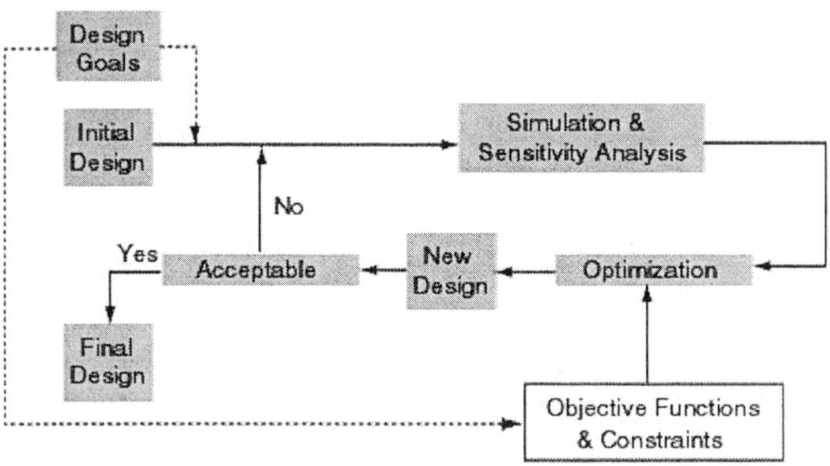

Figure 6. Flow chart for optimal design.

3.1 Simulation of the driving force

As indicated in Eqs. (7) and (8), the driving force exerted on the moving electrode due to the electrostatic field is related to the charge density distributed on this electrode. Thus, to calculate the driving force, we need to compute the charge density first. This can be done by solving the exterior Laplace equation, i.e., Eq. (6). Typically, analytical solutions to the Laplace equation with the prescribed boundary conditions are not possible except for a few simple cases. For structures with arbitrary shapes, a numerical method is often required. There are several methods that can be employed to solve this problem, for example, the Finite Element Method (FEM), the Finite Difference Method (FDM), the Boundary Element Method (BEM), etc. Among them, the BEM is the most efficient method for handling the exterior

Laplace problem. One main advantage of the BEM compared with domain methods such as FEM, FDM, is that only a surface mesh of the structure is required for this method. The boundary condition at infinity is embedded in the governing equation and therefore there is no need to "discretize infinity". For a detailed introduction to the BEM, readers are referred to, for example, [11]. Here we give a very brief description of the method.

In the BEM, instead of solving the Laplace equation in its differential form, an equivalent integral equation is solved. For a two-dimensional problem, the integral formulation of the Laplace equation is [12]

$$\begin{aligned} \phi_i(\mathbf{x}) &= \sum_{j=1}^{m} \int_{\partial s_j} \frac{\partial \phi(\mathbf{y})}{\partial n} G(\mathbf{x},\mathbf{y}) ds(\mathbf{y}) + \hat{C}, \qquad i = 1,2,\cdots,N \\ Q &= \sum_{j=1}^{m} \int_{\partial s_j} q(\mathbf{y}) ds(\mathbf{y}) = \varepsilon \sum_{j=1}^{m} \int_{\partial s_j} \frac{\partial \phi}{\partial n}(\mathbf{y}) ds(\mathbf{y}) \end{aligned} \tag{10}$$

where $\mathbf{x}$ is the position vector of the evaluation point located on conductor i, $\mathbf{y}$ is the position vector of a field point located on conductor j, G is the Green's function which is equal to $\dfrac{1}{4\pi\|\mathbf{x}-\mathbf{y}\|}$ in 3-D and $\dfrac{1}{2\pi}\ln\|\mathbf{x}-\mathbf{y}\|$ in 2-D, ∂s_j is the surface of conductor j, Q is the total charge of the system and $\hat{C}$ is the unknown constant. The integral formulation of the Laplace equation in the 3-D space is simply

$$\phi_i(\mathbf{x}) = \sum_{j=1}^{m} \int_{\partial s_j} \frac{\partial \phi(\mathbf{y})}{\partial n} G(\mathbf{x},\mathbf{y}) ds(\mathbf{y}), \qquad i = 1,2,\cdots,N \tag{11}$$

These integral equations can be solved either by collocation, Galerkin or qualocation schemes. In a collocation scheme, the surfaces of conductors are discretized into l elements. On each element, the charge density is assumed to be a function of its node values. By insisting that the integral equation is satisfied at each node, a linear system of equations in unknowns (charge density) can be derived as follows.

$$\mathbf{A}\mathbf{q} = \phi \tag{12}$$

where $\mathbf{q}$ and ϕ are vectors that contain the unknown charge density and the known potential at each node respectively, and $\mathbf{A}$ is the matrix that relates $\mathbf{q}$

with ϕ. This linear system can then be solved by either direct methods such as Gaussian elimination or iterative methods such as GMRES [13]. To reduce the cost, iterative methods are highly recommended.

3.2 Sensitivity analysis

Design sensitivity coefficients (DSCs) are the derivatives of physical quantities, for example forces, stress, temperature etc., with respect to design variables such as geometrical parameters that determine the shape of a structure. In optimization problems, they are used as a guide to find the best search direction in nonlinear programming algorithms. These algorithms typically iterate on the design variables along these directions until an optimal design is obtained. Accurate determination of the DSCs typically leads to fast convergence, and thus to more efficient designs.

There are several methods for computing DSCs. Among them, the finite difference method (FDM) is the easiest one. It calculates two functions obtained from two slightly different design variables, and takes the difference of these functions, divided by the difference of the design variables as the DSC. This method is very easy to use but may not be accurate, in that the accuracy of the DSCs of certain physical quantities obtained from the FDM, is one-order lower than that of the physical quantities. An alternative and more accurate method is the direct differential approach (DDA). In this approach, DSCs are calculated by differentiating the integral equations for the functions with respect to the design variables, and solving the resultant equations. One of the advantages of the DDA is high accuracy. The computed DSCs are typically obtained with the same accuracy as the physical quantities. The other advantage is computing efficiency. After discretizing the resultant integral equations, the linear system obtained has the same coefficient matrix **A** as the one obtained from the calculation of the physical quantities. Only the right-hand-side vector must be recalculated.

In design of actuators, the design variables are the shape parameters of the actuators and the physical quantity is the driving force. Let c be one of the parameters that determine the shape of the actuator. According to Eq. (8), the derivative of the driving force with respect to c is

$$\dot{\mathbf{F}} = \int_{\Gamma} \dot{\mathbf{f}} ds + \int_{\Gamma} \mathbf{f} d\dot{s} \tag{13}$$

where $\dot{(*)} = \partial/\partial c$. From Eq. (7), the sensitivity of **f** is

$$\dot{\mathbf{f}} = -\frac{q\dot{q}}{\varepsilon}\mathbf{n} - \frac{1}{2}\frac{q^2}{\varepsilon}\dot{\mathbf{n}} \tag{14}$$

The sensitivity of the charge density q can be obtained by solving the following equations. In 2-D:

$$\begin{aligned}
&\sum_{j=1}^{m} \int_{\partial s_j} \dot{q}(\mathbf{y})G(\mathbf{x},\mathbf{y})ds(\mathbf{y}) + \dot{C} = \\
&-\sum_{j=1}^{m} \int_{\partial s_j} q(\mathbf{y})\dot{G}(\mathbf{x},\mathbf{y})ds(\mathbf{y}) - \sum_{j=1}^{m} \int_{\partial s_j} q(\mathbf{y})G(\mathbf{x},\mathbf{y})d\dot{s}(\mathbf{y}). \qquad (15)\\
&\sum_{j=1}^{m} \int_{\partial s_j} \dot{q}(\mathbf{y})ds(\mathbf{y}) = -\sum_{j=1}^{m} \int_{\partial s_j} q(\mathbf{y})d\dot{s}(\mathbf{y})
\end{aligned}$$

Here, $C = \varepsilon\hat{C}$ is a constant. And, in 3-D:

$$\begin{aligned}
&\sum_{j=1}^{m} \int_{\partial s_j} \dot{q}(\mathbf{y})G(\mathbf{x},\mathbf{y})ds(\mathbf{y}) = \\
&\sum_{j=1}^{m} \int_{\partial s_j} q(\mathbf{y})\dot{G}(\mathbf{x},\mathbf{y})ds(\mathbf{y}) - \sum_{j=1}^{m} \int_{\partial s_j} q(\mathbf{y})G(\mathbf{x},\mathbf{y})d\dot{s}(\mathbf{y})
\end{aligned} \tag{16}$$

Equations (14) and (15) are obtained by differentiating Eqs. (10) and (11) with respect to c. Finally, the sensitivity of the objective function Ψ with respect to a design variable c is

$$\dot{\Psi} = \int_{l_1}^{l_2} 2\left[F(c_i,\delta) - F_e(\delta)\right]\dot{F}(\delta)d\delta \tag{17}$$

Once the sensitivity $\dot{q}$ is known, the sensitivity of the objective function can be obtained from Eqs. (13), (14) and (17).

3.3 Optimization

The optimization problem, i.e., minimizing the objective function Ψ subject to certain constraints $a_j(c_i) \leq 0$, can be solved by several techniques. An excellent discussion of various optimization methods can be found in [14]. There are also many existing software packages that supply optimization codes for various problems. Among them, code "E04UCF" from the NAG Fortran Library was used to solve the optimization problem in the comb-drive example discussed in the next section. This code is designed for solving nonlinear programming problems – minimization of a smooth nonlinear function subject to a set of constraints on the variables, either linear or nonlinear. It implements a sequential quadratic programming (SQP) method [14]. The basic structure of E04UCF involves major and minor iterations. The major iterations generate a sequence of iterates $\{\mathbf{c}_k\}$ that converge to $\mathbf{c}^*$, a first-order Kuhn-Tucker point. At a typical major iteration, the new iterate $\bar{\mathbf{c}}$ is defined by

$$\bar{\mathbf{c}} = \mathbf{c} + \alpha \mathbf{p} \tag{18}$$

where $\mathbf{c}$ is the current iterate, the non-negative scalar α is the step length and $\mathbf{p}$ is the search direction. Also associated with each major iteration are estimates of the Langrange multipliers and a prediction of the active set (the set in which all the constraints are active, i.e., satisfied exactly).

The search direction $\mathbf{p}$ in Eq. (18) is the solution of a quadratic programming subproblem of the form

$$\begin{aligned} &\text{minimize} && \mathbf{g}^T\mathbf{p} + \frac{1}{2}\mathbf{p}^T\mathbf{H}\mathbf{p} \\ &\text{subject to} && a_j(\mathbf{p}) \le -a_j(\mathbf{c}) \end{aligned} \tag{19}$$

where $\mathbf{g}$ is the gradient of Ψ at $\mathbf{c}$ and the matrix $\mathbf{H}$ is a positive-definite quasi-Newton approximation to the Hessian of the Lagrangian function. The BFGS formula is used to update this matrix $\mathbf{H}$ [15].

This quadratic programming subproblem is solved by a two-phase quadratic programming method. The two phases of the method are: finding an initial feasible point by minimizing the sum of infeasibilities (the feasibility phase), and minimizing the quadratic objective function within the feasible region (the optimality phase). In general, this must be done iteratively. These iterations are the minor iterations in E04UCF.

Once $\mathbf{p}$ has been computed, the major iteration proceeds by determining the Lagrangian merit function, which is the objective function Ψ itself in linear constraint cases. Finally, the approximation to the transformed Hessian matrix is updated using a modified BFGS quasi-Newton update to incorporate new curvature information obtained in the move from $\mathbf{c}$ to $\bar{\mathbf{c}}$.

4. AN EXAMPLE: VARIABLE COMB-DRIVE ACTUATORS

The goal here is to synthesize a comb-drive actuator with desired force profile such as linear, quadratic or cubic. As mentioned in Section 1, such comb drives can be used to tune the linear or cubic stiffness coefficients of the beam in a resonator or be used as a low-power actuator with large travel range.

The design variables in this problem are the parameters that determine the shapes of the fingers in a comb drive. According to Eq. (4), the driving force depends on both the gap between the fingers and the height of the fingers. This observation leads to at least two approaches for solving this problem, namely a 2-D approach [16] and a 3-D approach [17]. In the 2-D approach, the height of the fingers is kept constant. Only the gap between the fingers is changeable. Since many comb drives have large aspect ratio between the height and the gap (for the purpose of obtaining large force), a two-dimensional model is sufficient to model such devices. This approach has the following advantages: (1) It is easy to model and design, and (2) the resultant design is easy to fabricate. However, the gap between the fingers in a variable comb drive is usually larger than that of a standard comb drive in order to achieve a variable force profile. Therefore, such an approach leads to a design that occupies larger area (compared to a standard comb drive). In order to overcome this drawback, a 3-D approach is proposed. In this approach, the gap is kept constant, i.e., the gap is equal to the minimum gap that one can fabricate. The variable force profile is achieved by changing the height profile. Such an approach requires a three-dimensional model and is much more difficult to fabricate using present fabrication technology. However, it occupies less area as compared to the 2-D approach.

In the section that follows, designs based on the 2-D approach are discussed. Next, we present the designs obtained from the 3-D approach. Finally, in the last section, we show a fabricated variable comb drive based on the 2-D design and its test results. A comparison between the test results and simulation results is also given in this section.

4.1 2-D Designs

First, it is assumed that the fixed fingers of a comb drive are of variable width while the moving fingers are uniform. In view of the approximate formula of Eq. (4), it is proposed that the gap profile between a fixed and a moving finger be an inverse polynomial

$$g(\delta) = \frac{1}{c_0 + c_1\delta + c_2\delta^2 + c_3\delta^3} \tag{20}$$

If $g(\delta)$ varies slowly, one would expect to have a driving force $F \approx \varepsilon h V^2 (c_0 + c_1\delta + c_2\delta^2 + c_3\delta^3)$, i.e., a combination of linear, quadratic and cubic functions. The design variables are $c_i, i = 1,2,3,4$.

4.1.1 Driving force

Given a set of c's, the driving force of the comb drive whose shape is determined by this set of c's is computed using the Boundary Element Method as described in Subsection 3.1. A standard comb drive is considered first to validate the developed code. The dimensions of this comb drive are shown in Fig. 7. The boundaries of the comb drive are discretized using quadratic elements. A total of 152 elements are used here. It is known that the solution of an exterior Laplace equation becomes unbounded at a corner. In this work, the corners are "rounded off" by Hermetian curves in order to avoid these singularities.

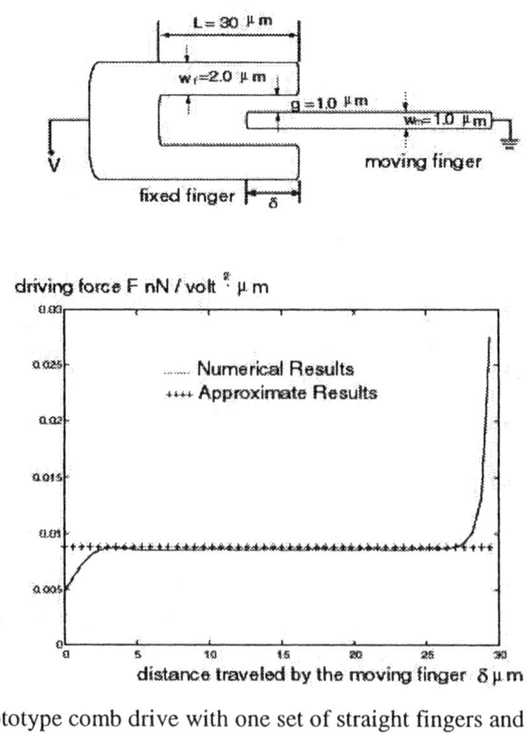

Figure 7. A prototype comb drive with one set of straight fingers and its force profile.

Numerical results for the driving force as a function of the distance traveled by the moving finger are shown in Fig. 7. Also, the results from the approximate formula in Eq. (4) are shown there for comparison. The results show that the driving force remains constant if the gap g and the height h remain constant. The difference between the two solutions is only 1.1%, which indicates that the capacitance model is a very good approximate formula for a standard comb drive with a uniform gap profile.

A comb drive with its fixed fingers having a linear width profile shown in Fig. 8 is considered next. The driving force of this "linear profile" comb drive, shown also in Fig. 8, is a sixth-degree polynomial function of the travel distance. Here, o's denote the sixth-degree polynomial approximation to the actual numerical simulation. This further supports the idea of by changing the gap profile to achieve variable force profile.

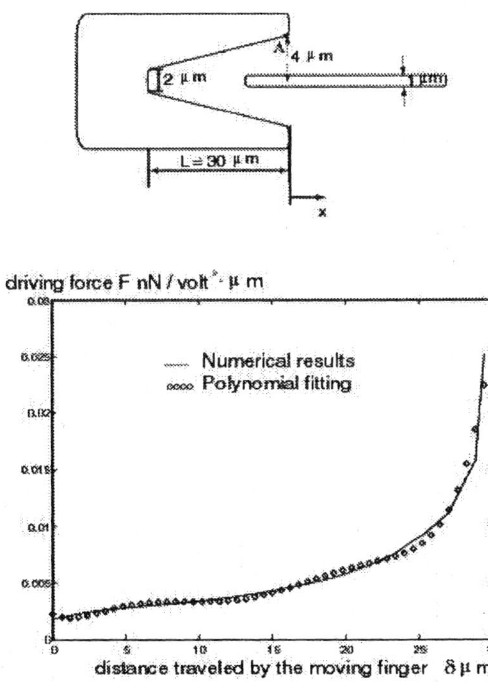

Figure 8. A comb drive with a linear gap profile and its force profile.

4.1.2 Sensitivity analysis

Numerical results for $q = \partial q/\partial c_i$ from the DDA of the BEM, are compared with those computed by the finite difference method (FDM) in

Fig. 9. They agree very well when the gradients are not very large. When the value of q is very sensitive to those of c_i, bigger discrepancies between the results are observed. However, the forces in this problem are mostly determined by the fringe field so that the values of the sensitivities on the tip of the moving finger (nodes 105 – 115) are of primary interest. In this region, the results from the DDA and the FDM agree.

One can also calculate the sensitivity of the maximum driving force (F) with respect to the gap profile. For example, one can pick a specific point, say $x = -2\mu m$ (which is point A in Fig. 8), and calculate the quantity $(|\Delta F|/F)/(|\Delta g|/g)$, keeping the slope g' and curvature g'' constant at that point. This calculation gives:

$$\frac{\frac{|\Delta F|}{F}}{\frac{|\Delta g|}{g}} = 0.7869 \tag{21}$$

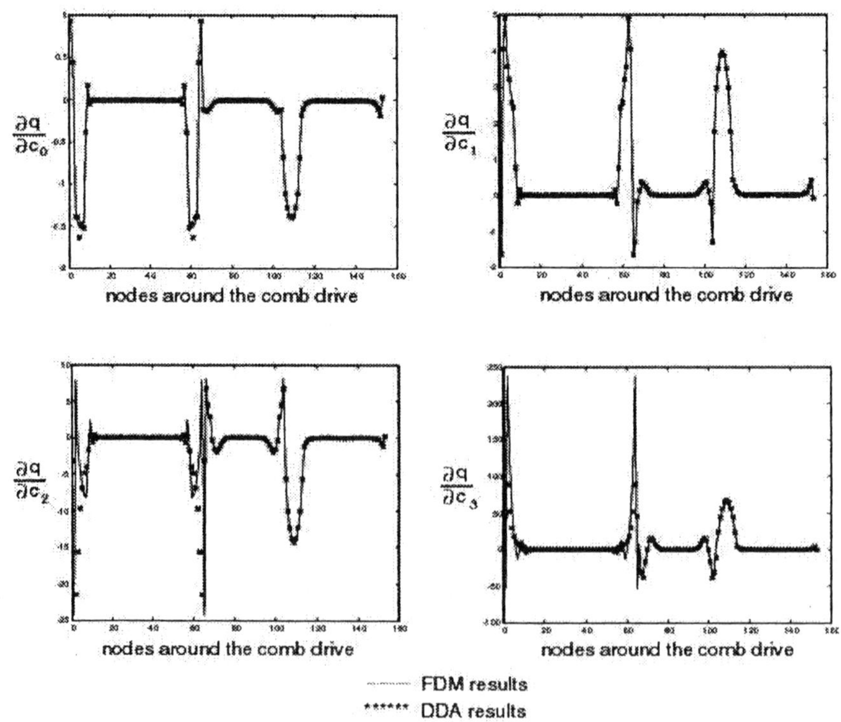

Figure 9. Numerical results for the gradients of charge density with respect to shape parameters.

Similarly, one can find the quantity $(|\Delta F|/F)/(|\Delta g''|/g'')$ keeping the gap g and the slope g' constant at that point. This time, one gets

$$\frac{\frac{|\Delta F|}{F}}{\frac{|\Delta g''|}{g''}} = 0.146 \tag{22}$$

These results indicate that the maximum driving force is not very sensitive to small changes in finger shape and dimensions. Any perturbations in shape or dimensions caused by manufacturing uncertainties would not deteriorate the driving force that much. In other words, these designs are robust.

4.1.3 The inverse problem

The goal here is to design three variable comb drives, which have linear, quadratic or cubic driving force profiles as functions of the distance traveled by the movable fingers. However, a comb drive with purely linear, quadratic, or cubic driving force profile usually occupies a large area. For example, for a comb drive with driving force $F(\delta) = \delta^3$, according to the simple formula from Eq. (4), the gap $g(\delta)$ will be roughly proportional to $1/\delta^3$. If δ, the position of the tip of the moving finger, varies between 5 – 25 μm, the gaps at the two ends will have a ratio of 125, i.e., if $g(\delta)$ at $\delta = 25\mu m$ is 1 μm, then $g(\delta)$ at $\delta = 5\mu m$ will be 125 μm! This will result in an unacceptably wide comb drive. Due to this practical design consideration, the range of δ in the desired function $h(\delta)$ is shifted by δ_o. Instead of having $h(\delta)$ proportional to δ, δ^2 or δ^3, it is taken to be proportional to $(\delta + \delta_o), (\delta + \delta_o)^2$ or $(\delta + \delta_o)^3$. By choosing suitable values of δ_o, the opening between the fixed fingers can be controlled. Also, in these examples, the design space is enlarged by adding another term $c_4\delta^4$ in the denominator of the expression for $g(\delta)$ in Eq. (20). The new objective function Ψ for the inverse problem is

$$\Psi(c_i) = \int_{l_1}^{l_2} [F(c_i, \delta) - h(\delta + \delta_o)]^2 \, d\delta \tag{23}$$

The constraints are:

$$1.5\mu m \leq g(\delta) \leq 10\mu m \tag{24}$$

To simplify this problem, instead of insisting that the constraints are satisfied at every point within the range of δ, we require that the constraints are satisfied at several sample points within the range of δ, i.e.,

$$1.5\mu m \leq g(\delta_i) \leq 10\mu m \qquad i = 1,2,\cdots m \tag{25}$$

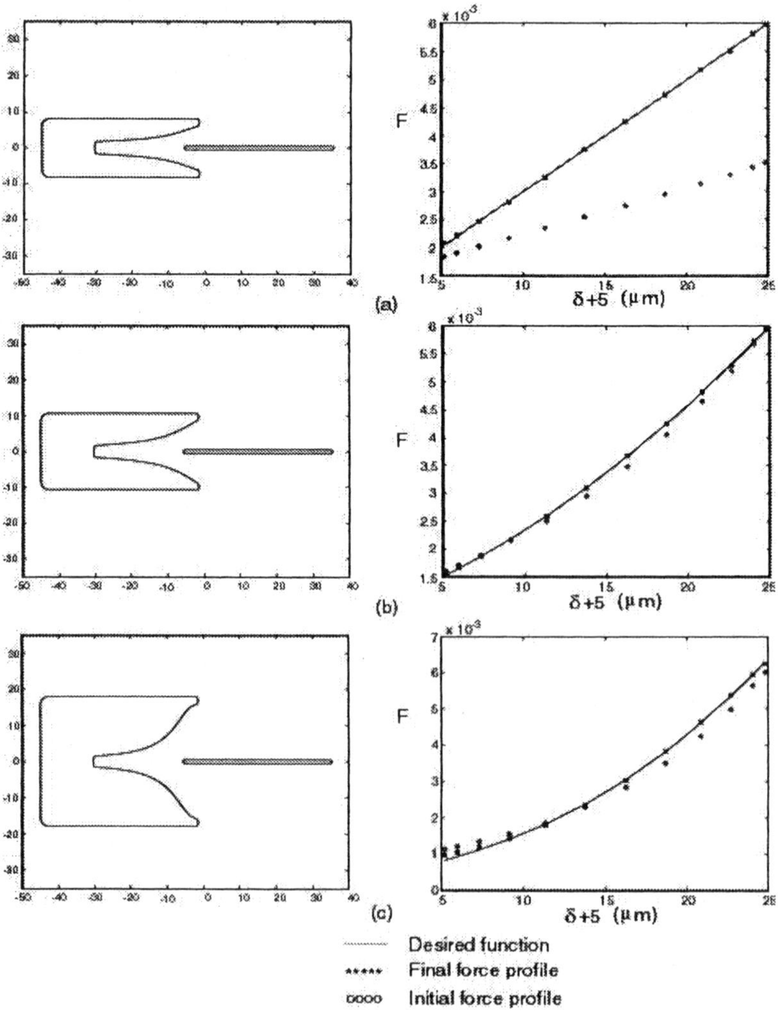

Figure 10. Designs of variable comb drives with uniform moving fingers, together with their force profiles. (a) linear motor, (b) quadratic motor, (c) cubic motor.

This simplification reduces the nonlinear constraints to linear constraints and thus reduces the complexity of the corresponding optimization problem.

The optimization problem is solved by the optimization code "dlcong" from the IMSL package. The method used in this code is similar to the one used in "E04UCF" described in Section 3. The stopping criterion is $|\partial\Psi/\partial c_i| \le 0.001$. Although this only guarantees a local minimum, it is usually sufficient from a practical design point of view.

The optimizer uses the function F and sensitivity $\dot{F}$ from the comb drive simulation approach presented above. The second derivatives are approximated by the BFGS formula. Three designs are shown in Fig. 10, along with their force profiles. All the shape parameters and the expressions for the driving forces are given in Table 1. These three, together with the standard comb drive, serve as the fundamental comb drives.

Table 1. Design parameters and driving forces for variable comb drives with uniform moving fingers.

Initial design parameters	Optimal design parameters	Driving force (nN/volt^2 x μm)
$c_0 = 0.1522$ $c_1 = -0.7246\times10^{-2}$ $c_2 = 0.0$ $c_3 = 0.0$ $c_4 = 0.0$	$c_0 = 0.1507$ $c_1 = -0.3135\times10^{-2}$ $c_2 = 0.1679\times10^{-2}$ $c_3 = 0.8051\times10^{-4}$ $c_4 = 0.1244\times10^{-5}$	Linear $F = 0.002 +$ 0.0002δ
$c_0 = 0.1067$ $c_1 = -0.8581\times10^{-2}$ $c_2 = 0.2860\times10^{-3}$ $c_3 = 0.0$ $c_4 = 0.0$	$c_0 = 0.1048$ $c_1 = -0.2795\times10^{-2}$ $c_2 = 0.1324\times10^{-2}$ $c_3 = 0.4863\times10^{-4}$ $c_4 = 0.6654\times10^{-6}$	Quadratic $F = 1.5\times10^{-3} +$ $1.5\times10^{-4}\delta$ $+0.375\times10^{-5}\delta^2$
$c_0 = 0.07249$ $c_1 = -0.4760\times10^{-2}$ $c_2 = 0.3173\times10^{-3}$ $c_3 = -0.7052\times10^{-5}$ $c_4 = 0.0$	$c_0 = 0.06897$ $c_1 = -0.5754\times10^{-2}$ $c_2 = 0.1419\times10^{-2}$ $c_3 = 0.1401\times10^{-4}$ $c_4 = -0.1870\times10^{-6}$	Cubic $F = 8\times10^{-4} +$ $1.2\times10^{-4}\delta$ $+0.6\times10^{-5}\delta^2 +$ $1.0\times10^{-7}\delta^3$

A problem with the above designs, however, is that they produce relatively small driving forces at the maximum gap, especially for the cubic motor. Since the moving finger is very flexible, it is very easy for it to become unstable for a large applied voltage. To address this problem, the moving finger is also shaped. The shape of the moving finger is chosen such that when it is fully inserted, the gap profile becomes uniform. The shape parameters for the fixed fingers are shown in Table 2. The choice of those parameters is guided by consideration of the minimum gap and maximum opening of the comb drive. For example, the linear motor in Table 2 has a minimum gap of 1.5 μm and a maximum gap of 3 μm.

Table 2. Design parameters of the fixed fingers in the comb drive in Figure 4.7.

	Design parameters
Linear	$c_0 = 0.3188, \quad c_1 = -0.7246 \times 10^{-2},$ $c_2 = 0.0, \quad c_3 = 0.0, \quad c_4 = 0.0$
Quadratic	$c_0 = 0.2378, \quad c_1 = -0.5721 \times 10^{-2},$ $c_2 = 0.1907 \times 10^{-3}, \quad c_3 = 0.0, \quad c_4 = 0.0$
Cubic	$c_0 = 0.1522, \quad c_1 = -0.7262 \times 10^{-2},$ $c_2 = 0.7469 \times 10^{-3}, \quad c_3 = 0.1893 \times 10^{-4}, \quad c_4 = 0.0$

Table 3. Force ranges of different comb drives

	Force profile	Force range ($nN/volt^2 \times \mu m$)
Standard comb drive	Constant	0.0059
Comb drive with uniform moving fingers	Linear	(0.002, 0.006)
	Quadratic	(0.0015, 0.006)
	Cubic	(0.001, 0.0063)
Comb drive with variable moving fingers	Linear	(0.0042, 0.011)
	Quadratic	(0.0033, 0.014)
	Cubic	(0.0023, 0.013)

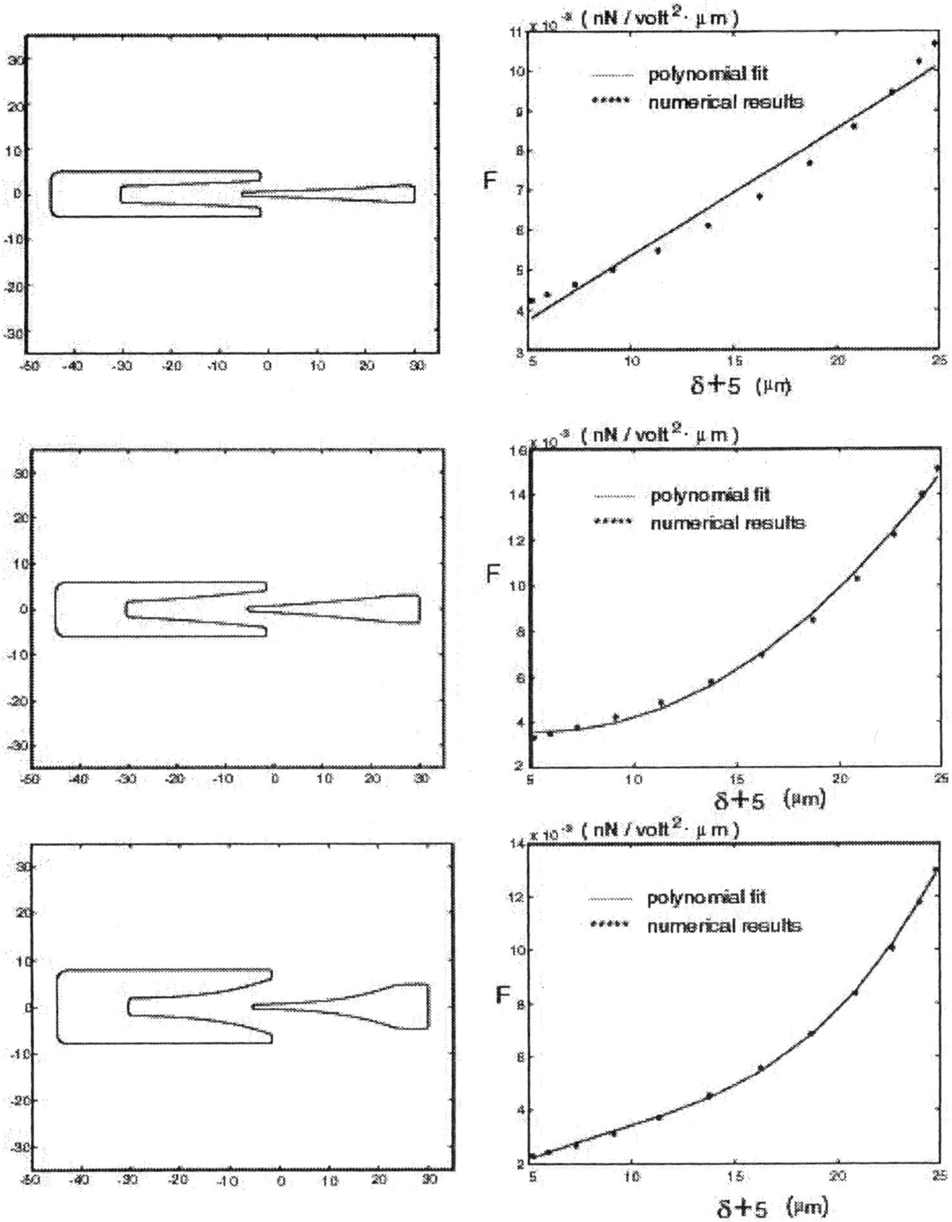

Figure 11. Designs of variable comb drives with variable moving fingers, together with their force profiles. (a) linear motor: $F(\delta) = 0.00032\delta + 0.0038$ (b) quadratic motor: $F(\delta) = 3.0\times10^{-5}\delta^2 - 3.2\times10^{-5}\delta + 0.0036$ (c) cubic motor: $F(\delta) = 1.57\times10^{-6}\delta^3 - 2.0\times10^{-5}\delta^2 + 3.2\times10^{-4}\delta + 0.00214$.

The resultant comb drives have several advantages. First, the maximum driving force increases dramatically. Second, the structure is more stable and therefore large voltages can be applied. Finally, the area occupied by a comb drive is much smaller. The new designs with variable widths of both the fixed and the moving fingers, and their force profiles, are shown in Fig. 11. Table 3 gives a comparison of a standard comb drive with the two kinds of variable shape. The minimum gap in each comb drive is 1.5 μm, and the maximum displacement of the moving fingers is 20 μm.

4.2 3-D Design

In this approach, the gap between the fingers is kept constant. A variable force profile is achieved by varying the height profile of the moving fingers.

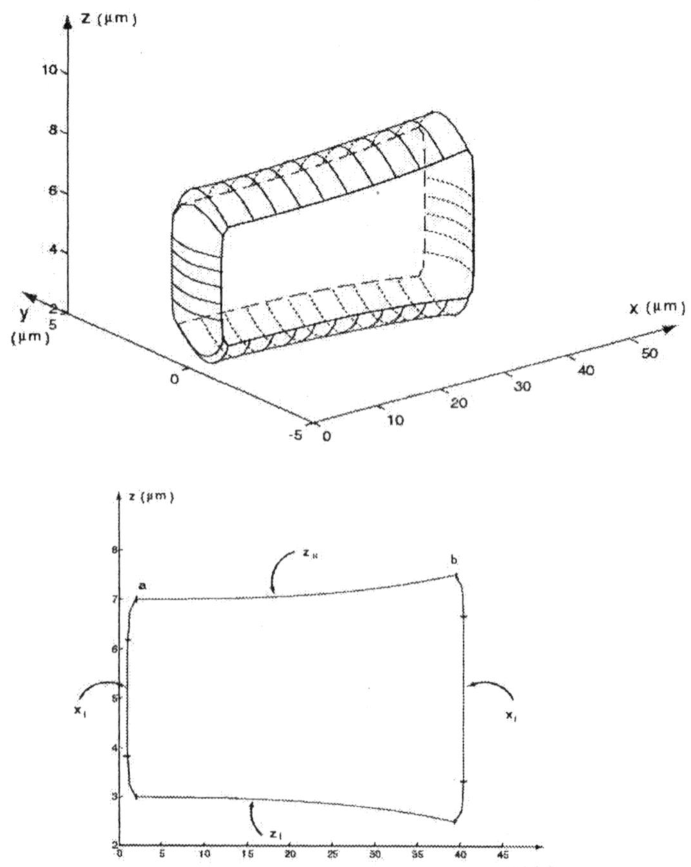

Figure 12. A moving finger with variable height profile.

Figure 12 shows the three-dimensional shape of a proposed moving finger and its projection onto the $y=-1$ plane. This structure is symmetric about $z=5$. The upper curve between *a* and *b* is a polynomial function of *x*:

$$z_u = c_0 + c_1 x + c_2 x^2 + c_3 x^3 \tag{26}$$

where c_0, c_1, c_2, c_3 are design variables. The lower curve is simply set as

$$z_l = 10 - z_u \tag{27}$$

to maintain symmetry about the plane $z=5$. The left and the right lines are $x_l = 1$ and $x_r = 40.5$, respectively. Between these lines, Hermetian curves are used to smooth the corners. This is done for the purpose of avoiding the corner singularities that exist in exterior potential problems. The edges in the lateral plane are rounded off by semi-circles because of the same reason. It is important that edges and corners remain rounded after subjecting a structure to geometric perturbations.

4.2.1 Driving force

A 3-D numerical code based on the Boundary Element Method has been developed to calculate the driving force for a given structure. A standard comb drive with uniform gap and height profiles is again used as the test example to validate the developed code. The dimensions of the comb drive are shown in Figs. 13 and 14 make the driving force a function of the distance traveled by the moving fingers. The driving force obtained from the two dimensional simulation is also shown there for comparison. As expected, the driving force remains constant during most of the travel range because the gap between the fixed fingers and the moving fingers is constant, and so are the heights of these fingers. When the moving finger is fully inserted into the fixed fingers, the electrostatic potential breaks down due to the contact between these two sets of fingers. The difference between these two results is around 5%. The driving force obtained from the three dimensional model is slightly higher than the driving force obtained from the two-dimensional model. This is because in the 3-D model, in addition to the local fringe field, the global fringe field is also accounted for.

4.2.2 Sensitivity analysis

Equation (16) is solved by the usual boundary element method. As mentioned before, the matrix obtained from Eq. (16) is the same as the one obtained from Eq. (11). Only the right-hand side involves the sensitivities of nodal coordinates (in the $\dot{G}$ function) that must be calculated carefully. Consistency must be maintained before and after perturbation of the geometry.

Table 4 shows the sensitivities of the objective function Ψ with respect to the design variables c_i obtained from the direct differentiation and the finite difference method, respectively, for the case where $c_0 = 7.0, c_1 = 0.0, c_2 = 0.0, c_3 = 0.0$ and $\delta = 20\mu m$. After carefully choosing the perturbation step sizes in the finite difference method, the difference between two sets of results obtained from the two different methods is very small.

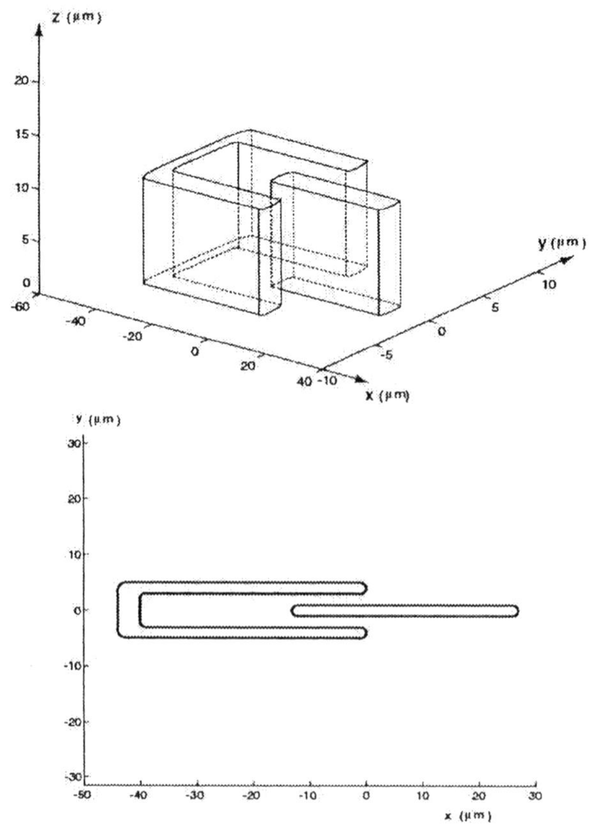

Figure 13. A 3D standard comb drive and its projection onto the $x-y$ plane.

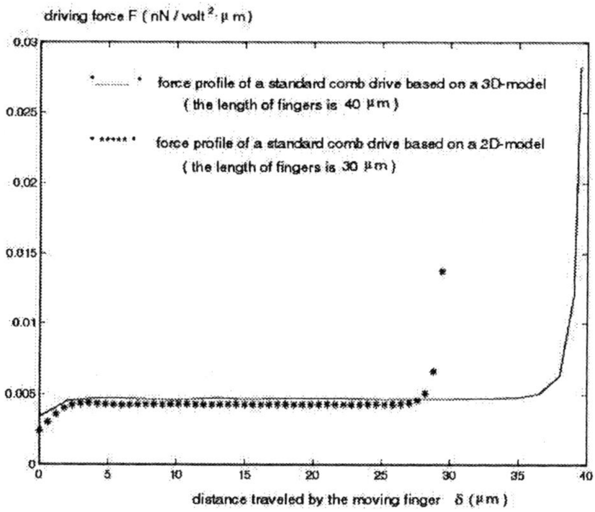

Figure 14. Driving force profiles of a standard comb drive based on the 3D and 2D models.

Table 4 The sensitivities of Ψ with respect to c_i, $i = 0, 1, 2, 3$ obtained from DDA and FDM.

	$\frac{\partial \Psi}{\partial c_0}$	$\frac{\partial \Psi}{\partial c_1}$	$\frac{\partial \Psi}{\partial c_2}$	$\frac{\partial \Psi}{\partial c_3}$
DDA	-1.24×10^{-3}	-0.0204	-0.349	-6.874
FDM	-1.24×10^{-3}	-0.0202	-0.3477	-6.867

4.2.3 The inverse problem

The goal here is to design a comb drive that has a polynomial force profile as a function of the distance traveled by its moving fingers. So the desired force profile $F_e(\delta)$ in Eq. (9) is set as a polynomial function of δ. For the purpose of illustration, only one example is shown in this section. In this example, a cubic force profile is chosen as the desired force profile.

$$F_e(\delta) = 5.0\times10^{-7}\,\delta^3 + 3.8\times10^{-2} \tag{28}$$

The initial guesses for the design variables c_i are:

$$c_0 = 7.0, \quad c_1 = 0.0, \quad c_2 = 0.0, \quad c_3 = 0.0 \tag{29}$$

Keeping in mind the manufacturing techniques accessible to these designers, the maximum height of the moving finger is set as 30 μm. This provides constraints on the design variables c_i, i.e.

$$c_0 + c_1 x + c_2 x^2 + c_3 x^3 \leq 30 \tag{30}$$

Also, at points *a* and *b* in Fig. 12, the minimum values of z_u are set as 6.5 μm and 7.5 μm respectively, to ensure that the height of the moving finger is at least 5 μm.

Figure 15 shows the projection in the *x-z* plane of the initial and final shape of the moving finger for a comb drive with a cubic force profile. The force profile of the comb drive at its initial shape, two intermediate shapes, and the final shape, are also shown there.

Each iteration needs 2 hours of CPU time on one node of an IBM SP2 computer. Also, convergence of the optimization problems would require a large number of iterations. Due to limitations of computing resources, these minima were not achieved – instead, the calculations were terminated when the magnitude of $Z^T \nabla \Psi$ became sufficiently small. Here Z is a matrix whose columns form an orthogonal basis for the set of vectors orthogonal to the space spanned by the active constraints. Table 5 shows $\left\| Z^T \nabla \Psi \right\|$ at the final design. Of course, $\left\| Z^T \nabla \Psi \right\|$ should ideally be zero when Ψ achieves its minimum. Also, the value of Ψ at the final design is shown in Table 5. From a practical point of view, the design shown in Fig. 15 is adequate.

Table 5. The values of the normal of the projected gradient and objective function at the initial and final designs.

		$\left\| Z^T \nabla \Psi \right\|$	Ψ
Cubic Motor	Initial design	0.12	1.62×10^{-4}
	Final design	0.0001	2.4×10^{-6}

Comb drives with linear, quadratic force, (or other) profiles can be designed by following exactly the same procedure. These comb drives, plus the one shown above, together with the standard comb drive, serve as fundamental comb drives. These fundamental comb drives can be arranged in parallel, with suitable bias voltages applied to each comb drive, in order to obtain any desired polynomial force profile.

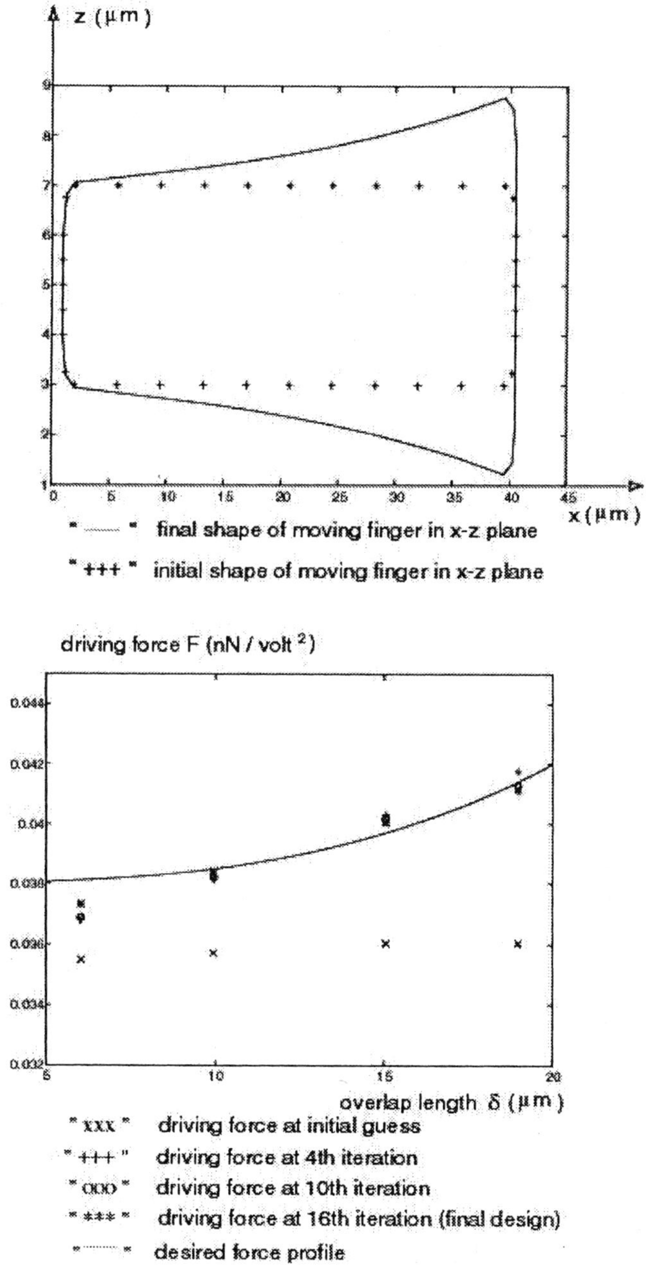

Figure 15. Cubic motor: Initial and final shapes of each moving finger in the *x-z* plane and the force profiles from this comb drive at different iterations.

4.3 Fabrication of a shaped motor – a demonstration

Fabrication of a shaped motor, with an approximately cubic force profile, is discussed in this section. SCREAM I process [18] is used to fabricate the device. SCREAM I is a single-mask bulk micromachining process. It uses Reactive Iron Etch (RIE) of a silicon substrate to fabricate suspended movable single-crystal silicon structures. Only one lithography step is needed to define beams and structures simultaneously as well as all necessary contact pads, electrical interconnects and lateral capacitors. A brief outline of the SCREAM I process is described in the next subsection.

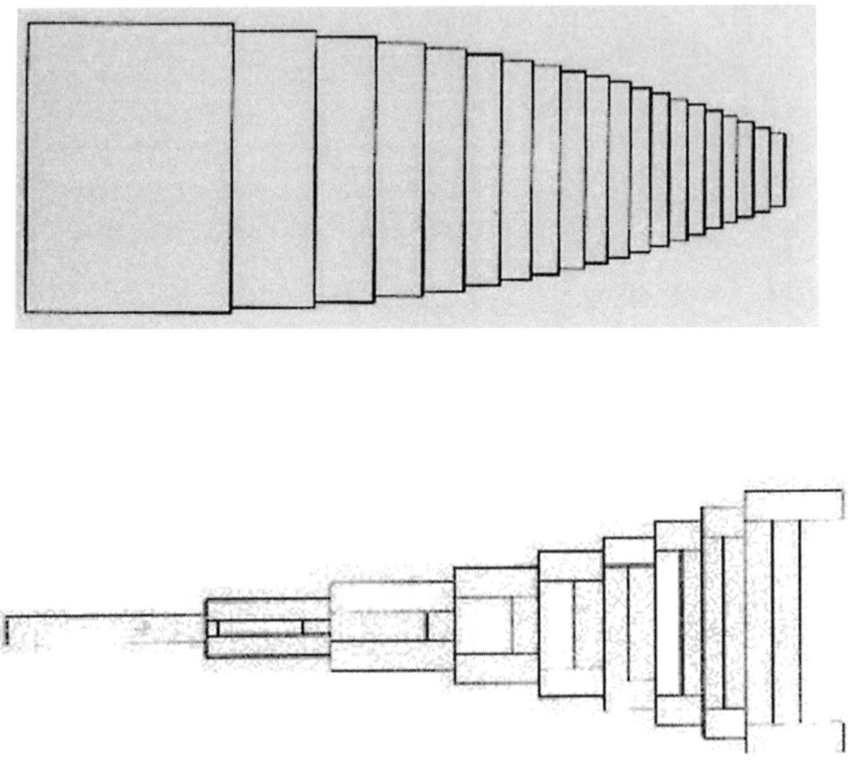

Figure 17. CAD design of the moving finger in a variable comb drive.

Due to limitations of photolithography, generation of the patterns of a comb drive with the precise shapes defined in Subsection 4.1.3 is very expensive and time consuming. Because of this fact, a comb drive with an approximate shape is used in our device. Figures 16 and 17 show the shapes of one fixed finger and one moving finger defined in the CAD file. This file is used to generate patterns on the mask. The shape of the fixed finger is approximated by a set of rectangles. The thicknesses of adjacent rectangles differ by only 0.2 μm. Since the resolution of photolithography is about 0.5 μm, the corners shown in Fig. 16 will be smoothed out after photolithography. Thus a fixed finger with smooth shape is created. Due to special requirements in the SCREAM I process on the thickness of released structures, a honey-comb structure is used in the moving finger. The smallest feature size is chosen as 0.5 μm. This is determined by considering the resolution of photolithography and the relatively thick oxide needed to be deposited on the wafer. The final shapes of the moving fingers are not as smooth as those of the fixed fingers because of all the manufacturing process limitations. In order to study the effects of stepped shaped fingers on the driving force profiles, a calculation of the driving force from such a comb drive, with the moving finger shown in Fig. 17, is performed. The driving force profile of this comb drive, together with the force profile of the cubic motor, is shown in Fig. 18. In the range between 5 – 20 μm, the maximum difference between them is around 10%. It should be noted that simulations of a comb drive, one with and without holes in the moving fingers, give essentially the same force profiles. Figure 19 shows the CAD design of the whole device. It consists of one beam and two comb drives. One is the standard comb drive and the other is the variable comb drive. The beam length is 800 μm. The initial overlap length is 3 μm.

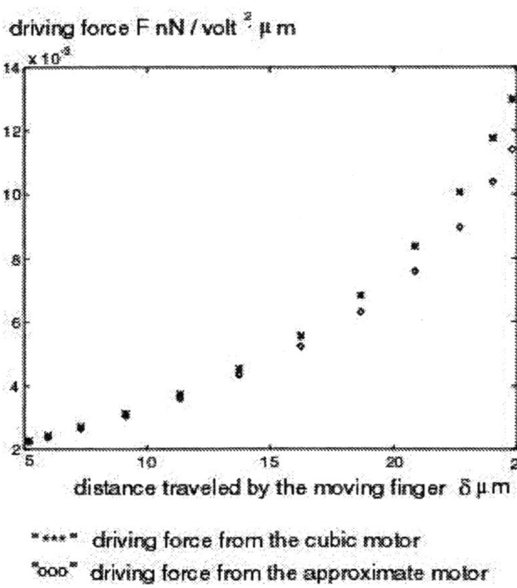

Figure 18. Driving force profiles from the approximate comb drive and the ideal cubic motor.

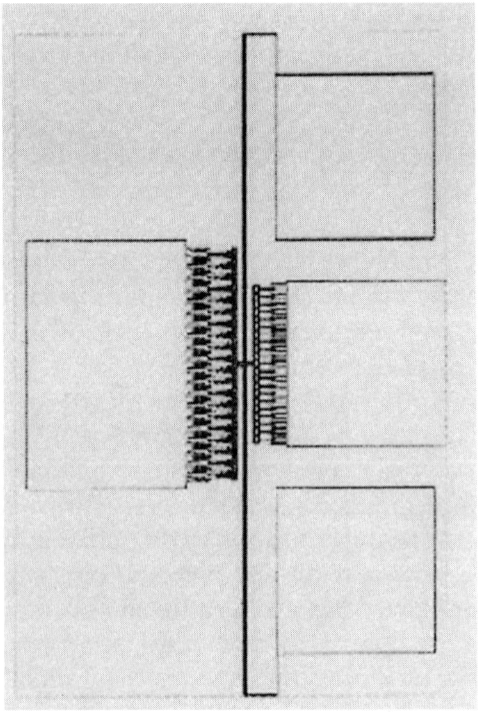

Figure 19. CAD design of a device that contains one standard comb drive and one variable comb drive.

4.3.1 SCREAM I process

An outline of the SCREAM I process is given in Fig. 20. A layer of 2.4 μm thick silicon oxide (SiO_2) is first deposited on the silicon wafer using the standard PECVD process. The thickness of silicon oxide is determined by the etch rate and etch time in the successive RIE steps. Next, photolithography is performed and patterns are transferred from mask to the photo resist spinned on the top of SiO_2. The CHF_3 RIE is used to transfer the patterns further from resist to silicon oxide. After this step, the resist is stripped by an O_2 plasma etch. The next step is to transfer the patterns into silicon. This is done by an anisotropic RIE etch. Another layer of silicon oxide with thickness of 368 nm is deposited using PECVD to form the sidewall oxide layer. The floor oxide is cleaned by CHF_3 RIE. In step 8, another deep silicon etch is performed using the same method described in step 5. The structure is then ready for release. A standard SF6 parallel plate RIE is used for isotropic release etch. Finally, in step 10, a layer of aluminum of 300 nm thick is sputtered on the released device. Figure 21 shows the SEM of the part of the device after release [19]. The height of the released structure is about 10 μm.

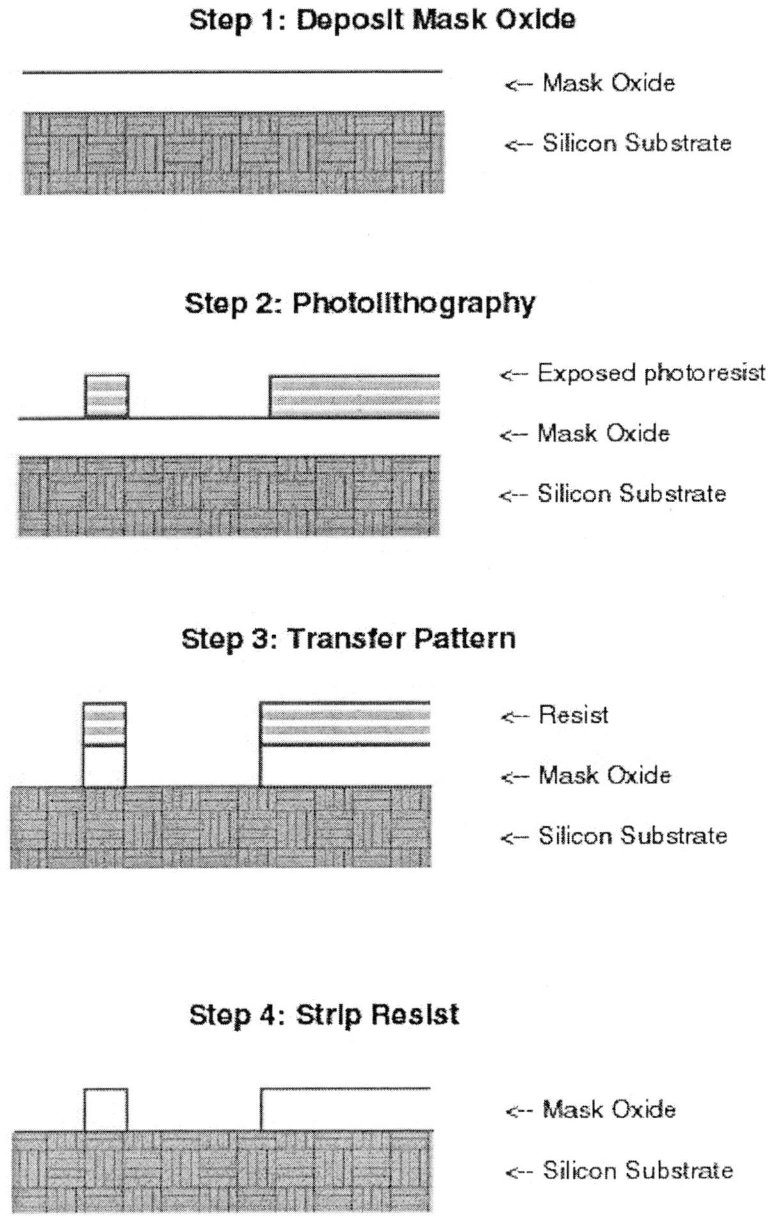

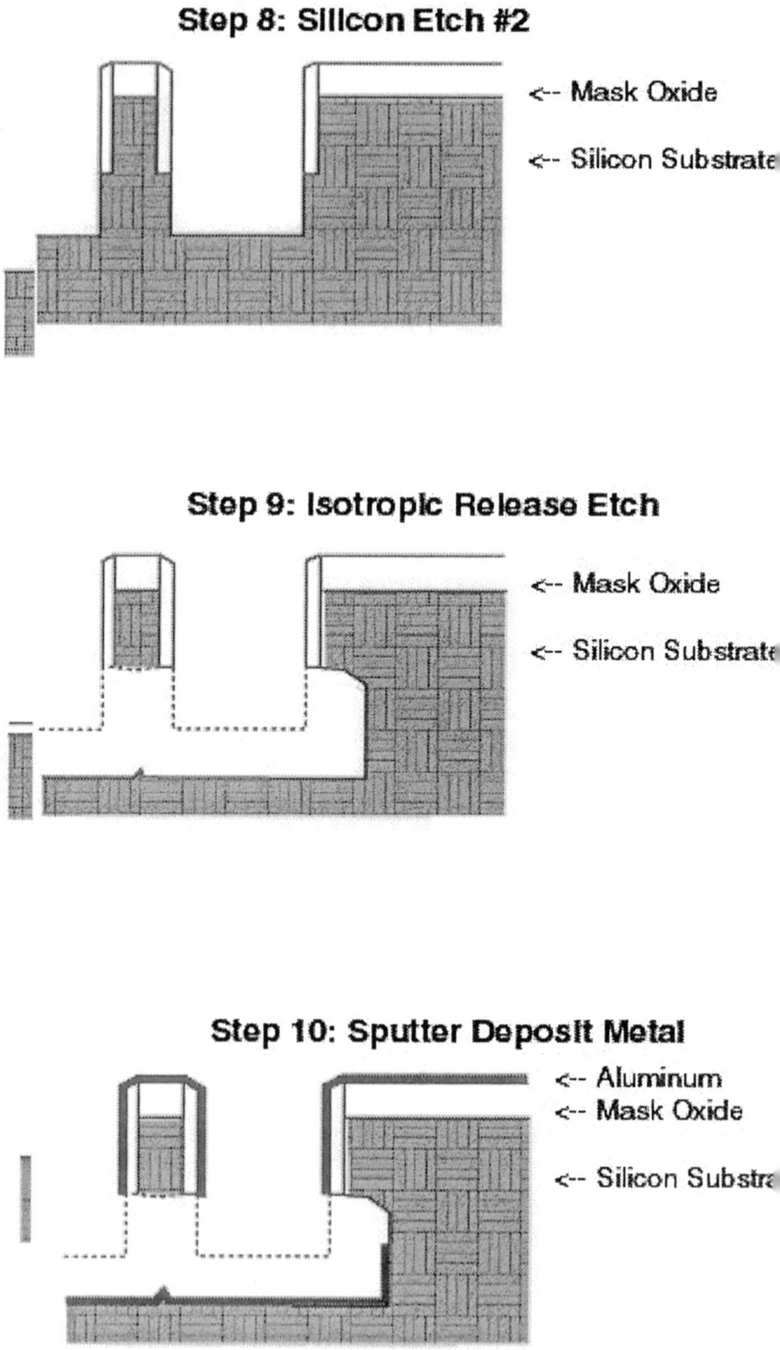

Figure 20. SCREAM I process outline (reprint from [18]).

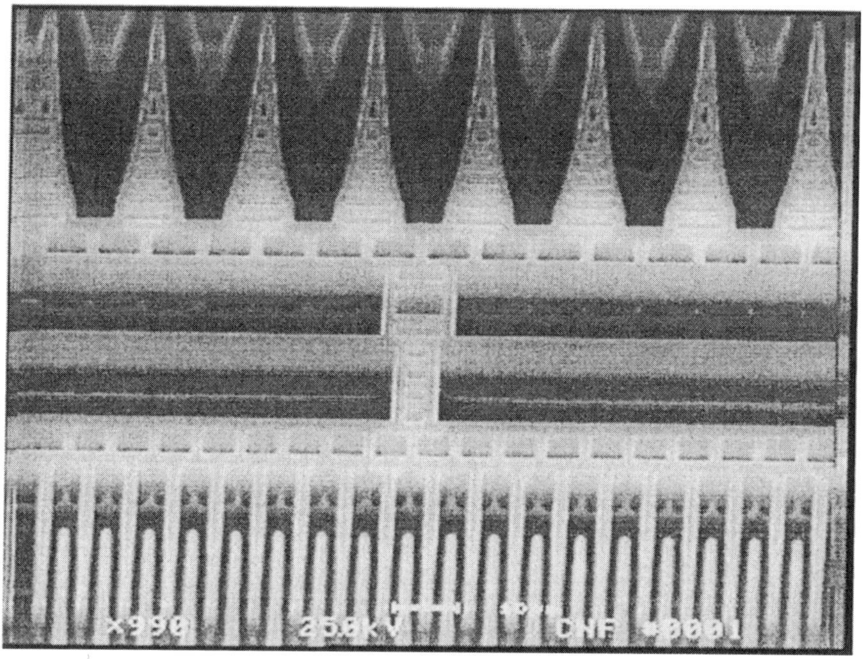

Figure 21. SEM of the released structure.

Table 6. Driving force from the variable comb drive shown in Figure 4.13.

	Driving force per finger (nN/volts^2 x μm)
Experimental result	1.729×10^{-3}
Simulation result	2.25×10^{-3}

4.3.2 Test results

The fabricated device is electrically tested by applying a bias DC voltage to the fixed fingers (the moving fingers are grounded). In the standard comb drive, the moving fingers start to move at 6 volts. At $V = 14$ volts, the device starts to have sideways motion and becomes unstable. In the variable comb drive, the moving fingers start to move at 5 volts. Up to 19 volts, the variable comb drive is stable. The displacements of the moving fingers in the variable comb drive, driven by different voltages, are measured using an optical microscope. The driving forces from the comb drive at this configuration are then calculated from the measured displacements. The experimental result is shown in Table 6, together with the simulation result. The experimental result is the average from three applied voltages – 5, 10 and 15 volts, respectively. The difference between the experimental and the computed results is about 23%.

5. CLOSURE

A synthesis methodology for electrostatically driven MEMS devices, with applications to variable shape comb drives, is presented in this chapter. This method is fairly general and can be applied to a wide range of electrostatically actuated MEMS. It has been shown in this chapter that with the systematic synthesis procedures presented here, structures that will deliver desired performance can be designed efficiently and accurately.

REFERENCES

1. G. Kovacs, *Micromachined Transducers Sourcebook*, McGraw-Hill, New York, 1998.
2. S. J. Sherman, W. K. Tsang, T. A. Core, R. S. Payne, D. E. Quinn, K. H.-L. Chau, J. A. Farash and S. K. Baum, "A low cost monolithic accelerometer; product/technology update," *Proceedings of IEEE International Electron Devices Meeting*, San Francisco, CA, Dec. 13 – 16, pp. 501-504, 1992.
3. M. Putty and K. Najafi, "A micromachined vibrating ring gyroscope", *Solid State Sensor and Actuator Workshop*, Hilton Head Island, SC, June 13-16, pp. 213 – 220, 1994.
4. W. C. Tang, T.-C. H. Nguyen, M. W. Judy and R. T. Howe, "Electrostatic-comb drive of lateral polysilicon resonators," *Sensors and Actuators*, Vol. A 21-23, pp. 328 – 381, 1990.
5. K. Wang, Y. Yu, A.-C. Wong and C. T.-C. Nguyen, "VHF free-free beam high-Q micromechanical resonators," *Technical Digest, IEEE MEMS Conf.*, Orlando, Florida, pp. 453 – 458, Jan. 17 – 21, 1999.
6. J. J. Yao and M. F. Chang, "A surface micromachined miniature switch for telecommunications applications with signal frequencies from DC up to 4 GHz," *Digest of Technical Papers, Transducers 95*, Stockholm, Sweden, June 25 – 29, vol. 2, pp. 257 – 264, 1995.
7. J. Drake, H. Jerman, B. Lutze and M. Stuber, "An electrostatically actuated micro-relay," *Proceedings of Transducers '95, the 8th International Conference on Solid-State Sensors and Actuators*, Stockholm, Sweden, June 25 – 29, vol. 2, pp. 380 – 383, 1995.
8. J. D. Jackson, *Classical Electrodynanmics*, John Wiley and Sons, Singapore, second edition, 1975.
9. W. Ye, *Optimal Design and Fabrication of Microelectromechanical Structures – with Applications to Comb Drives*, Ph. D Thesis, Cornell University, 1998.

10. S. Adams, F. Bertsch and N. C. MacDonald, "Independent tuning of the linear and nonlinear stiffness coefficients of a micromechanical device," *Proc. of IEEE the 9th Annual International Workshop on MEMS*, San Diego, CA, pp. 32 – 37, 1996.
11. P. Banerjee, *The Boundary Element Methods in Engineering*, MaGraw-Hill, London, 1994.
12. F. Shi, P. Ramesh, and S. Mukherjee, "On the application of 2-D potential theory to electrostatic simulation," *Communications in Numerical Methods in Engineering*, Vol. 11, pp. 691 – 701, 1995.
13. Y. Saad, and M. Schultz, *"GMRES: A generalized minimal residual algorithm for solving symmetric linear systems"*, SIAM, J. Sci. Statis. Comput., Vol. 7, pp. 856 – 869, 1986.
14. R. Fletcher, *Practical Methods for Optimization*, John Wiley and Sons, Chichester, 2nd edition, 1987.
15. J. E. Dennis, Jr. and J. J. More, "Quasi-Newton methods, motivation and theory," *SIAM Review*, vol. 19, pp. 46 – 89, 1977.
16. W. Ye, S. Mukherjee and N. C. MacDonald, "Optimal shape design of an electrostatic comb drive in Microelectromechanical systems," *Journal of MEMS*, Vol. 7, No. 1, pp. 16 – 26, 1998.
17. W. Ye and S. Mukherjee, "Optimal design of three-dimensional MEMS with applications to electrostatic comb drives," *International Journal for Numerical Methods in Engineering*, Vol. 45, pp. 175-194, 1999.
18. K. A. Shaw, Z. L. Zhang and N. C. MacDonald, "SCREAM I: A single mask, single-crystal silicon, reactive ion etching process for microelectromechanical structures," *Sensors and Actuators A*, Vol. 40, pp. 63 – 70, 1994.
19. W. Ye and S. Mukherjee, "Design and fabrication of an electrostatic variable gap drive in micro-electro-mechanical systems," Computer Modeling in Engineering and Sciences, Vol. 1, pp. 111-120, 2000.

Chapter 4

SYNTHESIS METHODS FOR ELECTRO-THERMAL ACTUATION

G. K. Ananthasuresh
Department of Mechanical Engineering and Applied Mechanics, University of Pennsylvania, Philadelphia, PA 19104, U. S. A.

Synopsis: Electro-thermal actuation has a number of advantages in minute mechanical devices such as the ones found in Micro-Electro-Mechanical Systems (MEMS). In this chapter, we describe synthesis methods for a general class of electro-thermal-compliant (ETC) actuators and mechanisms. The ETC devices are similar to the widely used bimorph actuators, but to achieve differential thermal expansion they rely upon their particular shape of a single-material structure rather than mismatched thermal expansion coefficients of two or more materials. These are *smart by design* because their functionality does not depend on the "smartness" of materials. Determination of the shape of the elastic structure for a specified behavior constitutes the synthesis problem. The synthesis procedure incorporates accurate thermal modeling including convection, which is important at the micro scale due to the familiar scaling law of surface to volume ratio. Design concepts and algorithms, microfabrication procedures, and prototype devices are described in the chapter.

1. INTRODUCTION

In the micrometer to millimeter size-range of Micro-Electro-Mechanical Systems (MEMS), actuators based on the Joule heating-induced thermal expansion provide large forces with moderate input power. When coupled with *compliant mechanisms*, which are single-piece elastic structures capable of force and motion transmission via elastic deformation, these actuators can also deliver large deflections. This type of *electro-thermal-compliant (ETC)* actuation generates much larger forces over large distances compared to the widely used electrostatic actuation. In terms of ease of fabrication, ETC actuators are better than electromagnetic actuators and actuators made of piezoelectric, shape memory alloy, ferroelectric and other smart materials. However, ETC actuation consumes more power than some of the other actuation methods such as electrostatic. Another disadvantage is the heat produced, which must be taken away from the workpiece.

An appropriately shaped single-layer, released structure made of a conducting material anchored to a fixed substrate that has the electrical connections is all that is required to realize in-plane ETC actuators in practice. The principle of operation of ETC devices is explained below.

Thermal actuation as in a bimorph is not new [1]. The bimorph effect using two or more materials with dissimilar thermal expansion coefficients has been used in MEMS [2-4, and others]. This type of actuation is more suitable for out-of-plane actuation in MEMS. For in-plane actuation, non-uniform electro-thermal expansion using a single material of appropriate shape has been demonstrated [5-20 and others]. A basic electro-thermal actuator, sometimes called a *heatuator*, proposed in [5, 6] is the basis for this type of non-bimorph electro-thermal actuators. As shown in Fig. 1a, a U-shaped folded beam consisting of a narrow arm, a wide arm, a connector, and a flexure for the wide arm, all made of a single material, behaves in the same way as a bimorph would. When a voltage potential is applied between the two fixed ends of the U-shaped beam, the same amount of current flows through the narrow and wide arms, which can be viewed as electrical resistors in series. Because of the small cross-section of the narrow arm, the current density in it is larger than that it is in the wide arm. Therefore, Joule heating per unit volume is larger, which makes the narrow arm get hotter than the wide arm. Then, the structure achieves equilibrium by bending towards the wide arm to accommodate larger expansion of the narrow arm compared to the wide arm. The flexure that enables the wide arm to bend is necessary because the latter's larger moment of inertia would not otherwise allow it to deform to any significant degree. Fig. 1b is a scanning electron microscope (SEM) image of the actuator made of single-crystal silicon using the PennSOIL [15-18] process. It is about 2 mm long and is one of the larger size devices of its type. Less than one tenth of this size is routinely possible in most microfabrication processes.

Generalization of the above simple yet powerful concept of shape-enabled electro-thermal actuation leads to a versatile microactuation technique. In Section 2, several variations of the U-shaped electro-thermal actuators are described. The concept of ETC actuation is introduced at the end of the section. The rest of the paper describes modeling (Section 3), a simple formulation and implementation of the synthesis procedure (Section 4), some sample problems that exemplify the synthesis algorithm including an example based on a more complicated modeling and constraints (Section 7), an alternative implementation of the synthesis procedure (Section 6), and microfabrication (Section 8). The chapter is concluded with some remarks and avenues for further research in Section 9.

(a)

(b)

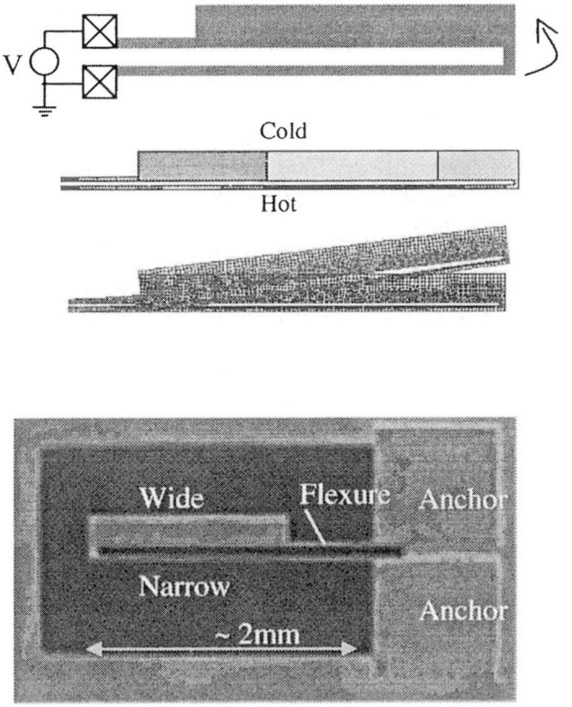

Figure 1. (a) Basic electro-thermal actuator: shape and boundary conditions; temperature distribution; deformation (b) fabricated prototype made of single-crystal silicon using PennSOIL [15-18].

2. GENERALIZATION OF THE BASIC ELECTRO-THERMAL ACTUATOR

The mode of deformation of the U-shaped actuator is not as simple as it was described above. That is, larger current density in the narrow arm does not necessarily give rise to deformation that makes the structures always bend towards the wide arm. A number of features, configurations, and environmental conditions make this simple structure bend in either direction. Four different variations are described below.

2.1 Changing dimensions

Widths and lengths of the wide arm, the narrow arm, the flexure, and the small segment that connects the narrow and wide arms, are the dimensions that are at the discretion of the designer in the U-shaped actuator. By changing these dimensions, the structure can be made to bend in either direction for the same voltage under identical conditions. For example, in [10], the same cross-section was used for narrow and wide beams but their lengths were different. The flexure was not necessary in that design. Fig. 2 shows another example with a longer and narrower flexure. As we change the length of the flexure, the direction of bending changes. The device shown in the figure bends towards the narrow arm rather than the wide arm.

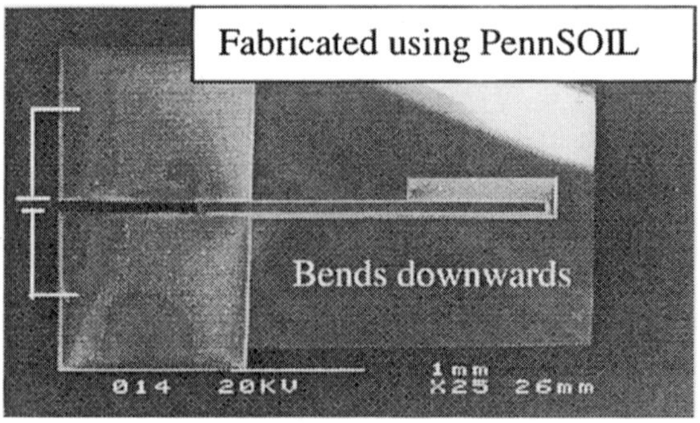

Figure 2. A U-shaped beam with a long flexure

2.2 Changing material properties

Electro-thermal actuators are, in principle, amenable for any conducting material at any size scale. When a semi-conducting material such as silicon is used at the micro scale, an additional freedom arises in easily changing the electrical resistivity of selected portions of the structure by a process called doping. This requires a mask in the doping process in which impurities, usually Boron or Phosphorous, are diffused or ion-implanted into silicon. When only one arm is doped, the structure still bends even though the two arms have the same width. This is because the electrical resistance is decreased for one arm due to doping, which changes the amount of Joule heating generated in that portion [18]. Figure 3 shows this effect. We exploit this in our design methodology as described in Section 4.

2.3 Changing thermal boundary conditions

The Joule heat generated in electro-thermal actuators gets transmitted from the device to its environment in all three modes of heat transfer viz. conduction, convection, and radiation to varying degrees depending on device's geometry, material properties, and the environment. The familiar scaling law states that surface related phenomena dominate over volume related ones when the size goes down because the surface area decreases by an order of magnitude less compared to that of the volume. Therefore, a careful analysis is necessary to see which modes of heat transfer are important for a given device. Figure 4 is an example where convection leads to markedly different behavior. That is, this device would deform down in vacuum and up in normal atmosphere where convection is present.

Conduction of the generated heat to the substrate also plays a significant role [21]. In many instances, the anchored portions are directly connected to the substrate (the wafer in the case of silicon) without any significant thermal insulation in between. In these cases, most of the heat gets conducted away into the substrate. This situation is modeled as a constant ambient temperature condition at the base pads as an essential boundary condition (EBC), also known as Dirichlet boundary condition. However, this is not true when the thermal mass of the substrate is not large enough to maintain the ambient temperature or when the anchors are insulated thermally from the substrate. Then, a natural boundary condition (NBC), also known as Neumann boundary condition, must be imposed which leads to a non-uniform temperature distribution at the base pad that will be known only after completing the thermal analysis.

Radiation becomes important when large voltages are applied to achieve larger forces or displacements. Red-hot temperatures are not uncommon in the usage of these actuators [6, 15, 22]. By varying thermal boundary conditions, the same actuator can be made to behave in different ways. Thus, this is also at the discretion of the designer. We take advantage of this in the design methodology described in Section 4.

2.4 Changing electrical boundary conditions

Yet another variation of the basic electro-thermal actuator is concerned with the way the voltage is applied. Figure 5 shows how the mode of deformation is changed by applying the voltage such that the wide and narrow arms are electrically in parallel. In this case, the wide arm draws larger current because its overall resistance, equal to resistivity times length divided by the area of cross-section, is less than that of the narrow arm. So, the wide arm gets hotter than the narrow one, and the structure deforms in a

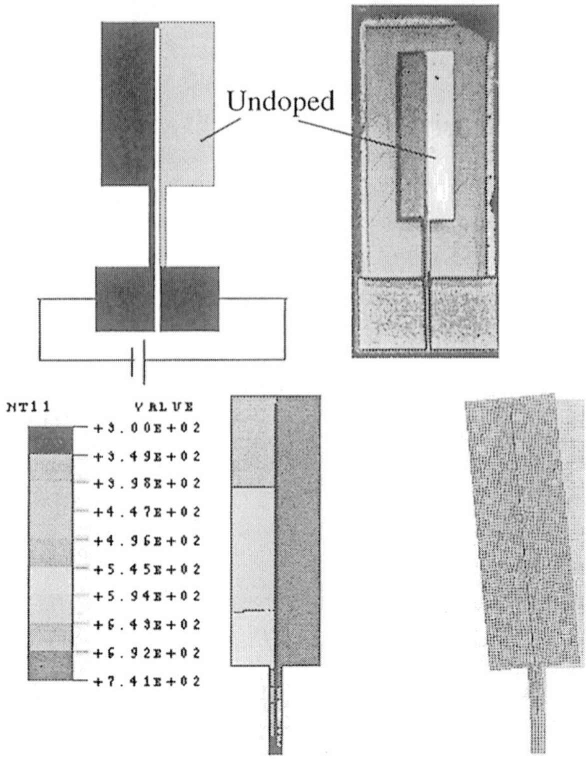

Figure 3. Basic electro-thermal actuator with changed electrical resistivity in one arm via masked doping: shape; prototype fabricated using PennSOIL; temperature distribution; and the deformed model [18]

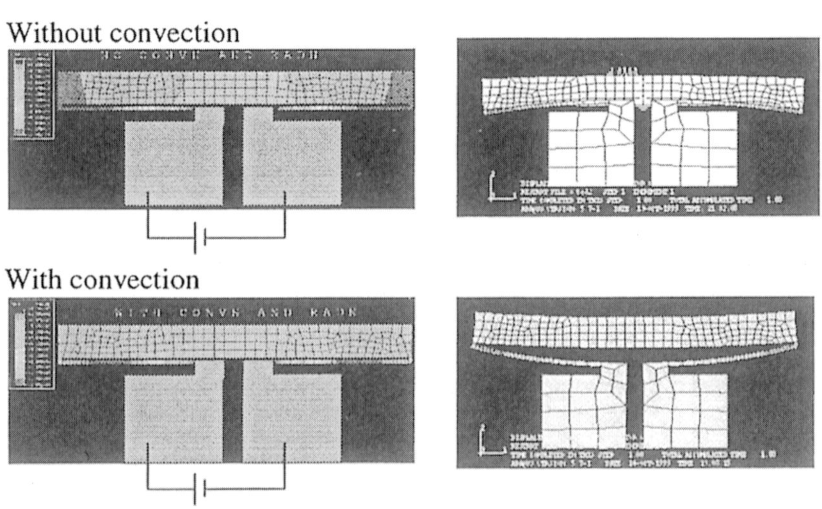

Figure 4. An example of an electro-thermal device in which the mode of deformation changes with and without convection [22]

direction opposite to that of the case shown in Fig. 1. In Fig. 5b, a 25 μm diameter gold wire is bonded using an epoxy to achieve the aforementioned parallel electrical connection. In practice, there are several other ways to achieve this type of parallel connection when a number of such structures are combined as can be seen in the examples presented later in the chapter.

(a)

(b)

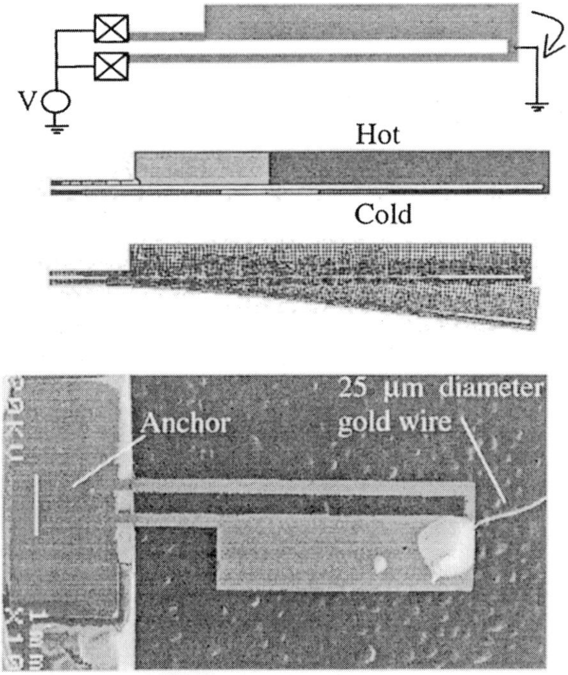

Figure 5. Parallel actuation of the basic electro-thermal actuator (a) shape; temperature distribution; and deformation (b) excimer laser micromachined, mm-size prototype

2.5 Electro-thermal-compliant designs

As seen above, there are a number of ways to modify the basic electro-thermal actuator to change its mode of deformation. The most important point to note about this actuator is the way axial thermal expansion (and therefore the load) is transformed into lateral deflection by virtue of a special shape of the structure. In addition to changing the direction of deflection, the magnitude is also increased to a significant degree. This is because the shape of the structure is such that it is very stiff in the axial direction but flexible in the lateral bending. Thus, a small deflection but large force in one direction is made to result in a large deflection in another direction. In a similar manner, an appropriately shaped elastic structure can be made to deform in a desired manner when subject to a voltage difference between two or more points. That is, we can embed actuator within the mechanism. As an example of an ETC mechanism with embedded electro-thermal actuation is shown in Fig. 6. The shape of the structure and the mechanical, electrical, and thermal boundary conditions enable the structure to deform as desired. Thus, all of these present themselves as variables in the design procedure. The shapes of the devices presented so far in this chapter are all intuitively conceived. In order to take full advantage of the ETC concept (i.e., redirecting and amplifying the thermal expansion by mechanical leveraging) for any desired deformation behavior, a systematic synthesis methodology is useful. Before presenting the synthesis algorithms, we need to understand the three-step analysis of ETC devices.

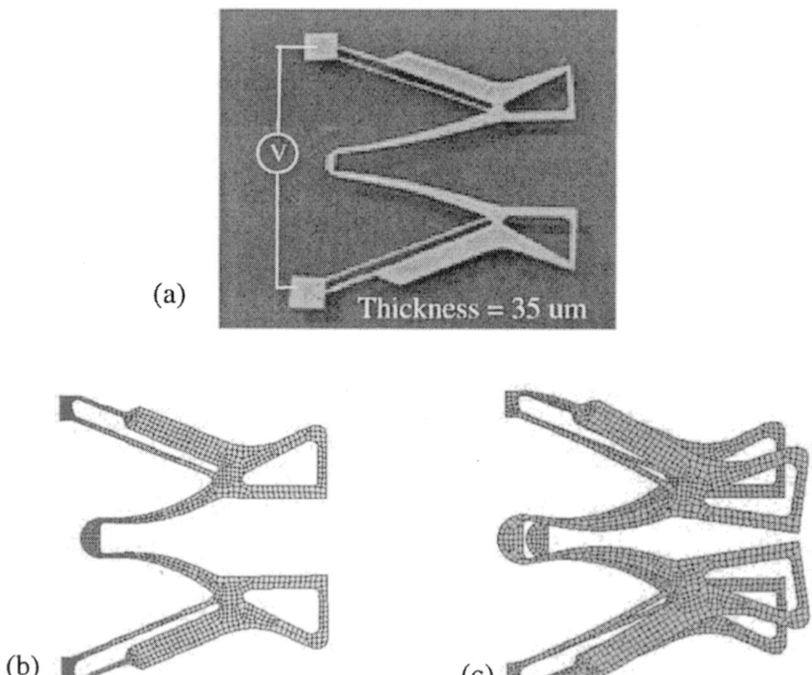

Figure 6. Electro-Thermal-Compliant device (a) micromachined aluminum prototype (b) temperature distribution (c) deformation

3. MODELING

As is typical in MEMS devices, ETC micro devices operate in multiple energy domains, electrical, thermal, and elastic in particular. Therefore, simulation of the behavior of an ETC device entails three analyses steps: electrical analysis to determine the current distribution, thermal analysis to determine the temperature distribution in the presence of Joule heating, and elastic deformation analysis in the presence of thermal loads. These three steps can be carried out sequentially. However, they may be coupled due to various phenomena depending on the material properties, which warrants coupled, self-consistent solution. One significant source of coupling comes from the temperature dependence of properties. This can be accounted for by considering the temperature dependence of electrical and thermal conductivities,

Table 1. Governing equations for the stead-state equilibrium analysis of ETC devices

In partial differential equation form	In weak (variational) form for the 2-D domain
Electrical	
$\nabla \cdot (k_e \nabla V) = 0$ in Ω $V = V_{\text{specified}}$ on Γ_{eE}	$\int_\Omega \nabla^T V\, k_e \nabla V_v\, t d\Omega = 0$
Thermal	
$\nabla \cdot (k_t \nabla T) + \nabla^T V\, k_e\, \nabla V = 0$ in Ω $T = T_{\text{specified}}$ on Γ_{eT} $\mathbf{n} \cdot (k_t \nabla T) = f_{nT}$ on Γ_{nT} where $\mathbf{n}$ is the unit normal to the surface and f_{nT} is the temperature-dependent boundary heat flux that accounts for convection, radiation, and conduction in some cases.	$\int_\Omega \nabla^T \phi\, k_t \nabla \phi_v\, t d\Omega -$ $\int_\Omega \nabla^T V\, k_e \nabla V\, \phi_v\, t d\Omega +$ $\int_\Omega 2h\phi\, \phi_v d\Omega +$ $\sum_e \left(\sum_{k=1}^{N_n} \int_{l_k} h(1-\gamma_k)\phi\phi_v\, t\, dl_k \right) = 0$
Elastic	
$\nabla \cdot \boldsymbol{\sigma} = 0$ $\boldsymbol{\sigma} = \mathbf{E}(\boldsymbol{\varepsilon} - \alpha(T - T_\infty)\mathbf{I})$ $\boldsymbol{\varepsilon} = (\nabla u + \nabla^T u)/2$ $u = u_{\text{specified}}$ on Γ_{eM}	$\int_\Omega \boldsymbol{\varepsilon}^T(u)\mathbf{E}\boldsymbol{\varepsilon}(u_v)\, t d\Omega -$ $\int_\Omega \boldsymbol{\varepsilon}_{th}^T\, \mathbf{E}\boldsymbol{\varepsilon}(u_v)\, t d\Omega = 0$ with $\boldsymbol{\varepsilon}_{th} = \alpha\phi\{1 \quad 1 \quad 0\}^T$ = thermal strain

thermal expansion coefficient, and convective heat transfer coefficients. We first examine the governing equations in all three domains, which are shown in Table 1 and are discussed separately below.

3.1 Electrical analysis

The objective of the electrical analysis is to obtain the current density distribution $k_e \nabla V$, within the structure for specified value of the voltage V on the boundary for a given electrical conductivity k_e, which is constant for a single-material design but will vary spatially in the general case. This is a boundary value problem as can be seen in Table 1. The region of interest is denoted as Ω and its boundary as Γ. The portion on which the voltage is specified is indicated by Γ_{eE}. In the weak (variational) form, V_v denotes the admissible trial function (or variation) for V. Since this is written for 2-D (i.e., planar) structures, t is used to denote the thickness, which is assumed to be uniform everywhere. The closed-form, analytical solutions to this problem exist only for the simplest geometries and boundary conditions of the structures. A finite element or finite difference discretization is necessary to solve this problem numerically. The current density distribution, which is the output of this problem, will appear as the Joule heat source term in the thermal analysis. When temperature dependence of k_e is not considered, this analysis can be done first. Otherwise, it must be done in conjunction with the thermal analysis, which is described next.

3.2 Thermal analysis

Taking the Joule heat term $\nabla^T V \, k_e \nabla V$, convective heat transfer terms from the boundaries (radiation terms are not shown in Table 1), and specified temperature boundary conditions, thermal analysis gives the temperature distribution T, in the entire structure. In the weak form, $\phi = T - T_\infty$ is used instead of T where T_∞ is the ambient temperature. The corresponding trial function is ϕ_v. The material property that is relevant to this analysis is the thermal conductivity k_t, which can either be uniform everywhere in Ω or may vary in the case of multiple materials. The third term in the weak form denotes the convective heat transfer from the top and bottom surfaces of a planar structure, which explains the factor of two in this term. In this term, h is the convective heat transfer coefficient for the interface between the structure and the ambient. The fourth term accounts for convection from the side surfaces. Usually, this applies only to some portion of Γ, and would have been denoted as $\int_{\Gamma_{eT}} h\phi\phi_v \, t \, d\Gamma$. However, the unconventional form, which also happens to be partly in the discretized finite element notation shown here applies "side convection" to all of Ω. Here, γ_k denotes a value associated with k^{th} element out of N_n neighbors of an element. This notation facilitates the synthesis algorithm where holes may be automatically created within the structure. This will be elaborated further in Section 4.1.

When temperature dependence of k_t and h are not considered, this problem can be numerically solved in one step. Otherwise, an iterative procedure is used. A detailed discussion of thermal modeling including the sources of data on material and physical properties is described in [22]. The output of this analysis T, is used as an input to the elastic analysis as discussed next.

3.3 Elastic analysis

The governing equations for the elastic analysis shown in Table 1 do not include the usual actuating mechanical force terms, as they usually do not exist in the ETC problem. Instead, they show thermal loading as an initial stress term, $\mathbf{E}\boldsymbol{\varepsilon}_{th}$ where $\mathbf{E}$ is the material-property matrix relating stress $\boldsymbol{\sigma}$, and strain $\boldsymbol{\varepsilon}$ vectors. The thermal expansion coefficient α, is the other material property in this analysis. The state variable in this problem is the deformation u whose trial function in the weak form is denoted by u_v. The elastic analysis will be the third single-step analysis in the sequential analysis following electrical and thermal analyses when there are no temperature dependent material (k_e, k_t, α) and thermo physical (h) properties. Otherwise, an iterative analysis is necessary. The reader may refer to [22] where a comprehensive modeling was presented.

For accurate modeling, large deformations must be considered in elastic analysis. While this offers no conceptual difficulties, the implementation becomes more involved than the linear modeling. In [19], it was noted that it is necessary to use geometrically nonlinear modeling in the synthesis procedure to get realistic model of the deformation. In what follows, we will first discuss the concepts in synthesis using linear (small deformations and small strains) modeling without considering the temperature-dependent properties and with only essential constraints. In Section 4 the continuum (i.e., differential equation) form is used, and in Section 6 the discretized (numerical) form is used. Later, an advanced example is presented (Section 7).

4. SYNTHESIS

A generalized version of the ETC design problem is shown in Fig. 7. The objective here is to determine the topology (i.e., the number of holes and connectivity) and shape of the structure that fits within a design region and deforms as desired when subject to electrical, thermal, and mechanical boundary conditions as shown. Aided by the analysis procedure outlined above and the material properties, one way to solve this problem is to perform numerous analyses by varying the design of the structure by trial and error and by using "designer's intuition". The key is to identify how many holes are needed, where they should be placed, what their shapes and those of the remaining structure should be to meet desired displacement requirement. Many problem specifications might not be within the reach of intuition of a human designer or they might require many trials before a satisfactory design is obtained. Alternatively, an efficient way to solve this problem is to employ tools of the design optimization, topology optimization [23, 24] in particular. To generate a solution, topology optimization does not need any more input from the user than what is shown in Fig. 7. Like all optimization problems, it needs an objective function to minimize and some constraints. We use the output displacement as the objective function to be maximized (or the negative of the output displacement as the function to be minimized). A constraint on the amount of material used is included. Additional constraints on maximum temperature, maximum stress, current drawn, etc., can be included as necessary. Before presenting a mathematical statement of the problem, we must define what the design variables are. This is done in the next section.

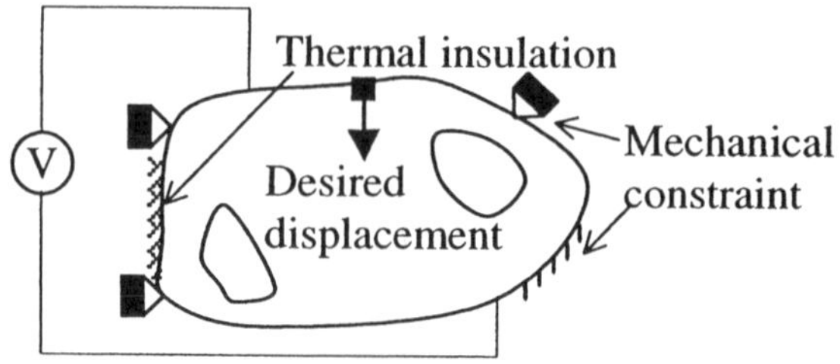

Figure 7. Problem specifications for the synthesis of electro-thermally actuated structures and mechanisms

4.1 Design parameterization

Topology optimization is essentially a "material distribution" problem in that we need to optimally distribute a certain amount of material in a larger design region. It is achieved with a fixed reference domain. That is, when the design region is discretized in the numerical implementation, the mesh is fixed throughout the optimization process. Material "re-distribution" on a fixed mesh is made possible by defining a design variable for each element in the mesh. This variable decides if material will be placed in that element or not. In order to use continuous, gradient-based optimization algorithms, this "binary" variable must be continuous and it should be able to smoothly change the element's status from being empty to being full with material. However, in view of manufacturability, this variable should reach a value such that the element is either empty or full but is not in between at the end of the optimization procedure. The homogenization method, the power law model, etc. are used to define such design variables. See [25] for a review. In this paper, we use a *peak-function material interpolation* model, as it is well suited for multiple-material distribution with fewer variables per element [26, 27].

The equation below shows how a variable ρ defined for an element can smoothly interpolate a material property, E in this case, to assume the status of being filled with material 1 of property value E_1 or material 2 of property value E_2, or being empty with property value E_{void}, which is a very small number typically of the order of 1E-4.

$$E = E_1 e^{-(\rho-\mu_1)^2/2\sigma_1^2} + E_2 e^{-(\rho-\mu_2)^2/2\sigma_2^2} + E_{void} \tag{1}$$

The above equation serves as a "material selector" because, as shown in Fig. 8, the value of ρ selects the material. Pre-defined parameters μ_1 and μ_2 "locate" the materials on the variable-axis, and σ_1 and σ_2 determine the sharpness of the material selection. In the material interpolation curves of Fig. 8, the values σ_1 and σ_2 decrease from the outermost curve to the innermost "peak function" curve. So, we begin optimization procedure with large values of σ_1 and σ_2, and decrease them steadily to finish with peaks at the end. Thus, when optimization converges, each element in the design region will contain material 1 ($\rho = \mu_1$) or material 2 ($\rho = \mu_1$), or will be empty (any other value of ρ). In a similar manner, we can interpolate any number of properties of any number of materials. The following are relevant for the ETC design problem for a two-material design. Note that convective heat transfer coefficient is also interpolated in the same way as the four material properties.

$$\begin{Bmatrix} k_e \\ k_t \\ \alpha \\ \mathbf{E} \end{Bmatrix} = \sum_{m=1,2} e^{-(\rho-\mu_m)^2/2\sigma_m^2} \begin{Bmatrix} k_e \\ k_t \\ \alpha \\ \mathbf{E} \end{Bmatrix}_m + \begin{Bmatrix} k_e \\ k_t \\ \alpha \\ \mathbf{E} \end{Bmatrix}_{void} \tag{2a}$$

$$h = \sum_{m=1,2} e^{-(\rho-\mu_m)^2/2\sigma_m^2} h_0 \tag{2b}$$

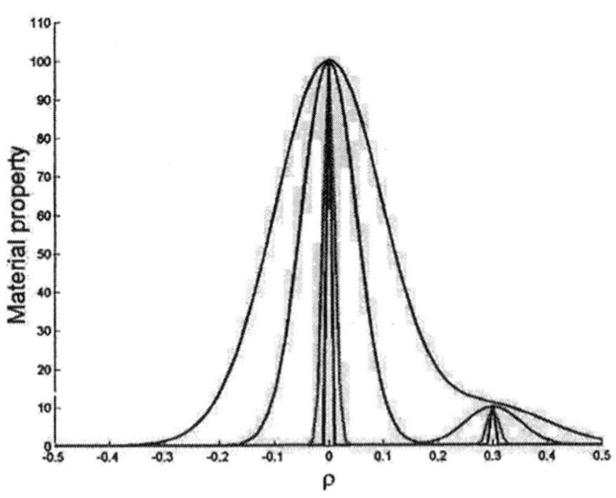

Figure 8. Material selection with a single variable, ρ

The interpolation of h needs additional explanation. Here, h_0 is the heat transfer coefficient at the face of an element when exposed to the ambient. If the element becomes empty, convection does not take place from that element. If it were filled with material 1 or 2, the element would convect. When the element is in between, such as the case for most elements until the end of optimization, the convection would also be *intermediate* for that element. Although one can explain intermediate convection by way of a porous material, such an explanation is not really necessary because, at the end, every element is essentially either there or not there but is not in between. Returning to the third and fourth terms in the weak form of the thermal equilibrium equations in Table 1, one can now understand how convection is handled in topology optimization where number, sizes, shapes, and locations of holes may vary freely. Convection from the top and bottom faces of an element (third term) is easy to model in the standard form. To model convection from the side surfaces of an element, we should look at its neighbors. If a neighboring element is "empty", then full side-convection takes place at that element but not otherwise. Therefore $(1-\gamma)$ is included where γ is the design variable ρ of the neighboring element.

4.2 Problem statement

The optimization problem can now be stated mathematically as follows.

Minimize $(-u_{out})$

With respect to

ρ_i, $i = 1,2,\ldots,$ number of elements

Subject to

Three equilibrium equations in the weak form as shown in Table 1

Volume constraint: $$\int_{\Omega}\left(\sum_{m=1}^{2} w_m\, e^{-(\rho-\mu_m)^2/2\sigma_m}\right)d\Omega \le \Omega^* \quad (3)$$

where w_m are pre-specified weights, and Ω^* is the upper limit on the amount of material to be used.

4.3 Solution procedure

By writing the Lagrangian of the constrained minimization problem in Eq. (3), and taking its first variation with respect to the design variable ρ, we obtain the first order necessary condition in the following form:

$$A + \Lambda D = 0 \quad (4)$$

where

$$A = \nabla^T V \frac{\partial k_e}{\partial \rho} \nabla V_v t + \nabla^T \phi \frac{\partial k_t}{\partial \rho} \nabla \phi_v t - \nabla^T V \frac{\partial k_e}{\partial \rho} \nabla V \phi_v t +$$

$$2\frac{\partial h}{\partial \rho} \phi \phi_v + \sum_{k=1}^{N_n} \int_{l_k} \frac{\partial h}{\partial \rho} (1-\gamma_k) \phi \phi_k t dl_k - \sum_{k=1}^{N_n} \int_{l_k} h_k \frac{\partial \gamma}{\partial \rho} \phi \phi_k t dl_k + \tag{4a}$$

$$\varepsilon^T(u) \frac{\partial E}{\partial \rho} \varepsilon(u_v) t - \varepsilon_{th}^T \frac{\partial \mathbf{E}}{\partial \rho} \varepsilon(u_v) t - \frac{\partial \varepsilon_{th}^T}{\partial \rho} \mathbf{E} \varepsilon(u_v) t$$

$$D = \sum_{m=1}^{n} -(\rho - \mu_m)/\sigma_m^2 \, w_m \exp\left(-\frac{(\rho - \mu_m)^2}{2\sigma_m^2} \right) \tag{4b}$$

and γ_k is the interpolated value in the neighborhood element k. The values of V_v, ϕ_v, and u_v in the above equation are computed by solving the following equations which arise when the first variation of the Lagragian of the problem is taken with respect to V , ϕ, and u , respectively.

$$\int_\Omega \nabla^T \delta V \, k_e \nabla V_v \, t \, d\Omega - 2\int_\Omega \nabla^T \delta V \, k_e \nabla V \, t \, d\Omega = 0 \tag{5a}$$

$$\int_\Omega \nabla^T \delta\phi k_t \nabla \phi_v t d\Omega + \int_\Omega 2h \delta\phi \phi_v d\Omega +$$

$$\sum_e \left(\sum_{k=1}^{N_n} \int_{l_k} h(1-\gamma_k) \delta\phi \phi_v t dl_k \right) \tag{5b}$$

$$- \int_\Omega \delta \varepsilon_{th}^T E \varepsilon(u_v) t d\Omega = 0$$

$$\int_\Omega \varepsilon^T(\delta u) E \varepsilon(u_v) t d\Omega - \delta u_{out} = 0 \tag{5c}$$

where δV , $\delta\phi$, δu are admissible arbitrary perturbations from the actual physical fields. The solution of the above equations proceeds in the following order. First, we solve the three physical equilibrium equations shown in Table 1, then we solve Eq. (5c) followed by Eq. (5b), and Eq. (5a) at the end. The Lagrange multiplier Λ of the resource constraint of Eq. (3) is obtained as described next. From the first order Taylor series approximation, we get

$$\int_{\Omega} \sum_{m=1}^{n} w_m \exp\left(-\frac{(\rho^{(k+1)} - \mu_m)^2}{2\sigma_m^2} \right) d\Omega = \int_{\Omega} B^{(k)} d\Omega + \int_{\Omega} D^{(k)} (\rho^{(k+1)} - \rho^{(k)}) d\Omega \leq \Omega^* \tag{6}$$

where

$$\sum_{m=1}^{n} w_m \exp\left(-\frac{(\rho - \mu_m)^2}{2\sigma_m^2} \right) = B^{(k)} \tag{6a}$$

Λ can now be obtained in the analytical form:

$$\Lambda = \frac{\Omega^* - \int_{\Omega} B^{(k)} d\Omega + \int_{\Omega} D^{(k)} \rho^{(k)} d\Omega - \int_{\Omega_l} D^{(k)} \rho_l d\Omega - \int_{\Omega_u} D^{(k)} \rho_u d\Omega}{\int_{\Omega_a} D^{(k)} D^{(k)} d\Omega} \tag{7}$$

where the domain Ω_l refers to the domain controlled by the lower moving limit, Ω_u to the domain controlled by the upper moving limit, and Ω_a to the active domain controlled by Eq. (3). As is usual in this method, an inner loop is needed to decide these domain partitions.

Using Eq. (3), we update the design variables (ρ for each element) iteratively using the optimality criteria method, which has been successfully used in the topology optimization.

$$\rho^{(k+1)} = \rho^{(k)} - \kappa \left\{ \rho^{(k)} + \Lambda^{(k)} D^{(k)} \right\} \tag{8}$$

where k refers to iteration number and κ is some constant to control the step-length of the update.

5. NUMERICAL EXAMPLES

First, two examples are presented to illustrate that different optimal topologies could result if we include or do not include convection from top, bottom, and side surfaces. In these two examples, a heat flux was specified rather than Joule heating but the observations made here are equally applicable to the ETC design problem. The subsequent four examples include convection and are the optimal solutions for the ETC design problem. The data for the examples assumed silicon as one material and SiO_2 as the other material. This material property data, primarily taken from [28] are shown in Table 2. In all the examples, the environmental temperature is assumed to exist at the mechanical anchors and it is imposed as an essential (Dirichlet type) boundary conditions in the thermal analysis problem.

Table 2. Material and thermo-physical properties used in the numerical examples

Property	*Silicon*	*Silicon dioxide*
Electrical resistivity	$1\ \Omega - cm$	Insulator
Thermal conductivity	0.157 W/mm-K	0.0014 W/mm-K
Young's modulus	190 GPa	73 GPa
Thernal expansion coefficient	2.33E-6	0.55E-6
Heat transfer coefficient (arbitrary)	0.00025 W/mm^2-K	0.00025 W/mm^2-K

The specifications for the first example are shown in Fig. 9a. As shown in the figure, heat flux is supplied from the two sides of a rectangular domain fixed mechanically at the four corners. The middle point of the top edge is prescribed to move upwards. The resulting optimal designs, as per the problem stated in Eq. 3, are shown for three cases in Figs. 9b-9d. The three cases are, respectively, without convection, with convection only from top and bottom, and with convection from all surfaces including the side surfaces. The filled squares represent silicon dioxide, and the center-filled ones silicon. It can be clearly seen that topologies and shapes are quite different when convection is included in the model and when it is not. When convection is not modeled, the only mode of heat transfer from the device is by conduction through the mechanical anchors that are specified to be at the environmental temperature. Therefore, the material with larger thermal conductivity connects to the corners. This is not the case when convection is included. As can be seen in Figs. 9c and 9d, lower conductivity material also connects to the corners. In all the cases, the heat flow into the output portion is shielded by placing lower conductivity material. This is good from the viewpoint of practical application because keeping that part at reasonably low temperature is essential in most applications. It should be noted that there is virtually no difference between the solutions obtained with and without side surface convection. The reason for this is that the thickness of the structure is much smaller (as is generally true in MEMS devices) and that leads to very little surface area on the sides. However, when temperatures get really high, this convection also becomes important (see [22]). An example where larger thickness is assumed is presented next.

As shown in Fig. 10a, heat flux is supplied at the bottom edge of the rectangular design domain anchored at the four corners. A non-intuitive displacement behavior is prescribed by asking that the middle point of the top edge move down as a result of heating. The principal use of topology optimization becomes apparent in these non-intuitive problems where it will be difficult for human designers to conceive a solution. This problem was solved by assuming that the thickness is 2 mm. As to be expected, the results without and with side surface convection are substantially different (see Figs. 10b-c). In both the designs it should noted that the output is prevented from getting excessive heat so that the central vertical bar does not expand much. This makes the output point move down due to the forces from the remainder of the structure rather than go up by its own free expansion. In the four ETC design examples that follow, we modeled convection from top, bottom, and side surfaces.

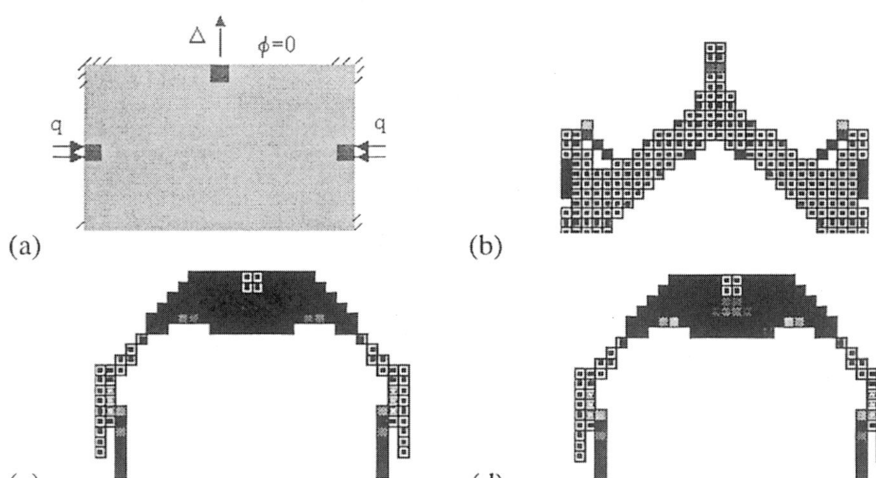

Figure 9. A synthesis example with heat-flux actuation (a) problem specification (b) solution without convection (c) with top and bottom convection (d) with side convection also included. Filled squares indicate SiO_2, and center-filled ones silicon. Side convection did not influence the design because the thickness was too small (from [27]).

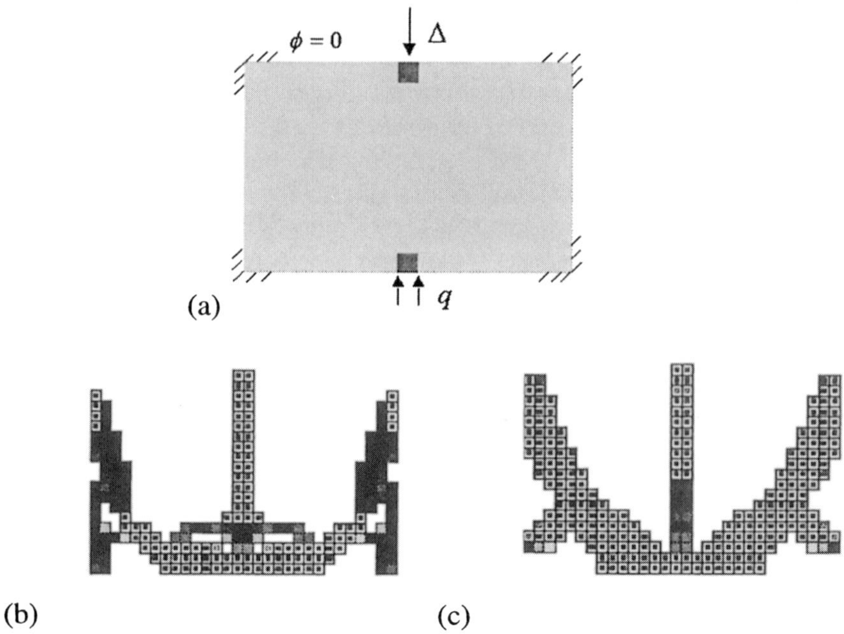

Figure 10. Example to illustrate that side convection will influence the result if thick structures are considered (thickness here was 2 mm) (Completely filled: SiO_2 and center-filled: Si) From [27].

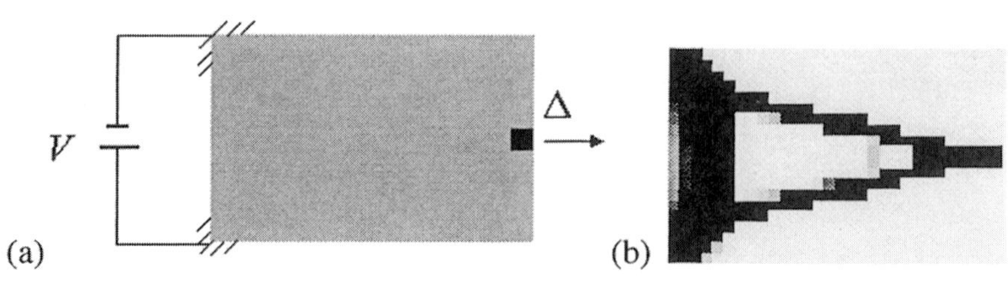

Figure 11. Electro-Thermal-Compliant (ETC) design example 1 with only one material (Si). From [27].

In all the ETC design examples, 30 V was applied between the points indicated in the figure showing the specifications of the problem. The first ETC design example is a straightforward and intuitive specification shown in Fig. 11a. The design solution shown in Fig. 11b has the shape that makes the output displace to the right by a large amount. This example was solved using a single material. The second example (Fig. 12a) is less straightforward. As can be seen in Fig. 12b, the topology is not completely clear and there is a lot of *intermediate material* in SiO_2. This happens however small the σ corresponding to this material is in the material interpolation function. That is, even if the distribution function becomes a sharp peak, the ρ value is ever so slightly different from μ value that the interpolated value falls in between full and empty. In other words, the algorithm insists on such a design. This behavior is not uncommon in body-force (like gravity and inertia forces) problems where more material means more force. Here, the thermal force gets larger when there is more material. So, the algorithm places only the intermediate material there. To check that the algorithm is going in the right direction, we manually sketched a design shown in Fig. 12c with only Si and SiO_2 and not intermediate material. By analyzing it, we see that the expected behavior is indeed obtained as can be seen in Fig. 12d where the output point moves up as desired. However, the deflection in the horizontal deflection is much more. To avoid this, further constraints are necessary as is done in [19] and [20]. Furthermore, designs with intermediate material are not practically feasible. Employing a technique to avoid intermediate material is quite common in topology optimization procedures. So, we modified the solution algorithm towards the end of the iterative process to eliminate the intermediate material as explained below.

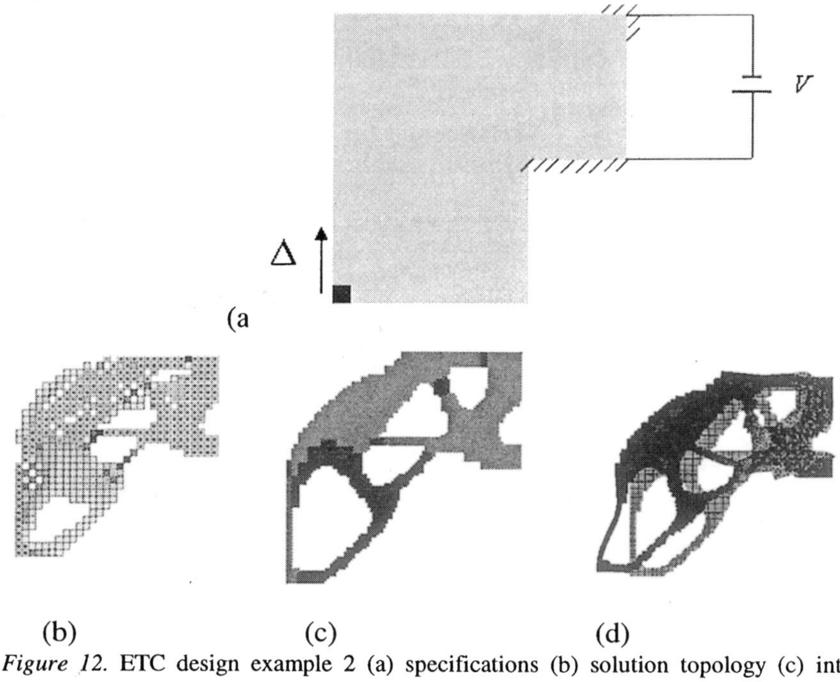

Figure 12. ETC design example 2 (a) specifications (b) solution topology (c) interpreted design (d) analysis of the interpreted design. From [27].

The modification entails adding a penalty term to the objective function towards the end after a problem converges and has some intermediate density. This term penalizes the intermediate values as can be seen in the equation below.

$$f_{obj}(new) = f_{obj}(old) + \alpha \sum_{e} \gamma(1-\gamma) \tag{10}$$

where γ is the interpolated value using the coefficient term in Eq. (2a) without. Since we minimize the objective function, the additional term will also be minimized. Therefore, if there is a value in between 1 and 0, it will be pushed to either extreme. The weighting factor α is chosen as 2E-3 based on some numerical experiments where we observed the objective function value to fall within a certain range.

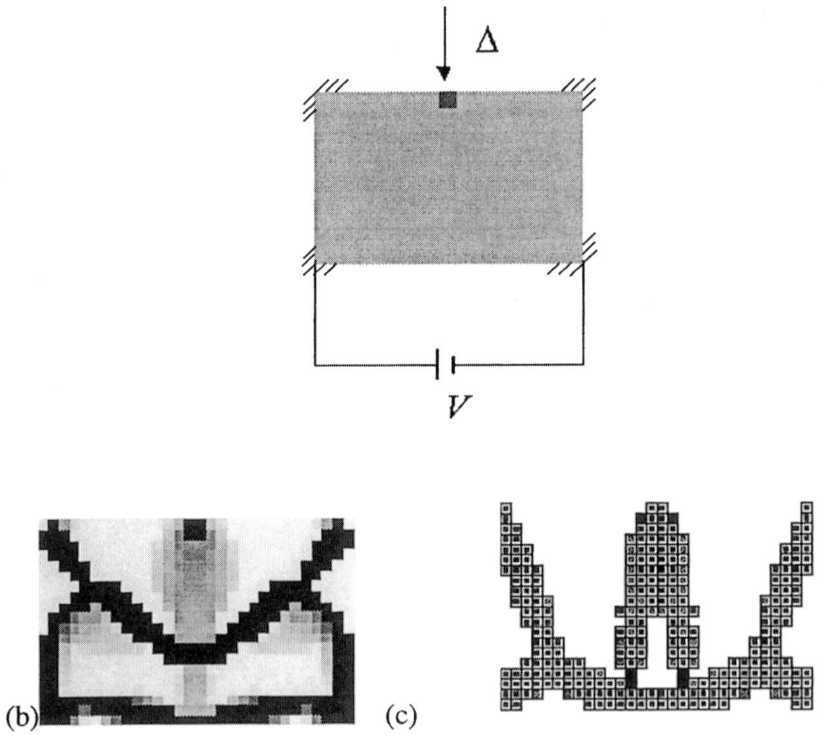

Figure 13. Third ETC synthesis example (a) Problem specification (b) Single material (Si) design (c) Two-material (Si (center-filled) and SiO_2 (completely filled)) design [27]

The third and fourth ETC design examples were solved using this modification as they both resulted in intermediate density with the original algorithm. As can be seen in Figs. 13 and 14, these examples were solved using a single material (Si) and two materials (Si and SiO_2). With a single material, the intermediate material could not be avoided even with this modification. This is perhaps due to the solution method used as [19, 20] report good results even with one material. In both the examples, the algorithm wants to make a structural connection to the output point but does not seem to want to transfer current or heat substantially. It achieves it with the intermediate material. However, when two materials are used, it can achieve this by placing the some SiO_2 so that no current passes through that segment, and thus less heat is given to that part. This is especially needed for the specification in Fig. 13 where the point at the top of the central vertical bar needs to move down. More heat in that bar would cause it to expand freely in the upward direction. Thus, although fabricating two-material structures is more cumbersome, there could be some advantages to heterogeneous structures. It is appropriate to note here that *functionally graded materials* and heterogeneous materials are quite common in biological designs. Topology optimization techniques can generate such designs, but for practical utility such designs are circumvented by additional techniques such as the one shown in Eq. (10).

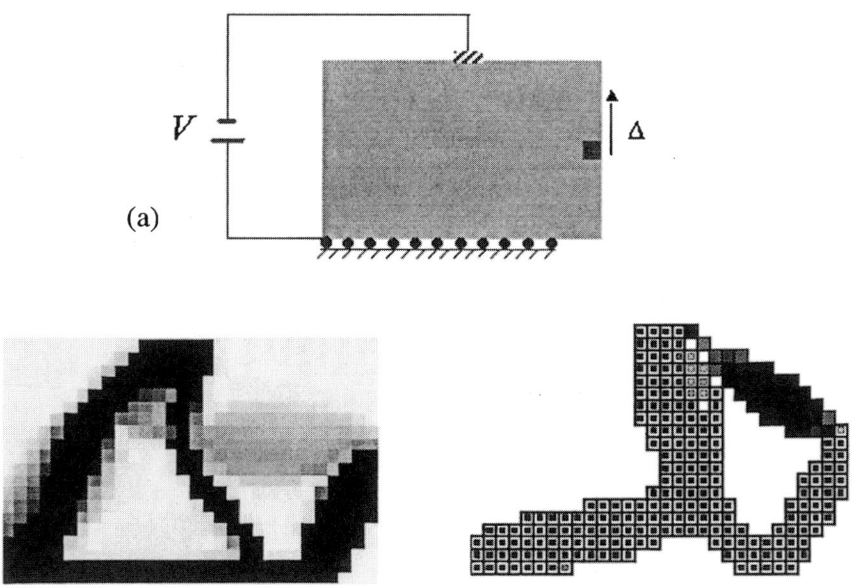

(b) (c)

Figure 14. ETC synthesis example 4. (a) specifications (b) solution with a single material (c) solution with Si (center-filled) and SiO_2 (fully filled) [27]

6. ALTERNATIVE IMPLEMENTATION USING "LINE ELEMENTS"

The topology synthesis formulation and its solution were presented in Section 4 without a detailed discussion of the implementation issues. To go from the equations of that section to a finite element implementation used to solve the examples of Section 5, there is considerable, albeit routine, work. Although the reader can refer to any standard textbook on finite element analysis to accomplish this, seeing a version of the synthesis problem in terms of the discrete finite element model may be beneficial at this point. In this section, we present an alternative implementation for the same synthesis problem using "line" finite elements. In addition to providing implementation details, this section also highlights the fact that a topology synthesis formulation is not unique and that any design parameterization can be used. Different formulations and design parameterizations have their own advantages and disadvantages. We compare and contrast the approach with planar continuum elements with that of the line elements. This section is adapted from [29].

6.1 Line elements

Most ETC devices contain slender or narrow segments, which more or less dictate the behavior of the device. For such segments a beam model may be just as appropriate as the continuum elements in elastic analysis, if not more. Similar beam-like models are suitable for electrical and thermal analyses. Towards this end, a "line element" (see Fig. 15) is defined here as a one-dimensional finite element model of the electrical, thermal, and elastic behaviors of a beam-like continuum. The design parameterization for line elements uses a variation of the *ground structure* [30] approach where the design region is divided into a mosaic of unit cells each containing eight line elements: two vertical, two horizontal, and four diagonal. Figure 16 shows a possible ground structure with specifications for a typical ETC synthesis problem. The design variables in the optimization problem are now the cross-section dimensions of the line elements. In the micromachined structures, since the thickness is usually the same everywhere, only the width of the line element becomes the design variable. The line elements have the following advantages over the continuum planar elements: 1) A coarse mesh for the design region could be used without significantly compromising on the resulting solutions. A coarse mesh reduces computational time in analysis and also in synthesis. 2) The resulting solutions adhere to the geometry of the mesh (e.g., orientation of segments as decided by the shape of the unit cell, widths of segments as decided by lower and upper bounds on them, etc.) so that a micromachining process can be chosen a priori. 3) Since the shape of an element is known, modeling surface heat transfer via convection and radiation becomes easy and accurate (see Fig. 15). For example, intra-device convection and sidewall convection can be easily incorporated [22]. The disadvantages include restricted geometries in the resulting solutions, and any modeling errors because of the one-dimensional abstraction in electrical, thermal, and elastic analyses. We will now proceed to write the governing equations using the line finite elements.

6.2 Finite element modeling with line elements

For the electrical analysis, each line element can be viewed as an electrical resistor. Voltage at each grid point is the nodal variable in the finite element formulation. The electrical resistivity (conductivity), the length of the element, and the cross-section area determine the electrical conductivity matrix for each element. By assembling the element conductivity matrices into a global matrix, we get the discretized version of the electrical boundary value problem of Table 1 as follows.

$$\mathbf{K}_e \mathbf{v} = 0, \qquad v = v_e \text{ on } \Gamma_{eE} \tag{11}$$

where $\mathbf{K}_e$ is the global electrical conductivity matrix, and $\mathbf{v}$ is the nodal voltage vector. The applied voltage at the nodes on the boundary Γ_{eE} is denoted as v_e. The above equation assumes that there are no current sources in the structure, which is a valid assumption for ETC devices. The design variables describing the cross-section of the line elements appear in $\mathbf{K}_e$ along with the electrical conductivity of the material k_e.

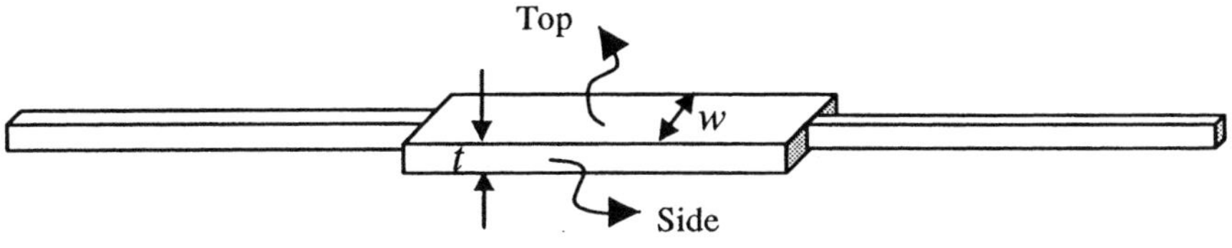

Figure 15. Three line elements with different widths. Line elements make accurate modeling of convection and radiation from top and side surfaces easy even in topology optimization.

In a similar way, the thermal boundary value problem can be shown in the discretized form as

$$\mathbf{K}_t\mathbf{T} = \mathbf{Q}, \qquad T = T_e \text{ on } \Gamma_{eT} \tag{12}$$

where $\mathbf{K}_t$ is the global thermal conductivity matrix, $\mathbf{T}$ is the nodal temperature vector, and $\mathbf{Q}$ is the thermal load vector comprising volumetric Joule heat generation and the convective heat loss from the surface. Joule heat is a function of $\mathbf{v}$ and the electrical conductivity of the material k_e. The imposed temperature at the nodes on the boundary Γ_{eT} is denoted as T_e. As in the electrical problem, the design variables describing the cross-section of the line elements appear in $\mathbf{K}_t$ along with the thermal conductivity of the material k_t.

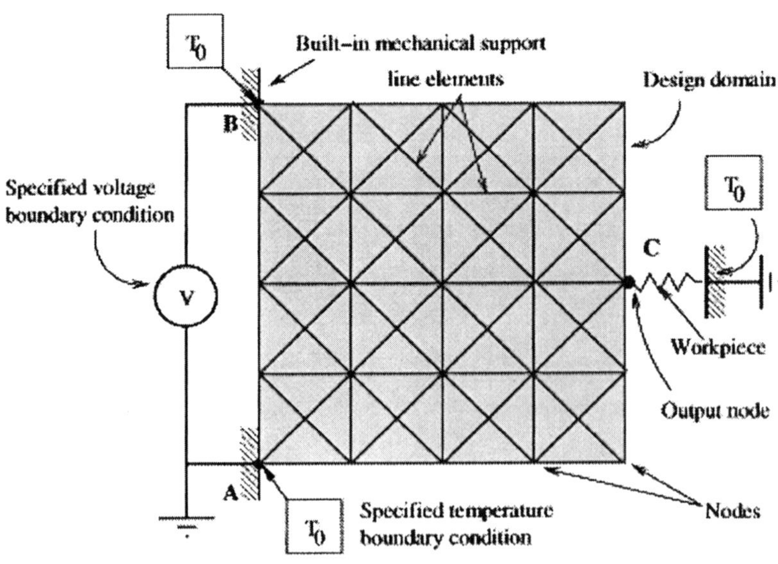

Figure 16. Ground structure with line elements. The width of each line element contributes a design variable in the synthesis procedure. The workpiece is shown as a linear spring.

The discretized equation for the boundary value problem of elastic equilibrium can now be written as

$$\mathbf{K}_m\mathbf{u} = \mathbf{f}_t, \qquad u = u_e \text{ on } \Gamma_{eM} \tag{13}$$

where $\mathbf{K}_m$ is the global stiffness matrix, $\mathbf{u}$ is the nodal displacement vector consisting displacements in the x and y directions, and $\mathbf{f}_t$ is the mechanical load vector due to induced thermal strains. The imposed displacement at the nodes on the boundary Γ_{eM} is denoted as u_e. The design variables appear in $\mathbf{K}_m$ along with Young's modulus E.

The details of computing element matrices and element load vectors and assembling into global matrices and vectors can be found in one of the many engineering textbooks on finite element analysis [e.g., 31]. A sequential solution of the above linear systems of equations is straightforward. If temperature-dependent material and physical properties are considered, then an iterative solution is necessary. If large deformations are included, elastic analysis by itself will be nonlinear and should be solved iteratively. In the examples presented in this section, small displacements are assumed. Therefore, the resulting solution will have the desired behavior only for a small range of motion.

6.3 Problem formulation

In Section 4.2, the problem statement implied that we want to maximize the displacement in the intended direction at an intended point subject to an upper limit on the volume of material. In most cases, there is also an output load against which the device must work. Towards this end, it is not sufficient to simply maximize the output displacement; the device must also be able to support the load and therefore, it must be reasonably stiff. This latter consideration can be met by minimizing the strain energy. Thus, the optimization problem may be formulated as follows:

$$\underset{\mathbf{w}}{Min} \quad f = -\frac{cu_{out}}{SE} \tag{14}$$

Subject to

Equilibrium equations: Eqs. (11) to (13)

Volume constraint: $\sum_{i=1}^{N_{elem}} l_i t w_i - V^* \leq 0$

where u_{out} is the output displacement, c is a factor to make the numerator and the denominator to be of the same order of magnitude in the objective function, SE is the strain energy, $\mathbf{w}$ is the array variable representing widths of N_{elem} line elements with w_i representing the width of the ith

element, t is the uniform thickness of all elements, and V^* is the available volume of material. Even though in MEMS devices, the economy of material is not an important consideration, limiting the available material helps the algorithm in forming the empty portions that define the functional character of the device.

As shown in Fig. 16, the workpiece from which an output load arises can be included in the model as a spring of given stiffness. In the linear modeling assumed here, constant spring rate may be assumed. If desired, one can also model the electrical and thermal resistances of the work-piece. This may be necessary if the work-piece is electrically and thermally conducting to a significant degree.

6.4 Sensitivity analysis and solution procedure

The sensitivity analysis and the solution procedure proceed along the same lines as it was done in Section 4.3 for continuous elements. It is presented again here to show the implementation details.

We begin the sensitivity analysis by taking the derivative of the objective function in Eq. (14) with respect to the ith design variable w_i.

$$\frac{\partial f}{\partial w_i} = -c\left(\frac{1}{SE}\frac{\partial u_{out}}{\partial w_i} - \frac{u_{out}}{SE^2}\frac{\partial SE}{\partial w_i}\right) \tag{15}$$

It is convenient to represent u_{out} as mutual strain energy (MSE) [30] where

$$MSE = \mathbf{u}^T \mathbf{K}_m \mathbf{u}_d \tag{16}$$

with $\mathbf{u}$ and $\mathbf{K}_m$ denoting the displacement and global stiffness matrix, respectively, as in Eq. (13), and $\mathbf{u}_d$ as the displacement obtained under a unit dummy load applied at the output. It satisfies

$$\mathbf{K}_m \mathbf{u}_d = \mathbf{f}_d \tag{17}$$

where $\mathbf{f}_d$ is the unit force vector containing 1 in the degree of freedom corresponding to the output and zeros everywhere else. The value of SE is given by

$$SE = \frac{1}{2} \mathbf{u}^{\mathrm{T}} \mathbf{K}_{\mathrm{m}} \mathbf{u} \tag{18}$$

Differentiation of Eqs. (16) and (18) gives the sensitivities of SE and MSE with respect to i^{th} variable w_i as follows.

$$\begin{aligned} \frac{\partial MSE}{\partial w_{\mathrm{i}}} &= \left(\frac{\partial \mathbf{u}_{\mathrm{d}}}{\partial \mathrm{w}_{\mathrm{i}}}\right)^{\mathrm{T}} \mathbf{f}_{\mathrm{t}} + \left(\frac{\partial \mathbf{f}_{\mathrm{t}}}{\partial w_i}\right)^{\mathrm{T}} \mathbf{u}_{\mathrm{d}} \\ \frac{\partial SE}{\partial w_i} &= \frac{1}{2}\left[\left(\frac{\partial \mathbf{u}}{\partial w_i}\right)^{\mathrm{T}} \mathbf{f}_{\mathrm{t}} + \left(\frac{\partial \mathbf{f}_{\mathrm{t}}}{\partial w_i}\right)^{\mathrm{T}} \mathbf{u}\right] \end{aligned} \tag{19}$$

The derivatives $\frac{\partial \mathbf{u}}{\partial w_i}$ and $\frac{\partial \mathbf{u}_d}{\partial w_i}$ can be computed by differentiating Eqs. (13) and (17) using the standard procedures [32]. The presence of $\frac{\partial \mathbf{f}_{\mathrm{t}}}{\partial w_i}$ in Eq. (19) couples the elastic sensitivity analysis with that of the thermal sensitivity analysis. The thermal load vector $\mathbf{f}_{\mathrm{t}}$ and its derivative with respect to w_i are given by:

$$\begin{aligned} \mathbf{f}_{\mathrm{t}} &= \sum_{j=1}^{N_{elem}} E_j w_{\mathrm{j}} t \,\varepsilon_{\mathrm{j}} \mathbf{M}_{\mathrm{j}} \\ \frac{\partial \mathbf{f}_{\mathrm{t}}}{\partial w_i} &= \sum_{j=1}^{N_{elem}} E_j \mathbf{M}_{\mathrm{j}} \left(t\,\varepsilon_{\mathrm{j}} \delta_{\mathrm{ij}} + w_{\mathrm{j}} t \frac{\partial \varepsilon_j}{\partial w_{\mathrm{i}}} \right) \\ \delta_{\mathrm{ij}} &= \begin{cases} 0 & i \neq j \\ 1 & i = j \end{cases} \end{aligned} \tag{20}$$

where $\mathbf{M}_j$ is a matrix for transforming between the local and global coordinate systems, E_j is the Young's modulus and ε_j is the thermal strain for the jth element. Next, the jth component of the thermal strain vector and its derivative are given as

$$\varepsilon_j = \frac{\alpha_j}{2}(T_1 + T_2)_j$$
$$\frac{\partial \varepsilon_j}{\partial w_i} = \frac{\alpha_j}{2}\left(\frac{\partial \mathrm{T}_1}{\partial w_i} + \frac{\partial \mathrm{T}_2}{\partial w_i}\right) \tag{21}$$

The temperatures at the two nodes of the jth element are denoted by T_1 and T_2, and α_j is the coefficient of linear thermal expansion for the element. The temperature derivatives in Eq. (21) are obtained by differentiating the thermal equilibrium equation, i.e., Eq. (12) as

$$\mathbf{K}_t \frac{\partial \mathbf{T}}{\partial w_i} = \frac{\partial \mathbf{Q}}{\partial w_i} - \frac{\partial \mathbf{K}}{\partial w_i}\mathbf{T} \tag{22}$$

The jth component of the heat flux load vector in the above equation, Q_j, and its derivative are obtained from

$$Q_j = \frac{l_j}{2}\left(\frac{(\mathrm{v}_2 - \mathrm{v}_1)^2}{R_j l_j} + 2h_j(w_j + t)\mathrm{T}_\infty\right)$$
$$\frac{\partial Q_j}{\partial w_i} = \frac{1}{R_j}(\mathrm{v}_2 - \mathrm{v}_1)\left(\frac{\partial \mathrm{v}_2}{\partial w_i} - \frac{\partial \mathrm{v}_1}{\partial w_i}\right) + \left(\frac{l_j(v_2 - v_1)^2}{2R_j w_j} + l_j h_j \mathrm{T}_\infty\right)\delta_{ij} \tag{23}$$
$$R_j = \frac{l_j \rho_j}{w_i t}$$

where v_2 and v_1 are the voltage potentials at the two nodes, R_j is the electrical resistance, T_∞ is the ambient temperature, l_j is the length, h_j is the film heat transfer coefficient and ρ_j is the electrical resistivity for the jth element. The voltage derivatives in Eq. (23) are obtained by differentiating the static electric potential equilibrium equation, i.e., Eq. (11) as

$$\mathbf{K}_e \frac{\partial \mathbf{v}}{\partial w_i} = -\frac{\partial \mathbf{K}_e}{\partial w_i}\mathbf{v} \tag{24}$$

With all the sensitivities computed as described above, the stage is now set to submit the optimization problem to a numerical optimization algorithm. The following examples were solved using Matlab's optimization toolbox, the function *fmincon* in particular. This function uses a combination of sequential quadratic programming and the truss region method.

6.5 Numerical examples with line elements

The material properties of silicon and film coefficients for MEMS scale used in solving the example problems are shown in Table 3. The specifications for the first example are in Fig. 16. A potential difference of 10 V was applied between points A and B. The obtained solution is shown in Fig. 17a. The widths of the lines indicate their relative sizes and the gray scale indicates the temperature, the light color implying high and the dark low values. The elements that have "vanished", i.e., reached their lowest limit (1E-6 μm), are shown as dotted lines. The initial guess was a uniform width value for all the elements such their total volume was 15% of the maximum possible value. It has an objective function value of 0.45. A positive value of f indicates that the output point was actually moving opposite to the intended direction. The optimized solution had a f value of –6.977. Furthermore, as shown in Figs. 17b and 17c, the topology obtained here agrees with an intuitively conceived design [33] and a design obtained using continuum elements [19], respectively. Of course, the ground structure approach used here restricts the orientation of the elements to 0^0, 45^0, or 90^0.

Table 3. Data used for the line element based synthesis examples

Electrical resistivity = 2.3E-5 $\Omega - m$
Thermal conductivity = 141.2 W/m-K
Coefficient of thermal expansion = 2.0E-6
Ambient temperature = 300 K
Film coefficient = 250 W/m^2 (estimated)
Young's modulus = 169E9 Pa
20 % volume constraint.
Work-piece:
Electrical resistance = 3.1E4 Ω
Thermal resistance = 9.5E6 K/W
Spring stiffness = 3.0E4 N/m

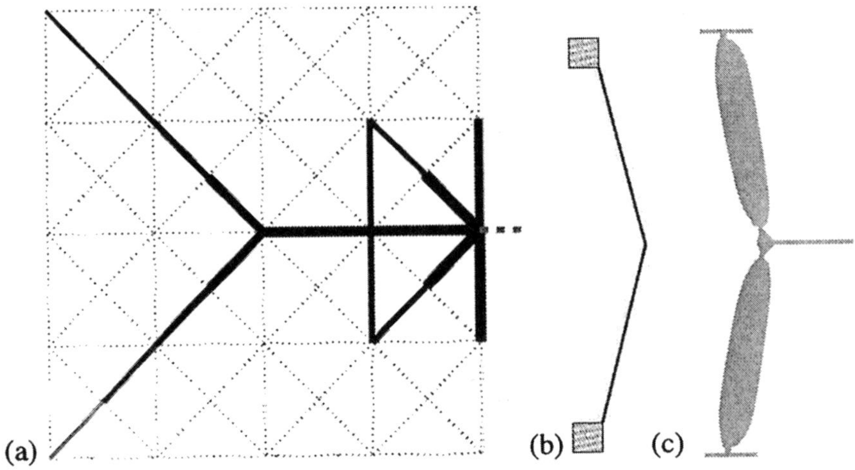

Figure 17. Example 1 with line elements. (a) Optimized solution. Widths show their relative values. Color shows temperature, light shade indicating high, and dark shade low values, respectively. "Vanished" elements are shown with dotted lines. The work-piece spring is the dashed line at the far right. (b) an intuitive "bent-beam" actuator [33] (c) a re-sketched design obtained with continuum elements [19].

In the second example, shown in Fig. 18, a potential difference of 10 V was applied. The work-piece spring was oriented at $\theta = 135^0$. Its initial guess had 6.67, and the optimized solution -17.7. Further studies on this example to see the effects of changing the angle and the electro-thermal properties of the work-piece spring are in [29].

7. ADVANCED EXAMPLE

This section is contributed by Ole Sigmund, Department of Mechanical Engineering, Solid Mechanics, Technical University of Denmark.

It is imperative that the large displacement theory is used for the modeling of the mechanisms used in real applications of the synthesis method. Furthermore, extra constraints on cross-sensitivity (i.e., the point moves precisely in the desired direction with little or no displacement in the perpendicular direction) and maximum current should be added to the design problem in order to obtain designs that can be readily used in practice. Several examples of such advanced applications can be found in [19, 20, 34]. The following example is taken from [19, 34].

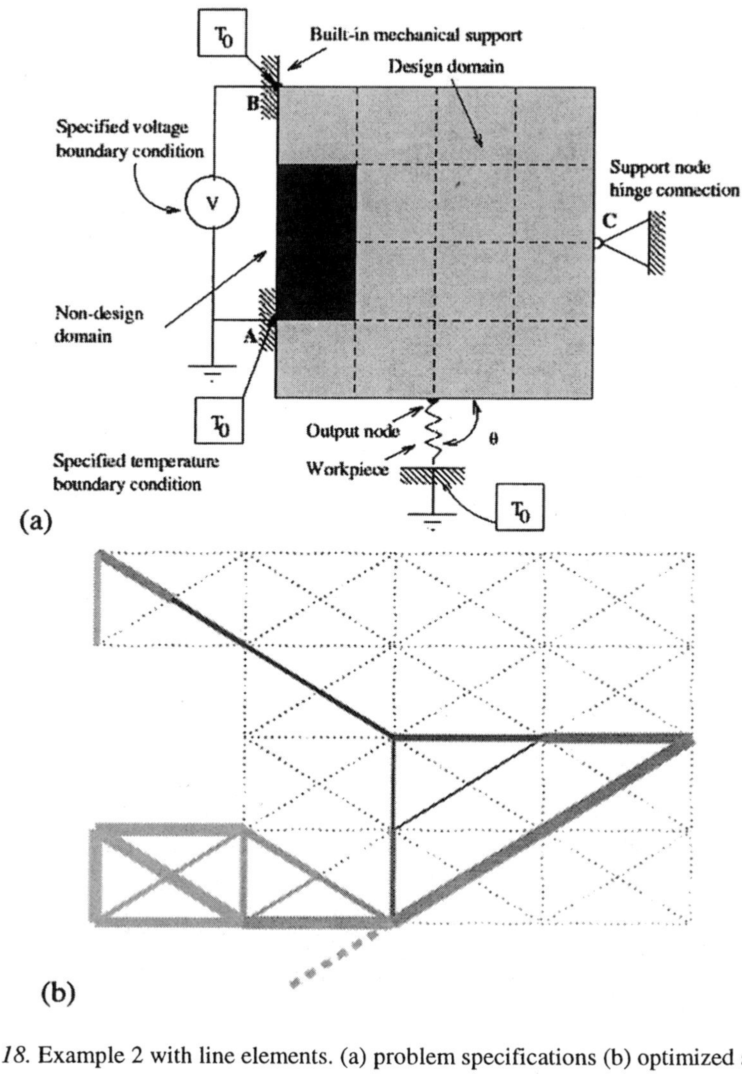

Figure 18. Example 2 with line elements. (a) problem specifications (b) optimized solution.

Consider the synthesis problem of a two-degree-of-freedom electro-thermal micro device with zero cross-axis sensitivity for scanning purposes. The specifications for this are sketched in Fig. 19a. The synthesis problem here consists of finding a scanning mechanism where the scanning head (the output point) moves in the horizontal direction for one electric input and in the vertical direction for another electrical input. The optimization problem consists of distributing a limited amount of material in the design domain such that the output displacements are maximized, the cross-sensitivity is smaller than a certain value, and the currents are also limited. The optimization problem involves 8 constraints and 8000 design variables (element densities). The iteration history is shown in Fig. 19b. The two modes of actuation are shown in Fig. 19c, and a micro-scale prototype of $500\mu m \times 400\mu m$ size laser-micromachined in Nickel is shown in Fig. 19d. This device was fabricated and tested at MIC, Technical University of Denmark.

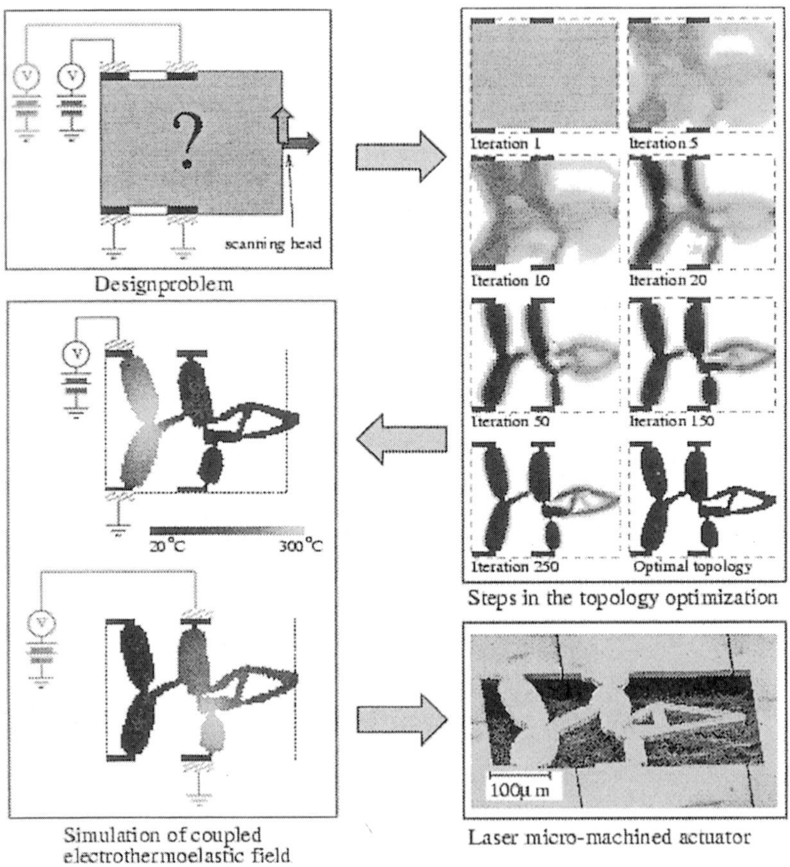

Figure 19. Two degree-of-freedom electrothermomechanical actuator synthesis. a) design problem specifications with two electrical inputs, b) iteration history, c) actuation modes and d) micro-fabricated actuator from [34].

8. MICROFABRICATION

Topology optimization is an excellent tool for conceptual design: it can provide designs for specifications that are difficult for human designers to conceive. Another of its strengths is providing solutions that can be readily transformed into manufacturing data such as masks for photolithography or instructions for a computer-controlled laser micromachining system. Three techniques described below can be used to make ETC devices at the micro scale.

8.1 PennSOIL

In the PennSOIL (University of Pennsylvania's Silicon-On-Insulator Layer) microfabrication process, we use commercially available SOI wafers, which consist of two silicon layers (525 μm thick *handle* layer, 20 μm thick *device* layer) bonded together with a 1 μm thick insulating oxide *box* layer between them. The process does not require any structural thin film depositions, and therefore, there is less concern about the process induced residual stresses. Also, the devices are suspended with nothing underneath eliminating the problem with conduction through air. Figure 20 illustrates the PennSOIL process sequence from [16, 17].

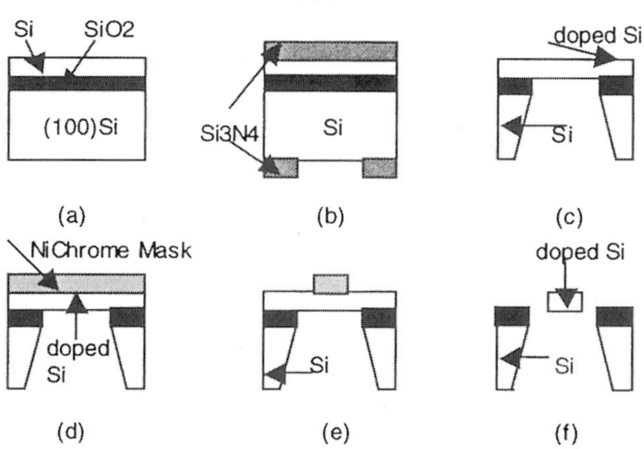

Figure 20. The PennSOIL process sequence [16, 17]

First, on a (100) SOI wafer (Fig. 20a), a 100nm thick silicon nitride layer is deposited by LPCVD on both sides. This serves as the mask for the KOH wet etch of the handle layer from the bottom. The wafer is then spin-coated with photoresist and patterned with the membrane pattern mask. The membrane pattern mask determines where the cavities will be etched in the handle layer from the bottom. The nitride layer is plasma-etched to leave windows where the bottom silicon is to be etched (Fig. 20b). Then the wafer is immersed in KOH solution to etch the handle silicon layer to leave a membrane of the oxide and the top silicon layers. The intermediate oxide layer works as an etch stop here. The simultaneous removal of the nitride mask layer and the exposed oxide layer with 50% HF follow this (Fig. 20c). Then the device layer is subsequently doped with phosphorus to reduce the electrical resistitivity. An intermediate step is needed to align the device layer to the back of the wafer. The photoresist is used as a mask for the shallow plasma etch for the purpose of alignment. Once the phosphorous diffusion is done, nichrome is evaporated as a mask for plasma etch of the top layer (Fig. 20d). The nichrome mask layer is patterned with the device mask (Fig. 20e). The device layer is then etched in the plasma chamber. Alternatively, a wet KOH is used if the devices have only straight edges and the resulting crystal-orientation dependent geometric features can be tolerated. The nichrome mask layer is stripped away in the end (Fig. 20f). The devices shown in Figs. 1c, 2, 3, and 4 were made using PennSOIL.

8.2 Excimer laser micromachining

The ETC devices can easily be made using a laser micromachining systems such as the Lambda physik EMG 100 excimer laser, which produces ultraviolet 248 nm wavelength laser pulses. Laser micromachining replaces the etching step of the deice layer in PennSOIL. The laser beam has the smallest spot size of 20 μm diameter. Its size can be adjusted using a mask and optics that exist between the laser beam that comes out of the EMG 100 and the XY-stage on which the wafer is positioned. The XY-stage has a resolution of 5 μm and can be programmed to cut intended shapes as in a Computer Numerically Controlled (CNC) milling machine.

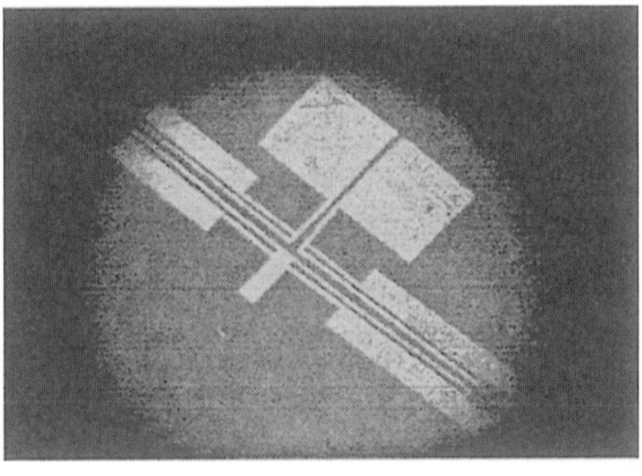

Figure 21. An excimer laser micromachined 35 μm thick aluminum ETC actuator with 20 μm wide beams

By using laser micromachining, ETC prototype devices were fabricated with a variety of metal foils including stainless steel, copper, aluminum and titanium. Various thicknesses of the metal foils were attempted. The metal ETC devices made using the laser were mounted on an insulating base using adhesives for the purpose of testing. Laser micromachining is a "serial" process and is slow compared to the photolithography based "batch" etching. However, recent advances in this area indicate that a powerful laser in conjunction with a mask can be used to machine the entire pattern of multiple devices at once. Figure 21 shows a linear actuator made in aluminum that has a thickness of 35 μm. The width of the smallest beam in this device is 20 μm. The device shown in Figs. 6a and 22 were also micromachined with the same laser. The device in Fig. 22 was a prototype, called a *pseudo-wheel* [35, 36] device with an intended application of reorienting micro-satellites.

Figure 22. A laser-micromachined *pseudo-wheel* [35, 36]

8.3 Electro-plating combined with photolithography

Microfabricated layered structures with multiple materials is very common, but two-material structures in which materials are next to each other in the same plane are not so common. When we further stipulate that they be released to allow elastic deformation, there are not many techniques to fabricate them at the micro level. But such a process is necessary to realize two-material designs presented in Section 5. So. We are currently developing a process by combining photolithography, etching, and electroplating. It is a modification of the PennSOIL process. Here, we first etch a cavity in the device layer up to the oxide layer using a mask. We then deposit a thin Nichrome and gold layers using electron beam evaporation. Nichrome is used for better adhesion with silicon. The thin gold layer acts as a seed layer for electrodeposition. This is followed by electroplating of gold in the cavity by placing it in an electrochemical bath consisting of an aqueous gold solution ($AU(CN_2)$). Electroplating is stopped when the required thickness of the gold layer is obtained so that silicon on the side is at the same level as the gold. The rest of the process is similar to PennSOIL in making the released structure. This technique is still in the experimental stage because controlling the morphology of electroplated structures is very challenging. Figures 23a and 23b show sample SEM pictures. Figure 23a shows silicon and gold side by side. In this, it can be seen that gold layer is rather rough. Controlling current density in the electrochemical bath helps to some degree as can be seen in Figure 23b where a slightly smoother surface obtained with larger current density can be seen at the top. At this point, it looks like a promising technique but much needs to be done further.

(a)

(b)

Figure 23. Fabrication of two-material (silicon-gold) planar structures by combining photolithography and electroplating (a) test structure showing gold and silicon side by side (b) control of surface roughness and porosity with different values of current density. From [37].

9. CLOSURE

Electro-thermal actuation of MEMS is advantageous from both the design and fabrication viewpoints. The microfabrication easy because only a single released and patterned layer is sufficient to get complex in-plane motions. The systematic synthesis is possible using topology optimization techniques as described in this chapter. The synthesis problem is also challenging because three sets of governing equations pertaining to electrical, thermal, and elastic domains need to be considered. It should be noted that proper thermal modeling, large deformation elastic modeling, and proper consideration of temperature-dependent material and thermo-physical properties are important to realize synthesized designs that can be readily used in practice. In this chapter, the description of the synthesis formulations was augmented with implementation details, a simple example, an advanced example, and relevant microfabrication techniques.

ACKNOWLEDGEMENTS

The work described in this chapter is a result of the synergistic effort of a number of current and past students and research associates (Timothy Moulton, Nilesh Mankame, Jun Li, Dr. Luzhong Yin and Luo Chen) in the Compliant and Micro Systems group at the University of Pennsylvania. This work was funded by Defense Advanced Research Projects Agency (contract #537640-52247), National Science Foundation (DMI-97-33196, DMR-98-00417), Pathway Technologies, Inc., and the Research Foundation at the University of Pennsylvania. This support is gratefully acknowledged.

REFERENCES

1. Timoshenko, S., "Analysis of Bi-metal Thermostats," *J. of the Optical Society of America*, Vol. 11, 1925, pp. 233-255.
2. Benecke, W. and Riethmuller, W, 1989, "Applications of Silicon Microactuators Based on Bimorph Structures," *Proceedings of the IEEE MEMS Workshop*, Salt Lake City, Utah, Feb. 1989, pp. 116-120.
3. Takeshima, N. and Fujita, H, 1990, "Polyimide Bimorph Actuators for a Ciliary Motion system," *Micromechanical Sensors, Actuators, and Systems*, 1991, pp. 203-209.
4. Chu, W.-H. and Mehregany, M., 1994, "MicrofabricatedTweezers with a Large Griping Force and a Large Range of Motion," Technical Digest of *Solid State Sensors and Actuators Workshop*, Hilton Head Island, SC, June 1994, pp. 107-100.

5. Guckel, H., Klein, J., Christenson, T., Skrobis, K., Laudon, M., and Lovell, E.G., 1992, 'Thermomagnetic Metal Flexure Actuators," *Technical Digest of Solid State Sensors and Actuators Workshop*, Hilton-Head Island, SC, 1992, p 73.
6. Comtois, J. and Bright, V, 1996, "Surface Micromachined Polysilicon Thermal Actuator Arrays and Applications," *Technical Digest of Solid-State Sensor and Actuator Workshop*, Hilton Head Island, SC, June 1996, pp. 174-177.
7. Lerch, P, Slimane, C.K., Romanwicz, B., and Renaud, P, 1996, "Modelization and characterization of asymmetrical thermal micro-actuators," *J. Micromechanics and Microengineering*, Vol. 6, 1996, pp. 134-137
8. Moulton, T., 1997, "Analysis and design of Electro-Thermal-Compliant micro devices" *Center for Sensor Technologies at the university of Pennsylvania technical report* #TR-CST31DEC97, pp.13-26.
9. Keller, C.G. and Howe, R.T., 1997, "Hexsil Tweezers for Teleoperated Micro-Assembly," *Proc. 10th Annual International Workshop on Micro-Electro-Mechanical Systems* (MEMS '97), Nagoya, Japan, January 26-30, 1997, pp. 72-77.
10. Pan, C. S. and Hsu, W., 1997, "An electro-thermally and laterally driven polysilicon microactuator," *J. Micromechanics and Microengineering*, 7 (1997), pp. 7-13.
11. Sigmund, O., "Topology Optimization in Multiphysics Problems," *Proceedings of the 7th AIAA/USAF/NASA/ISSMO Symposium*, Vol. 3, St Louis, August 1998, pp. 1492-1500.
12. Cragun, R. and Howell, L.L., "A Constrained Thermal Expansion Micro-Actuator," *Proceedings of the Micro-Electro-Mechanical Systems (MEMS) Symposium* at the International Mechanical Engineering Congress and Exhibition, DSC-Vol. 66, pp. 365-371.
13. Huang, Q. and Lee, N., "Analysis and Design of Polysilicon Thermal Flexure Actuator," *J. Micromechics and Microengineering*., Vol. 9, 1998, pp. 64-70.
14. Comtois, J. H., Michalicek, M.A., and Barron, C.C., " Electrothermal actuators fabricated in four-level planarized surface micromachined polycrystalline silicon." *Sensors and Actuators A Physical*, Vol. 70, 1998, pp 23-31.
15. Moulton, T. and Ananthasuresh, G.K., " Micromechanical devices with embedded electro-thermal-compliant actuation," *Proceedings of the Micro-Electro-Mechanical Systems (MEMS) Symposium*, International Mechanical Engineering Congress and Exposition, Nashville, TN, Nov., 1999, Vol. 1, pp. 553-560.
16. Moulton, T., "Micromechanical Devices with Embedded Electro-Thermal Actuation," *Masters thesis*, Mechanical Engineering and Applied Mechanics, University of Pennsylvania, 2000.
17. Li, J. and Ananthasuresh, G.K., "Microfabrication and Characterization of Electro-Thermal-Compliant Devices," *CD-ROM proceedings of the ASME 2000 Design Engineering Technical Conferences*, Baltimore, Maryland, Sep. 10-13, 2000, paper no. DETC2000/MECH14118.
18. Moulton, T. and Ananthasuresh, G.K., "Design and Manufacture of Electro-Thermal-Compliant Micro Devices," *Sensors and Actuators, Physical*, **90** (2001), pp. 38-48.
19. Sigmund, O., "Design of Multiphysics Actuators Using Topology Optimization – Part I: One-material structures," *Computer Methods in Applied Mechanics and Engineering*, Vol. 190, 2001, pp. 6577-6604.
20. Sigmund, O., "Design of Multiphysics Actuators Using Topology Optimization – Part II: Two-material structures," *Computer Methods in Applied Mechanics and Engineering*, Vol. 190, 2001, pp. 6605-6627.

21. Lin, L. and Chiao, M., "Electrothermal Responses of Lineshape Microstructures," *Sensors and Actuators, A Physical*, Vol. 55, pp. 35-41.
22. Mankame, N. and Ananthasuresh, G. K., "Comprehensive Thermal Modelling and Characterization of an Electro-Thermal-Compliant Microactuator," *J. Micromechanics and Microengineering*, Vol. 11, 2001, pp. 452-462.
23. Rozvany, G.I.N., Bendsoe, M.P., and Kikuchi, N., 1995, "Layout Optimization of Structures," *ASME Applied Mechanics reviews*, Vol. 48, 1995, pp. 41-119.
24. Bendsφe, M.P. and Sigmund, O., *Topology Optimization: Theory, Methods, and Applications*, Springer, Berlin, 2002.
25. Bendsφe, M.P. and Sigmund, O, 1999, "Material Interpolations in Topology Optimization," *Arch. Appl. Mech.*, Vol. 69, pp. 635-654.
26. Yin, L. and Ananthasuresh, G.K., "Topology Optimization of Compliant Mechanisms with Multiple Materials Using a Peak Function Material Interpolation Scheme," *Structural and Multidisciplinary Optimization*, Vol. 23, No. 1, 2001, pp. 49-62.
27. Yin, L. and Ananthasuresh, G.K., "Novel Design Technique for Electro-Thermally Actuated Compliant Micromechanisms," *Sensors and Actuators, A Physical,* Vol. 97-98, 2002, pp. 599-609.
28. Kovacs, G., *Micromachined Transducers*, WC McGraw-Hill, New York, 1998.
29. Mankame, N. and Ananthasuresh, G.K., "Topology Synthesis of Electrothermal Compliant (ETC) Mechanisms Using Line Elements," *CD-ROM Proceedings of 2001 ASME Design Engineering Technical Conferences*, Design Automation Conference, Pittsburgh, PA, Sep. 9-11, 2001, paper #. DETC2001/DAC-21019. 12 pages. To also appear in *Structural and Multidsciplinary Optimization* in 2003.
30. Saxena, A. and Ananthasuresh, G.K., "On an Optimal Property of Compliant Topologies," *Structural and Multidisciplinary Optimization*, Vol. 19, No. 1, 2000, pp. 36-49.
31. Cook, R.D., Malkus, D.S., and Plesha, M.E., *Concepts and Applications of Finite Element Analysis*, Third edition, John Wiley & Sons, New York, 1989.
32. Hafkta, R.T. and Gurdal, Z., *Elements of Structural Optimization*, Kluwer Academic Publishers, New York, 1992.
33. Que, L., Park, J., and Gianchandani, Y., "Bent-beam Electro-thermal Actuators for High-Force Applications," *Proceedings of IEEE Microelectromechanical Systems conference*, 1999, pp. 31-36.
34. Jonsmann, J., Sigmund, O., and Bouwstra, S., "Multi Degrees of Freedom Electro-Thermal Microactuators," *Proceedings of Transducers*, 1999, pp. 1372-1375.
35. Li, J., Koh, S. K., Ananthasuresh, G. K., Ayyaswamy, P. S., and Ananthakrishnan, P., 2001, "A Novel Attitude Control Technique for Miniature Spacecraft," *Proc. of MEMS symposium* at 2001 ASME International Mechanical Engineering Conference and Exposition, Nov. 2001, New York.
36. Koh, S.K., Ostrowski, J.P., and Ananthasuresh, G.K. "Control of Micro-satellite Orientation Using Bounded-Input, Fully-Reversed, MEMS Actuators." *International Journal of Robotics Research*, Vol. 21, No. 5-6, 2002, pp.591-605.
37. Chen, L., 2001, "Microfabrication of Heterogeneous Optimized Compliant Mechanisms," *SUNFEST Technical Report*, Center for Sensor Technology, University of Pennsylvania, Dec. 2001.

Chapter 5

SYNTHESIS WITH PIEZOELECTRIC ACTUATION

Emílio Carlos Nelli Silva
Department of Mechatronics and Mechanical Systems Engineering. Polytechnic School from University of São Paulo, Brazil.

Synopsis: Synthesis methods, such as topology optimization method, have been widely applied to design mechanical structures. Therefore, it is natural that they can also be applied to design smart materials and MEMS structures whose current development has been based on the use of simple analytical models, test of prototypes, and analysis using the finite element method (FEM), usually limiting the problem to a parametric optimization. In contrast to those approaches, this chapter presents systematic synthesis methods for MEMS actuated by piezoelectric materials by using topology optimization combined with homogenization technique. In this method, the topology of a flexible structure coupled to a piezoceramic (flextensional actuator) is designed to maximize the output displacement (or force) in some specified direction. Beginning with a brief introduction to the piezoelectric constitutive equations and the FEM theory applied to the piezoelectricity, the chapter provides a self-contained description of the synthesis method for piezoelectrically actuated MEMS. The examples presented show that the synthesis method is indeed a promising tool to design smart materials and structures.

1. INTRODUCTION

Piezoelectric materials have the property of converting electrical energy (electrical field and electrical charge) into mechanical energy (strain and stress) and vice versa. Examples of piezoelectric materials include quartz, ceramic (e.g., PZT), and polymer (e.g., PVDF). An overview of the many applications of piezoelectric materials is given by Newnham and Ruschau [1]. These materials are most widely used in electromechanical sensors and actuators, as resonators in electronic equipment, and in acoustic applications for generating and detecting sound waves represented by devices such as ultrasonic transducers, hydrophones, and naval sonars. Ultrasonic transducers are widely used in medical imaging and non-destructive evaluation (NDE), and typically operate in the low megahertz frequency range. Piezoelectric materials have been used in Microelectromechanical Systems (MEMS) as actuators [2-4] as well as sensors [5] in a number of applications including microphones, valves, pumps, liquid ejectors, fluid mixers, and motors.

The literature on piezoelectric transducers (sensors and actuators) is very extensive as they are available in many configurations [6]. In typical piezoelectric transducer applications, the usual objective functions are to obtain the maximum output displacement or force (in the case of a piezoelectric actuators) or to maximize energy conversion (in the case of hydrophones). These goals are not easy to achieve by trial and error analysis-based approaches; therefore, it is a natural idea to apply synthesis methods to design piezoelectric transducers. However, very little has been done in this direction [7-9].

In the past work, the design and optimization of piezoelectric transducers were limited to the optimization of a specific dimension of a transducer part (such as thickness of the piezoceramic, for example), based on analytical models used to describe the behavior of the piezoelectric transducer. These analytical models were usually one-dimensional and could be applied only to transducers with very simple geometry, failing to represent transducers with complex geometries. Actually, even simple geometries such as disks and parallelepiped shapes could not be modeled with the accuracy required for designing [10]. Therefore, engineers were limited to simple transducer geometries.

A great improvement in piezoelectric transducer design came with the development of the finite element theory for piezoelectricity developed by Allik and Hughes [11]. Since then, many researchers have discussed the implementation of FEM in piezoelectricity. We can cite, among others, [12-15] for an overview of this subject. As the modeling of transducers with complex geometry became possible, transducer engineers began to optimize transducer performance by exploring new configurations. Prototypes were needed only for the verification of FEM results.

However, most studies do not use a systematic approach to synthesize the transducer. Essentially, the design is accomplished by constructing response curves for some transducer performance as a function of a parameter, usually a transducer dimension, which limits the problem to a sizing optimization [16]. In other approaches, the transducer was designed via experimental testing [17] or simulating via FEM [18] of various *guessed* configurations, until satisfactory improvement was obtained.

In this chapter the potential of applying a powerful optimization method to synthesize piezoelectric transducers is presented. The chosen optimization method is topology optimization because it is a powerful and general method that has been widely applied recently in industry and academia. To illustrate its application in the design of piezoeletric transducers, the design of piezocomposite transducers and flextensional actuators are discussed.

A *piezocomposite transducer* is essentially a piece of piezocomposite material, made large enough to avoid the influence of boundary conditions in the transducer operation. In this case, the performance of the transducer depends mostly on the effective properties of the composite rather than the transducer shape. In the piezocomposite transducer design, topology optimization is applied to design the microstruture of piezocomposite material. Figure 1 schematically illustrates the idea. This is the topic of Chapter 6.

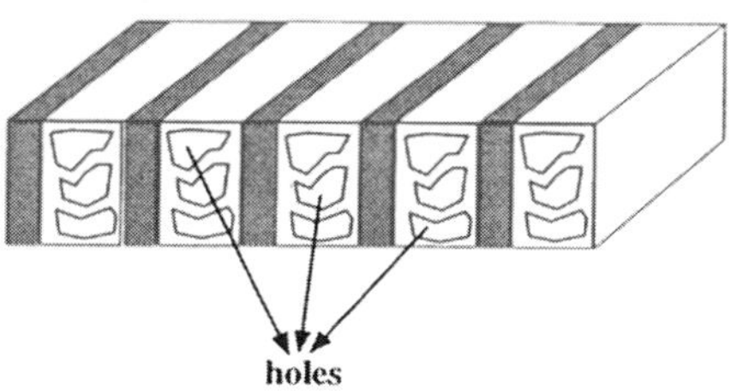

Figure 1. Piezocomposite material design.

A flextensional actuator is essentially a single ceramic element (disk or rectangular bar) connected to some coupling mechanical structure, which amplifies and changes the direction of displacements generated by the piezoceramic. They are also used as hydrophones and sonars [19] since both operate in a low-frequency range and require similar characteristics as actuators. In the flextensional actuator design, the topology optimization method is applied to design the transducer coupling structure (Fig. 2) so that it will increase its performance (i.e., increasing the output displacement or the blocking force).

The chapter is organized as follows. First, a background in piezoelectricity is given. Then, since finite element method is an important tool to design piezoelectric transducers, not only through an analysis approach but also by combining it with optimization methods, it is briefly described.

While in the design of piezocomposite materials the topology optimization is applied to design the microstructure (see Chapter 6), in the design of flextensional actuators the design of actuator itself is the objective. Prototypes manufactured by using Electro Discharge Machining (EDM) machine are also included in the chapter.

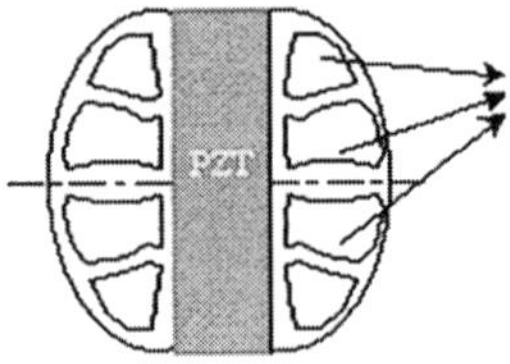

Figure 2. Flextensional actuator design.

2. BACKGROUND THEORY FOR PIEZOELECTRICITY

Piezoelectricity is a property that certain materials possess, which consists of generating an electrical polarization (generation of bound charges of opposite polarity) when the material is subjected to mechanical stress, and also generating strain in the material when it is subjected to an electric field. In the first case, it is termed the *direct piezoelectric effect*, and in second case it is termed the *inverse piezoelectric effect.* Piezoelectricity arises because of an interaction between Coulomb forces and the elastic restoring forces in a unit cell subject to either an electric field or a mechanical force. Therefore, piezoelectricity is intimately related to crystallographic properties of materials, existing only in materials that do not possess a central symmetry.

Consider the following didactic example where the behavior of the internal material structure is analyzed. It will help us to understand better the piezoelectric concept as well as the piezoelectric constitutive equations (described ahead) which relate mechanical stress (T), strain (S), electric field (E), and electric displacement (D). It consists of an electric dipole shown in Fig. 3a.

It is well known that the electric field due to an electric dipole on the axis of dipole, for $r>>L$, is given by $E=p/(2\pi\varepsilon_0 r^3)$, where p is the electrode dipole moment, $p=ql$. Consider now a hypothetical solid unit cell (Fig. 3b) where charged molecules are held by elastic forces represented by springs. This unit cell has no center of symmetry, i.e., a line connecting the negative charged molecule B and the center O will intersect a positively charged molecule E. Essentially, the unit cell consists of three dipoles AB, HC and

ED (Fig. 3b). At equilibrium, the electric field of the unit cell at point W for $r>>L$, is given by:

$$E = E_{AB} - E_{HC} + E_{ED} \cong \frac{(ql - q2l + ql)}{2\pi\varepsilon_0 r^3} = 0 \qquad (1)$$

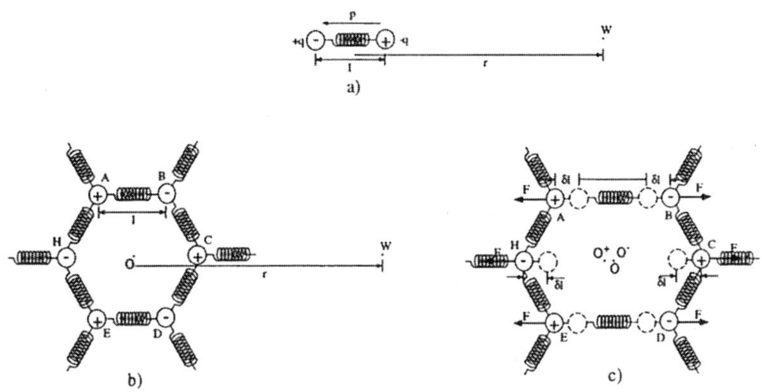

Figure 3. Piezoelectric unit cell model.

Thus, the total electric dipole moment of the cell is zero. If a symmetric force F is applied to the solid, the unit cell will deform as shown in Fig. 3b. For simplicity, it is assumed that the displacements of points B, C, and D are equal, i.e., $\delta l_1 = \delta l_2 = \delta l$. The electric field of the deformed cell at point W will now be

$$E = \frac{2q(2\delta l_1 - \delta l_2)}{2\pi\varepsilon_0 r^3} = \frac{2q\delta l}{2\pi\varepsilon_0 r^3} \qquad (2)$$

where the dipole moment can be written as $p=2q\delta l=2qlS$ and $S = \delta l / l$ is the longitudinal strain. Assuming that the body contains N molecules per unit volume, it follows that the electric polarization P, i.e., the dipole moment per unit volume, is given by $P=Np=2qlNS=eS$, where e is by definition the piezoelectric stress constant $e=2qlN$. Taking $N \cong 5.10^{27}\ \mathrm{m}^{-3}$, $q=1.610^{-19}\ C$, and $l = 5.10^{-10}$ m it is found that $e=0.8\ C/m^2$, which is comparable, for example, to the piezoelectric stress constant of ZnO, which is equal to $0.5\ C/m^2$.

The presence of the polarization P is taken into account by modifying the relation $D = \varepsilon E$ to $D = \varepsilon E + eS$. Notice that due to the application of force F, centers of positive and negative charges, initially both at point O (Fig. 3b), have been displaced to points O^+ and O^-, respectively, thus producing a displacement current, modifying *D* (see Fig. 3c). Therefore, electric displacement (*D*) in a piezoelectric material depends on the strain and electric field applied to the material. In an analogous way, when an external electric field *E* is applied, it can be shown that there is an additional stress term *-eE*. This is taken into account by modifying Hooke's law, *T=cS*, to *T=cS-eE*. If the charges of the positive and negative molecules were different in magnitude, the unit cell would have exhibited polarization even with no applied forces, an example of which can be found in quartz material.

Piezoelectric effect is a nonlinear effect that may be linearized as piezoelectric transducers mostly operate in the linear range of piezoelectric materials. Under this linear hypothesis, the governing equations described herein model response to linear changes in the electric field, electric displacement, or mechanical stress and strain. Furthermore, variations in temperature and magnetic field are considered unimportant. These assumptions are compatible with the piezoelectric ceramics, polymers, and composites in current use. Therefore, the behavior of the piezoelectric medium is described by the following piezoelectric constitutive equations, which relate the stress (T_{ij}), strain (S_{kl}), electric field (E_k), and electric displacement (D_i) [20]:

$$\begin{cases} T_{ij} = c^{E}{}_{ijkl} S_{kl} - e_{kij} E_k \\ D_i = e_{ikl} S_{kl} + \varepsilon^{S}{}_{ik} E_k \end{cases} \tag{3}$$

where $c^{E}{}_{ijkl}$ is a fourth-order elasticity tensor under short-circuit boundary conditions, $\varepsilon^{S}{}_{ik}$ is a second-order free-body dielectric tensor, and e_{kij} is a third-order piezoelectric strain tensor. The superscript refers to the transducer's electromechanical boundary conditions when these properties are measured. Strain S_{ij} is given by

$$S_{ij} = \frac{1}{2}\left(\frac{\partial u_j}{\partial x_k} + \frac{\partial u_k}{\partial x_j} \right) \tag{4}$$

where u_i are the displacement components.

Due to the symmetry of the tensors T_{ij}, S_{ij}, $c^{E}{}_{ijkl}$, and $\varepsilon^{S}{}_{ik}$, Eq. (3) can be written in a compact notation [20]:

$$\begin{cases} T_P = c^{E}{}_{PQ} S_Q - e_{Pk} E_k \\ D_i = e_{iQ} S_Q + \varepsilon^{S}{}_{ik} E_k \end{cases} \text{ or } \begin{cases} \mathbf{T} = \mathbf{c}^{E}\mathbf{S} - \mathbf{e}\mathbf{E} \\ \mathbf{D} = \mathbf{e}^{t}\mathbf{S} + \varepsilon^{S}\mathbf{E} \end{cases} \tag{5}$$

where the superscript "t" denotes a transposed matrix. The constitutive equations can also be represented in an alternate form:

$$\begin{cases} \mathbf{S} = \mathbf{s}^{E}\mathbf{T} + \mathbf{d}\mathbf{E} \\ \mathbf{D} = \mathbf{d}^{t}\mathbf{T} + \varepsilon^{T}\mathbf{E} \end{cases} \tag{6}$$

where $\mathbf{s}^{E}$ is the compliance tensor under short-circuit conditions, ε^{T} is the clamped body dielectric tensor, and **d** is the piezoelectric stress tensor. The relations among the properties in equations (5) and (6) are [20]:

$$\mathbf{s}^{E} = \left(\mathbf{c}^{E}\right)^{-1} \qquad \varepsilon^{T} = \varepsilon^{S} + \mathbf{d}^{t}(\mathbf{s}^{E})^{-1}\mathbf{d} \qquad \mathbf{d} = \left(\mathbf{s}^{E}\right)\mathbf{e} \tag{7}$$

In addition to the constitutive equations, piezoelectric medium behavior is also described by electrical and dynamic equations. The electric behavior is described by Maxwell equations by considering the piezoelectric medium an insulator (no free charges). Therefore, from Gauss theorem:

$$\nabla.\mathbf{D} = 0 \tag{8}$$

The electric field is related to the electric potential ϕ by:

$$\mathbf{E} = -\nabla\phi \tag{9}$$

The dynamic behavior is described by the Second Newton's law:

$$\rho\frac{\partial^{2}\mathbf{u}}{\partial t^{2}} = \nabla.\mathbf{T} \tag{10}$$

where $(\nabla.)$ is the tensor dyadic operator. In dynamic applications, we have $\mathbf{u} = \bar{\mathbf{u}}e^{j\omega t}$, $\mathbf{t} = \bar{\mathbf{t}}e^{j\omega t}$, $d = \bar{d}e^{j\omega t}$, and $\phi = \bar{\phi}e^{j\omega t}$ where ω stands for an excitation frequency, t stands for time, and $\bar{\mathbf{u}}, \bar{\mathbf{t}}, \bar{d}$ and $\bar{\phi}$ stand for amplitudes of $\mathbf{u}, \mathbf{t}, d$, and ϕ, respectively. If no damping is considered, there is no phase difference among these quantities.

By applying variational calculus to the energy expression of piezoelectric medium [21], the weak formulation of the equilibrium equations for piezoelectric medium is obtained, as described in Eqs. (11). The dependence on $e^{j\omega t}$ is not represented because it appears on both sides of the equation and cancels out. In addition, for the sake of simplicity of notation, the amplitudes are represented without the over bar.

$$\int_{\Omega}\mathbf{S}(\mathbf{u})^t\mathbf{c}^E\mathbf{S}(\mathbf{v})d\Omega-\omega^2\int_{\Omega}\rho\mathbf{u}^t\mathbf{v}d\Omega+\int_{\Omega}(\nabla\phi)^t\mathbf{e}^t\mathbf{S}(\mathbf{v})d\Omega=\int_{\Gamma_t}\mathbf{t}.\mathbf{v}d\Gamma$$
$$\int_{\Omega}\mathbf{S}(\mathbf{u})^t\mathbf{e}(\nabla\varphi)d\Omega-\int_{\Omega}(\nabla\phi)^t\varepsilon^S\nabla\varphi d\Omega=\int_{\Gamma_d}d\varphi d\Gamma \tag{11}$$

for $\mathbf{u},\phi\in$ V and $\forall\mathbf{v}$, $\forall\varphi\in$ V, where **t=T.n** and d=**D.n**, and **n** is the normal vector to the surface.

$$V=\{\mathbf{v}=v_i\bar{\mathbf{e}}_i,\ \varphi: v_i,\varphi\in H^1(\Omega)\ \text{with}\ \mathbf{v}=0\ \text{on}\ \Gamma_u\ \text{nd}\ \varphi=0\ \text{on}\ \Gamma_\phi, i=1\ \text{or}\ 3\}$$

where Ω is the domain of the piezoelectric medium (but it may contain non-piezoelectric materials also), ∇ is the gradient operator, and $\mathbf{c}^E$, **e**, and ε^S are the elastic, piezoelectric, and dielectric properties, respectively, of the medium, and ρ is the density. The superscript "t" denotes transpose, and $S_{ij}(\mathbf{u})$ is given by Eq. (4) and **v** and φ are virtual displacements and electric potential, respectively. In the case of a bidimensional analysis, the index i assumes value *1* or *3* because the problem is usually considered in the plane *1-3*. The piezoceramic is polarized in the *#3* direction. To write Eq. (11) in a more *compact form*, the following operators are defined:

$$A(\mathbf{u},\mathbf{v})=\int_{\Omega}\mathbf{S}(\mathbf{u})^t\mathbf{c}^E\mathbf{S}(\mathbf{v})d\Omega;\quad B(\phi,\mathbf{v})=\int_{\Omega}(\nabla\phi)^t\mathbf{e}^t\mathbf{S}(\mathbf{v})d\Omega;$$
$$C(\phi,\varphi)=\int_{\Omega}(\nabla\phi)^t\varepsilon^S\nabla\varphi d\Omega;\quad M(\mathbf{u},\mathbf{v})=\int_{\Omega}\rho\mathbf{u}^t\mathbf{v}d\Omega; \tag{12}$$
$$L_t(\mathbf{t},\mathbf{v})=\int_{\Gamma_t}\mathbf{t}.\mathbf{v}d\Gamma;\quad L_d(d,\varphi)=\int_{\Gamma_d}d\varphi d\Gamma$$

Therefore, Eq. (11) become:

$$A(\mathbf{u},\mathbf{v})-\omega^2M(\mathbf{u},\mathbf{v})+B(\phi,\mathbf{v})=L_t(\mathbf{t},\mathbf{v})$$
$$B(\varphi,\mathbf{u})-C(\phi,\varphi)=L_d(d,\varphi) \tag{13}$$

3. FEM APPLIED TO PIEZOELECTRICITY

It is important to state that even though most of the current piezoelectric transducers have simple shapes such as solid or hollow cylinders, parallepipeds, etc., they present a complex behavior due to the coupling nature of piezoelectric equations. These simple transducers may present strong coupling vibration mode behavior which makes it very hard to predict their operation by using analytical solutions derived from one dimensional equations. Therefore, FEM helps to model piezoelectric transducers easily and accurately. A good reference for this is [22]. Commercial FEM software programs often include a module for piezoelectric analysis. There are also some freeware [23]. In the particular case of topology optimization since complex topologies are expected, and a general method such as FEM is necessary for the transducer analysis. Therefore, in this section the finite element formulation applied to piezoelectricity is briefly described.

The basis of FEM consists of discretizing a domain into subdomains called *finite elements*. Then, scalar field variables are interpolated in each finite element by using polynomial-type functions calculated based on the values of these variables at the finite element vertices called nodes. These functions are known as *shape functions*. Boundary conditions must be specified, i.e., points where the structure is supported or fixed and applied loads on the structure. In the case of piezoelectric analysis the scalar field variables are electric potentials (voltage) and components of displacements in orthogonal directions at each node in the discretized finite element model. Therefore, electrical and mechanical boundary conditions must be considered.

The finite element equations for modeling a linear piezoelectric medium can be obtained by applying FEM approximation to equations (11). Therefore, considering dynamic excitation, these equations may be written as [12,13]:

$$\begin{bmatrix} \mathbf{M} & \mathbf{0} \\ \mathbf{0} & \mathbf{0} \end{bmatrix} \begin{Bmatrix} \ddot{\mathbf{U}} \\ \ddot{\boldsymbol{\Phi}} \end{Bmatrix} + \begin{bmatrix} \mathbf{C}_{uu} & \mathbf{0} \\ \mathbf{0} & \mathbf{C}^{d}{}_{\phi\phi} \end{bmatrix} \begin{Bmatrix} \dot{\mathbf{U}} \\ \dot{\boldsymbol{\Phi}} \end{Bmatrix} + \begin{bmatrix} \mathbf{K}_{uu} & \mathbf{K}_{u\phi} \\ \mathbf{K}^{t}{}_{u\phi} & -\mathbf{K}_{\phi\phi} \end{bmatrix} \begin{Bmatrix} \mathbf{U} \\ \boldsymbol{\Phi} \end{Bmatrix} = \begin{Bmatrix} \mathbf{F} \\ \mathbf{Q} \end{Bmatrix} \quad (14)$$

where $\mathbf{M}$, $\mathbf{C}$, $\mathbf{K}_{uu}$, $\mathbf{K}_{u\phi}$, and $\mathbf{K}_{\phi\phi}$ are the mass, damping, stiffness, piezoelectric, and dielectric matrices, respectively, and $\mathbf{F}$, $\mathbf{Q}, \mathbf{U}, \dot{\mathbf{U}}, \ddot{\mathbf{U}}$ and $\boldsymbol{\Phi}$ are the nodal mechanical force, nodal electrical charge, nodal displacements, nodal velocities, nodal accelerations, and nodal electric potential vectors, respectively. Since the piezoelectric medium is considered to be an insulator, $\mathbf{Q}$ has non-zero terms only in the piezoceramic electrodes. Considering N_d the number of displacements degrees of freedom, and N_v the number of electric degrees of freedom in the FEM mesh, matrices $\mathbf{M}$, $\mathbf{C}$, $\mathbf{K}_{uu}$ have size N_d x N_d, matrix $\mathbf{K}_{u\phi}$ has size N_d x N_v, matrix $\mathbf{K}_{\phi\phi}$ has size N_v x N_v, vectors $\mathbf{F}, \mathbf{U}, \dot{\mathbf{U}}, \ddot{\mathbf{U}}$ have size N_d x 1, and vector $\mathbf{Q}$ has size N_v x 1. Matrices $\mathbf{M}$, $\mathbf{C}$, $\mathbf{K}_{uu}$, $\mathbf{K}_{u\phi}$, and $\mathbf{K}_{\phi\phi}$ depend on the geometry and material properties of piezoelectric medium and are given by:

$$\mathbf{M} = \int_V \rho \mathbf{N}^t \mathbf{N} dV \qquad \mathbf{K}_{uu} = \int_V \mathbf{B}_u^t \mathbf{c} \mathbf{B}_u dV$$

$$\mathbf{K}_{u\phi} = \int_V \mathbf{B}_u^t \mathbf{e} \mathbf{B}_\phi dV \qquad \mathbf{K}_{\phi\phi} = -\int_V \mathbf{B}_\phi^t \varepsilon \mathbf{B}_\phi dV \tag{15}$$

and

$$\mathbf{u} = \mathbf{N}\mathbf{U},\ \phi = \mathbf{N}\Phi,\ \mathbf{S} = \mathbf{B}_u\mathbf{U},\ \mathbf{E} = -\nabla\phi = -\nabla(\mathbf{N}\Phi) = -\mathbf{B}_\phi\Phi$$

where **N** is the shape function (usually same for displacement and electric potential), u and ϕ are the displacement and electric potential functions, respectively, and $\mathbf{B}_u$ and $\mathbf{B}_\phi$ are functions of the derivatives of shape functions [24]. The damping matrix $\mathbf{C}_{uu}$ is given by Rayleigh's model:

$$\mathbf{C}_{uu} = \alpha\mathbf{M} + \beta\mathbf{K} \tag{16}$$

where α, β are constants defined in [24]. In addition to the mechanical damping matrix $\mathbf{C}_{uu}$, the electrical damping matrix $\mathbf{C}^d_{\phi\phi}$ related to dielectric losses can also be considered and is given by:

$$\mathbf{C}^d_{\phi\phi} = \frac{h}{\omega}\mathbf{K}_{\phi\phi} \tag{17}$$

where $h = -\tan\delta_E$ is the dielectric loss factor. This electrical damping causes a phase change between voltage and current in the piezoceramic.

This system of equations assumes that the piezoelectric structure is free in air, that is, it is not coupled to any propagation medium. Depending on the application of piezoelectric transducers, such as actuators for examples, operation frequency is lower than the first transducer resonance frequency, and thus the inertia effects are negligible. Therefore, they can be modeled through a static analysis, and Eq. (14) may be written as:

$$\begin{bmatrix} \mathbf{K}_{uu} & \mathbf{K}_{u\phi} \\ \mathbf{K}^t_{u\phi} & -\mathbf{K}_{\phi\phi} \end{bmatrix} \begin{Bmatrix} \mathbf{U} \\ \Phi \end{Bmatrix} = \begin{Bmatrix} \mathbf{F} \\ \mathbf{Q} \end{Bmatrix} \tag{18}$$

Notice from these static equations that if there is no piezoelectric effect ($\mathbf{K}_{u\phi}=\mathbf{0}$), then elastostatic and electrostatic equations are recovered:

$$\mathbf{K}_{uu}\mathbf{U} = \mathbf{F} \text{ and } \mathbf{K}_{\phi\phi}\Phi = \mathbf{Q} \tag{19}$$

The same idea is valid for dynamic equations (Eq. 14). Therefore, the matrix $\mathbf{K}_{u\phi}$ represents the coupling between electric and mechanical equations. This coupling introduces complexities in the behavior of piezoelectric materials making sometimes our intuition to fail when trying to understand these materials behavior especially in the dynamic case. This is usually not the case when modeling elastic materials.

Regarding the numerical solution of the matrix system of Eq. 18, the total matrix involving the sub-matrices $\mathbf{K}_{uu}$, $\mathbf{K}_{u\phi}$, and $\mathbf{K}_{\phi\phi}$ is symmetric and indefinite (contains negative and positive eigenvalues) due to the negative diagonal terms in the submatrix $-\mathbf{K}_{\phi\phi}$. This means that for the numerical solution of this matrix system, methods based on Cholesky decomposition cannot be used. Therefore, an active column solver (LDL^T factorization) using skyline storage [24] is used instead. Additionally, to avoid a large matrix band, the total matrix, composed of stiffness, piezoelectric, and dielectric matrices, must be stored using the following sequence of degrees of freedom: displacement in *x*, *y*, and *z* directions, and electrical potential.

For more details related on FEM applied to piezoelectricity, the reader may refer to Naillon *et al.* [12] and Lerch [13].

4. FLEXTENSIONAL ACTUATOR DESIGN

In this section, the design of flextensional transducers will be described. Essentially, a flextensional transducer consists of a piezoceramic (or a "stack" of piezoceramic - multilayer transducer) connected to a flexible mechanical structure that converts and amplifies the output displacement of the piezoceramic. Flextensional transducers are used for applications, such as low-frequency, large displacement actuators, high sensitivity hydrophones, and low-frequency underwater sound sources ("fish finders") [25,26]. We will consider static and low-frequency actuator applications of flextensional actuators. They can also be manufactured at the microscale by using MEMS fabrication techniques.

A well-known type of flextensional actuator is the "moonie" transducer, which consists of a piezoceramic disk sandwiched between two metal end caps, each having a shallow cavity on its surface (hence the name "moonie") [19,27] as shown in Fig. 4a. It can also be used for generating sound and as a hydrophone. Two important goals in the design of this kind of flextensional actuator are output displacement and generative (or "blocking") force [18]. Generative force is the maximum force supported by the transducer without deforming, or the necessary force to restore the undeformed shape of the transducer, for a certain applied voltage. The requirements for these goals change for each application. The moonie presents a combination of high output displacement and moderate generative force, filling the gap between the multilayer actuators (that have high generative force and small displacement) and bimorph actuators (that exhibit large displacement, but very small generative force). Another example of a flextensional actuator is the "cymbal" actuator (see Fig. 4c), developed by Dogan *et al.* [18], that presents a large displacement and relatively high generative (or blocking) force. Other important parameter in the flextensional actuator design is the first resonance frequency f_r corresponding to the flextensional mode. This frequency must be high to guarantee a linear operation of the transducer in a wide range of low frequencies, and a small time response ($t_{(sec)}=1/f_r$).

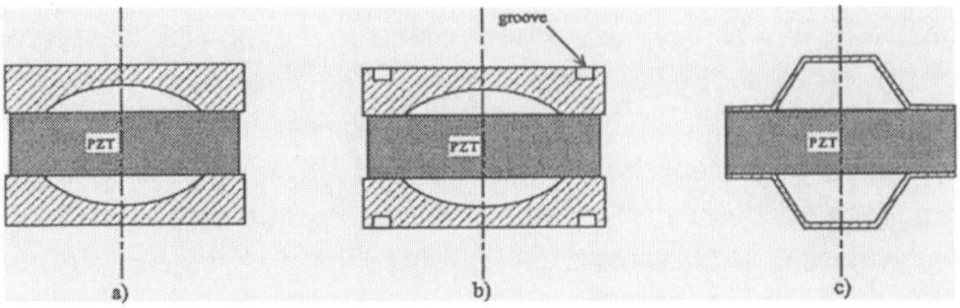

Figure 4. a) Moonie actuator; b) Groove moonie actuator; c) Cymbal actuator [48].

These flextensional actuators have been developed in the past by using simple analytical models and experimental techniques [17,27,28], and recently, some authors have applied the Finite Element Method (FEM) to optimize transducer design [19,29]. In this case, the optimization of the transducer is obtained by using FEM to build curves of output displacement and generative force as a function of some dimension of a specific topology chosen for the coupling structure which limits the problem to a sizing optimization. This choice is usually based on the physical intuition of the problem (for example, the end caps with shallow cavities, in the case of the moonie). In this way, a variation of the moonie transducer, called a "groove moonie", and the cymbal actuator were developed [18]. The groove moonie is essentially a moonie with a ring-shaped groove (transducer is circular) on the exterior surface of the end caps as shown in Fig. 4b. The shallow cavities in the common moonie transducers provide the end caps with a flexible structural behavior, that is, they become mechanical transformers that convert and amplify the piezoceramic lateral output displacement into an axial motion normal to the plates. The groove increases the flexibility of the exterior part of the end caps, giving larger output displacements than the common moonie. In the cymbal design, the output displacement is created by the combination of flexural and rotational motions, rather than only flexural motion, as in the case of the moonie. This yields larger output displacements than the previous transducers and also large generative force due to the high axial stiffness. Therefore, the development of these transducers shows that the output displacements and the generative force depend on the topology of the coupled structure connected to the piezoceramic (or a stack of piezoceramics). These two goals are to be traded-off since a high output displacement is obtained by increasing the coupling structure flexibility (or increase the structure compliance), while a high generative force requires an increase of coupling structure stiffness (or decrease the structure compliance). This means that the design of the coupling structure is related to the distribution of stiffness and flexibility in the structural domain, which is similar to a compliant mechanism design. Compliant mechanisms are structures that have mobility due to the flexibility of some or all of their members [30,31]. The design of compliant mechanisms with distributed flexibility and low weight can be achieved by using topology optimization [31,32,33,34,35]. By designing other types of flexible structures connected to the piezoceramic (or a multilayer transducer), we can obtain other types of flextensional actuators that produce high output displacements (and generative forces) in different directions, according to a specific application.

Therefore, flextensional actuators can be designed by applying topology optimization [36,37,38,39]. Considering the discussion presented above, the problem of designing a flextensional actuator is posed as the design of a flexible structure coupled to a piezoceramic (or a stack of piezoceramics) that maximizes the output displacement and generative force in some specified direction.

In some other efforts on flextensional actuator design [40,41] the flexible structure (or the compliant mechanism) is designed separately from the piezoceramic which is represented only by a force acting on the flexible structure. However, it is important to state that the force generated by the piezoceramic depends on the characteristics (stiffness and mass) of flexible structure attached to it, which means that if the flexible structure topology changes, then the piezoceramic force may also change considering the same applied electrical charge. This compromises the design of flexible structure alone. Therefore, the flexible structure must be designed together with the attached piezoceramic to obtain an optimized design in the flextensional actuator case, as described below.

4.1 Mean Transduction

To take into account the piezoceramic influence in the flexible structure design, the concept of *mean transduction* is introduced. This concept is obtained by considering the reciprocal theorem applied to the piezoeclectric medium, which is briefly described below [38,39].

The reciprocal theorem (or Betti's Theorem), well-known from elasticity theory [42], states that if the linear elastic body is kinematically supported and subjected to a traction $\mathbf{t}_1$ in a region $\Gamma\mathbf{t}_1$ that produces a displacement field $\mathbf{u}_1$, and a traction $\mathbf{t}_2$ in region $\Gamma\mathbf{t}_2$ that produces a displacement field $\mathbf{u}_2$, then:

$$\int_{\Gamma_{t_1}} \mathbf{t}_1.\mathbf{u}_2 d\Gamma = \int_{\Gamma_{t_2}} \mathbf{t}_2.\mathbf{u}_1 d\Gamma \tag{20}$$

where

$$\int_{\Gamma_t} \mathbf{t}.\mathbf{u} d\Gamma = \int_{\Gamma_t} t_i u_i d\Gamma ,$$

and $\mathbf{t}=\mathbf{T}.\mathbf{n}$, $\mathbf{T}$ is the stress tensor, and $\mathbf{n}$ is the normal vector to the surface. Now, considering the equilibrium equation for the elastic medium:

$$\int_\Omega \mathbf{S}(\mathbf{u})^t \mathbf{c}\mathbf{S}(\mathbf{v}) d\Omega = \int_{\Gamma_t} \mathbf{t}.\mathbf{v} d\Gamma \ \text{ for } \mathbf{u} \in \hat{V} \text{ and } \forall \mathbf{v} \in \hat{V} \tag{21}$$

where $\ddot{V} = \{ \mathbf{v} = v_i e_i : v_i \in H^1(\Omega)$ with $\mathbf{v}=\mathbf{0}$ *on* $\Gamma_\mathbf{u}\}$, Ω is the domain of the elastic medium, $\mathbf{c}$ is the elastic property matrix, and the superscript “t” denotes transpose. We have:

$$\int_\Omega \mathbf{S}(\mathbf{u}_1)^t \mathbf{c}\mathbf{S}(\mathbf{u}_2) d\Omega = \int_{\Gamma_{t_1}} \mathbf{t}_1.\mathbf{u}_2 d\Gamma$$

and

$$\int_\Omega \mathbf{S}(\mathbf{u}_2)^t \mathbf{c}\mathbf{S}(\mathbf{u}_1) d\Omega = \int_{\Gamma_{t_2}} \mathbf{t}_2.\mathbf{u}_1 d\Gamma \tag{22}$$

Therefore, we can write equation (20) in the form:

$$L_1(\mathbf{u}_2) = \int_{\Gamma_{t_1}} \mathbf{t}_1.\mathbf{u}_2 d\Gamma = \int_\Omega \mathbf{S}(\mathbf{u}_1)^t \mathbf{c}\mathbf{S}(\mathbf{u}_2) d\Omega = \int_\Omega \mathbf{S}(\mathbf{u}_2)^t \mathbf{c}\mathbf{S}(\mathbf{u}_1) d\Omega =$$
$$= \int_{\Gamma_{t_2}} \mathbf{t}_2.\mathbf{u}_1 d\Gamma = L_2(\mathbf{u}_1) \tag{23}$$

where L_1 (or L_2) is defined as the *mutual mean compliance* of the system, between $\Gamma\mathbf{t}_1$ and $\Gamma\mathbf{t}_2$. If the dummy load applied to generate the auxiliary solution discussed is a unit traction in a certain region of the body, the displacement at any point of the body due to some other load applied to another region will be equal to the mutual mean compliance between these two regions.

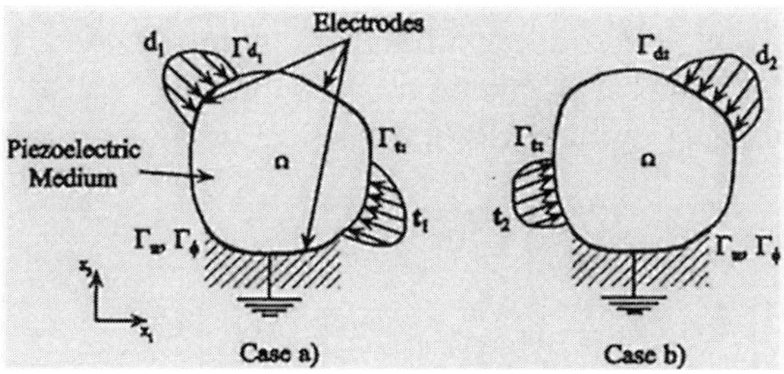

Figure 5. Reciprocal theorem to piezoelectricity [48].

An extension of the reciprocal theorem for the piezoelectric medium follows the same steps. Consider a piezoelectric body shown in Fig. 5, subjected to electrical charges d_1, d_2 applied to the boundary Γd_1 and Γd_2, respectively, and tractions $\mathbf{t}_1$, $\mathbf{t}_2$ applied to the boundaries $\Gamma \mathbf{t}_1$ and $\Gamma \mathbf{t}_2$. respectively. Since tractions and surface charges are applied gradually, the work done will be stored as strain and electric energy in the body. Therefore, it can be shown that [39]:

$$\mathrm{L}_1(\mathbf{u}_2,\phi_2) = \int_{\Gamma_{t1}} \mathbf{t}_1.\mathbf{u}_2 d\Gamma + \int_{\Gamma_{d1}} d_1.\phi_2 d\Gamma = \int_{\Gamma_{t2}} \mathbf{t}_2.\mathbf{u}_1 d\Gamma + \int_{\Gamma_{d2}} d_2.\phi_1 d\Gamma = \\ = \mathrm{L}_2(\mathbf{u}_1,\phi_1) \tag{24}$$

where $d = \mathbf{D}.\mathbf{n}$, $\mathbf{D}$ is the electric displacement vector, and $\mathbf{n}$ is the normal vector to the surface. This is the extension of reciprocal theorem for the piezoelectric medium. In analogy to the mutual mean compliance quantity, $\mathrm{L}_1(\mathbf{u}_2,\phi_2)$ (or $\mathrm{L}_2(\mathbf{u}_1, \phi_1)$) was denoted by *mean transduction* since mechanical and electrical quantities are involved (see Fig. 5).

By substituting piezoelectric equilibrium equations (11) into equation (24), they can be written in the form:

$$\mathrm{L}_1(\mathbf{u}_2,\phi_2) = \mathrm{A}(\mathbf{u}_1,\mathbf{u}_2) - \omega^2 \mathrm{M}(\mathbf{u}_1,\mathbf{u}_2) + \mathrm{B}(\phi_1,\mathbf{u}_2) + \mathrm{B}(\phi_2,\mathbf{u}_1) + \\ -\mathrm{C}(\phi_1,\phi_2) = \mathrm{A}(\mathbf{u}_2,\mathbf{u}_1) - \omega^2 \mathrm{M}(\mathbf{u}_1,\mathbf{u}_2) + \mathrm{B}(\phi_2,\mathbf{u}_1) + \mathrm{B}(\phi_1,\mathbf{u}_2) + \\ -\mathrm{C}(\phi_2,\phi_1) = \mathrm{L}_2(\mathbf{u}_1,\phi_1) \tag{25}$$

where operators $A(\mathbf{u},\mathbf{v}), M(\mathbf{u},\mathbf{v}), B(\phi,\mathbf{u})$, and $C(\phi,\varphi)$ were defined in equations (12).

Now, applying the FEM formulation by discretizing the domain Ω (displacements and electric potentials are approximated by using shape functions), the reciprocal theorem and mean transduction for the piezoelectric medium can be written in the following matrix form [38]:

$$\begin{aligned}
&L_1(\mathbf{U}_2,\mathbf{\Phi}_2)=\{\mathbf{U}_2\}^t\{\mathbf{F}_1\}+\{\mathbf{\Phi}_2\}^t\{\mathbf{Q}_1\}= \\
&=\begin{Bmatrix}\mathbf{U}'_2\\ \mathbf{\Phi}'_2\end{Bmatrix}^t\left\{\begin{bmatrix}\mathbf{K}_{uu} & \mathbf{K}_{u\phi}\\ \mathbf{K}^t_{u\phi} & -\mathbf{K}_{\phi\phi}\end{bmatrix}-\omega^2\begin{bmatrix}\mathbf{M} & 0\\ 0 & 0\end{bmatrix}\right\}\begin{Bmatrix}\mathbf{U}_1\\ \mathbf{\Phi}_1\end{Bmatrix}= \\
&=\begin{Bmatrix}\mathbf{U}'_1\\ \mathbf{\Phi}'_1\end{Bmatrix}^t\left\{\begin{bmatrix}\mathbf{K}_{uu} & \mathbf{K}_{u\phi}\\ \mathbf{K}^t_{u\phi} & -\mathbf{K}_{\phi\phi}\end{bmatrix}-\omega^2\begin{bmatrix}\mathbf{M} & 0\\ 0 & 0\end{bmatrix}\right\}\begin{Bmatrix}\mathbf{U}_2\\ \mathbf{\Phi}_2\end{Bmatrix}= \\
&=\{\mathbf{U}_1\}^t\{\mathbf{F}_2\}+\{\mathbf{\Phi}_1\}^t\{\mathbf{Q}_2\}=L_2(\mathbf{U}_1,\mathbf{\Phi}_1)
\end{aligned} \tag{26}$$

where $\mathbf{U}_i$, $\mathbf{\Phi}_i$, $\mathbf{F}_i$, and $\mathbf{Q}_i$, (i=1,2) are the nodal displacements, nodal electric potentials, nodal mechanical forces, and nodal electric charges, respectively. The matrices $\mathbf{K}_{uu}$, $\mathbf{K}_{u\phi}$, and $\mathbf{K}_{\phi\phi}$ were defined in Section 3.

Equation (26) allows us to implement the concept of mean transduction for the piezoelectric medium by using FEM formulated in a discretized domain. Thus, we can express the displacement or electric potential in any region of the piezoelectric medium as a function of a known displacement and electric fields caused by an external applied dummy traction or surface charge (analogous to the case of elastic medium).

4.2 Material Model

In the design of piezoelectric actuators using topology optimization the piezoceramic domain is kept fixed during the optimization and the design domain is only the coupling structure domain as shown in Fig. 6.

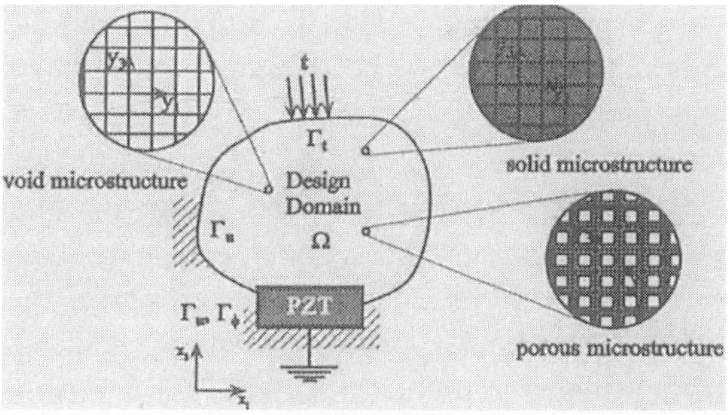

Figure 6. Distribution of material along the design domain

The material model for topology optimization applied consists of defining at each point of the domain, a microstructure [43] as described in Fig. 7 and consists of a unit cell with a rectangular hole inside whose dimensions are defined by design variables *a* and *b*, and the orientation θ. Therefore, in each point of the domain there is a composite material defined by the periodic repetition of the microstructure corresponding to that point. This composite can vary from void (*a*=*b*=1 or a maximum value) to full material (*a*=*b*=0). In this sense, the problem consists of optimizing the material distribution in a perforated domain with infinite microscale voids. The homogenized (or effective) elasticity properties ($\mathbf{c}^H$) of this composite material in each point of the design domain is obtained using the homogenization method (see Chapter 6) [34].

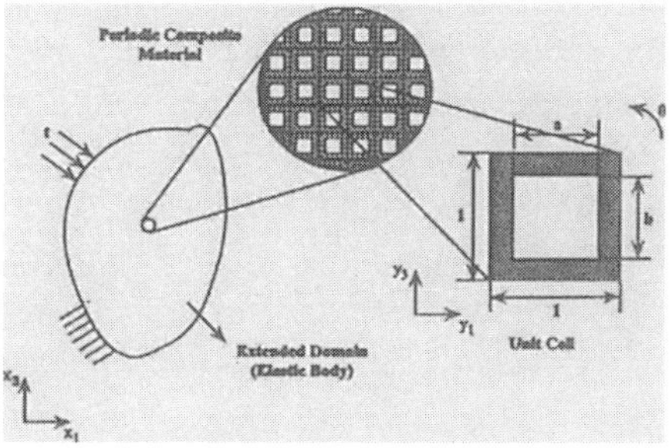

Figure 7. Microstructure used in the relaxation of the design domain.

During the optimization process, it is necessary to calculate the gradients of the elastic properties in relation to *a* and *b*. They are calculated by building a table that contains the values of homogenized properties for a large amount of combinations of discrete values of *a* and *b* [43]. The orientation θ is determined by considering the local principal stresses direction in each point of the domain [44].

The optimization problem is defined in a continuous form. However since the domain is discretized in finite elements, variables a, b, and θ which theoretically are a continuous function of coordinate axes, became sets of continuous design variables a_n, b_n, and θ_n, defined for the n finite element subdomain in the numerical problem. The upper-bounds $a_{upp}=0.995$ and $b_{upp}=0.995$ specified for a and b, respectively, are necessary to avoid numerical problems such as singularity of the stiffness matrix in the finite element formulation. Numerically, regions with $a_n=b_n=0.995$ have practically no structural significance and can be considered void regions.

4.3 Formulation of Optimization Problem

In the flextensional actuator design the objective is to maximize the output displacement (or generative force) in a domain region due to an input electrical charge in the transducer. Two different objective functions are required for the design: one is related to the electromechanical function and the other due to the structural function.

The electromechanical function provides electromechanical conversion between two regions (see Fig. 8), Γd_1 (electrical charges) and $\Gamma \mathbf{t}_2$ (mechanical tractions), of the design domain. The larger this function, the larger the displacement generated in a certain direction of the region $\Gamma \mathbf{t}_2$ due to an input electrical charge in the Γd_1 region. Its maximization is achieved by maximizing the mean transduction defined in equation (26) considering two load cases shown in case (a) of Fig. 8: the first one is related to the actuator response due to the application of a surface charge on surface Γd_1 of the transducer; and the second one is related to an applied dummy traction $\mathbf{t}_2$ to region $\Gamma \mathbf{t}_2$, in the same desired output displacement direction.

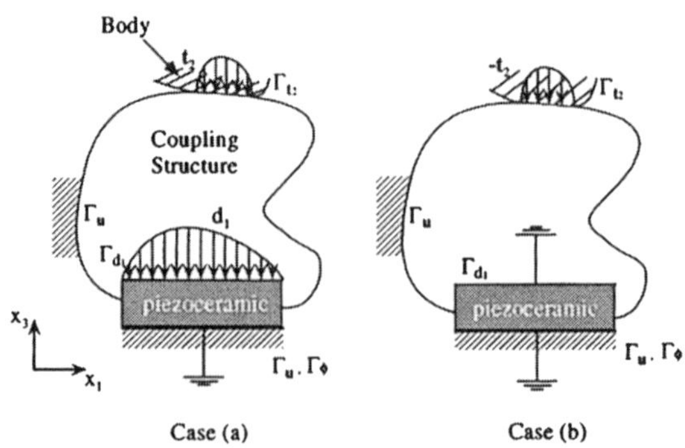

Figure 8. Load cases for calculation of mean transduction (case a) and mean compliance (case b) [48]

Considering equation (26), since the mechanical traction is null in the first load case ($\mathbf{t}_1$=0) and the surface charges are null in the second load case (d_2=0), the mean transduction in dynamic applications between regions Γd_1 and $\Gamma \mathbf{t}_2$ is given by (using FEM matrix formulation defined) [38]:

$$
\begin{aligned}
&L_2(\mathbf{U}_1,\boldsymbol{\Phi}_1)=\{\mathbf{U}_1\}^t\{\mathbf{F}_2\}=\{\boldsymbol{\Phi}_2\}^t\{\mathbf{Q}_1\}= \\
&=\begin{Bmatrix}\mathbf{U}'_2\\ \boldsymbol{\Phi}'_2\end{Bmatrix}^t\left\{\begin{bmatrix}\mathbf{K}_{uu} & \mathbf{K}_{u\phi}\\ \mathbf{K}^t{}_{u\phi} & -\mathbf{K}_{\phi\phi}\end{bmatrix}-\omega^2\begin{bmatrix}\mathbf{M} & 0\\ 0 & 0\end{bmatrix}\right\}\begin{Bmatrix}\mathbf{U}_1\\ \boldsymbol{\Phi}_1\end{Bmatrix}= \qquad (27)\\
&=\{\mathbf{U}_1\}^t[\mathbf{K}_{u\phi}]\{\boldsymbol{\Phi}_2\}-\{\boldsymbol{\Phi}_1\}^t[\mathbf{K}_{\phi\phi}]\{\boldsymbol{\Phi}_2\}
\end{aligned}
$$

since $\{\boldsymbol{\Phi}_1\}^t\{\mathbf{Q}_2\}=0$.

Notice that in the dynamic case, both the maximization or minimization of $L_2(\mathbf{U}_1,\phi_1)$ provide a high electromechanical conversion between regions Γd_1 (electrical charges) and $\Gamma \mathbf{t}_2$ (mechanical tractions), and so a high output displacement, because under the harmonic excitation, the deformation in the direction of $\mathbf{t}_2$, when the phase angle is ωt, is identical to the deformation when the phase angle is $\omega t+\pi$. Therefore, we need to consider the maximization of the absolute value of the mean transduction.

However, if we consider only the maximization of the mean transduction, the optimum solution obtained for this function may be a structure with very small stiffness. Therefore, a structural function must be defined to provide sufficient stiffness between $\Gamma \mathbf{t}_2$ and Γd_1, which is obtained by minimizing the mean compliance between $\Gamma \mathbf{t}_2$ and Γd_1. This also guarantees a large generative force, and also that the actuator will resist the reaction forces generated by the workpiece that the actuator is trying to hold or move. The mean compliance is calculated considering a *static* load case described in case 2 of Fig. 8 where the electrode surface Γd_1 is electrically grounded. Therefore, the mean compliance is given by the expression [39]:

$$
\begin{aligned}
&L_3(\mathbf{U}_3,\boldsymbol{\Phi}_3)=\{\mathbf{U}_3\}^t\{\mathbf{F}_3\}=\{\mathbf{U}_3\}^t[\mathbf{K}_{uu}]\{\mathbf{U}_3\}+\{\mathbf{U}_3\}^t[\mathbf{K}^t{}_{u\phi}]\{\boldsymbol{\Phi}_3\}= \\
&=\begin{Bmatrix}\mathbf{U}'_3\\ \boldsymbol{\Phi}'_3\end{Bmatrix}^t\left\{\begin{bmatrix}\mathbf{K}_{uu} & \mathbf{K}_{u\phi}\\ \mathbf{K}^t{}_{u\phi} & -\mathbf{K}_{\phi\phi}\end{bmatrix}\right\}\begin{Bmatrix}\mathbf{U}_3\\ \boldsymbol{\Phi}_3\end{Bmatrix} \qquad (28)
\end{aligned}
$$

since $\{\boldsymbol{\Phi}_3\}^t\{\mathbf{Q}_3\}=0$. By changing the mean compliance value we can control the generative force.

To combine both optimization problems the following objective function is proposed:

$$F = w * \ln(|L_2(U_1, \Phi_1)|) - (1-w) * \ln(L_3(U_3, \Phi_3)) \quad (29)$$
$$0 \le w \le 1$$

where w is a weight coefficient. This objective function allows us to control the contributions of the mean transduction (46) and the mean compliance (48) in the design. Therefore, the optimization problem is stated as:

Maximize F
a, b and θ
subject to:

$$t_3 = -t_2 \quad (\Gamma_{t_3} = \Gamma_{t_2})$$

equilibrium equations for all load cases

$$0 \le a \le a_{upp} \le 1$$
$$0 \le b \le b_{upp} \le 1$$
Volume constraint

A constraint related to symmetry conditions can be used to reduce the computational cost.

Therefore, the objective function can be calculated by solving three load cases shown in cases 1 and 2 of Fig. 8. These load cases are solved separately. The first and second load cases are used for the calculation of the mean transduction and the last one for the calculation of the mean compliance. Variations of the previous objective functions can be derived, depending on the design requirements of the flextensional actuator.

A flow chart of the optimization algorithm describing the steps involved is shown in Fig. 9. The initial domain is discretized by finite elements and the design variables are the values of a_n, b_n, and θ_n (defined above) in each finite element, defined only in the coupling structure (elastic material).

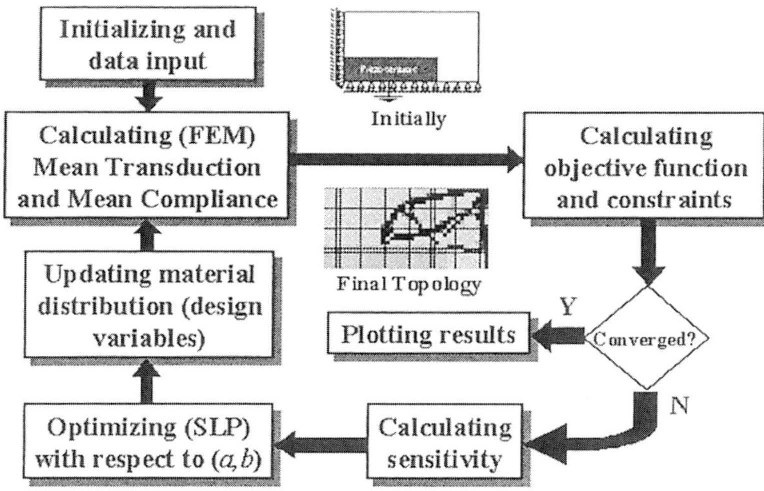

Figure 9. Flow chart of the optimization procedure

4.4 Sensitivity Analysis

The linearization of the problem (Taylor series) in each iteration of sequential linear programming requires the sensitivities (gradients) of the objective function and constraints relative to design variables a_n and b_n. The sensitivity of the objective function is obtained by differentiating Eq. (29). This will result in an expression which is a function of the sensitivities of the mean transduction $L_2(\mathbf{U}_1,\phi_1)$ and mean compliance $L_3(\mathbf{U}_3,\phi_3)$.

It is shown that the sensitivity of $L_2(\mathbf{U}_1,\phi_1)$ is given by (in a matrix form) [39]:

$$\frac{\partial L_2(\mathbf{U}_1,\mathbf{\Phi}_1)}{\partial A_n} = -\begin{Bmatrix} \mathbf{U}'_2 \\ \mathbf{\Phi}'_2 \end{Bmatrix}^t \begin{bmatrix} \dfrac{\partial \mathbf{K}_{uu}}{\partial A_n} & \dfrac{\partial \mathbf{K}_{u\phi}}{\partial A_n} \\ \dfrac{\partial \mathbf{K}^t_{u\phi}}{\partial A_n} & -\dfrac{\partial \mathbf{K}_{\phi\phi}}{\partial A_n} \end{bmatrix} \begin{Bmatrix} \mathbf{U}_1 \\ \mathbf{\Phi}_1 \end{Bmatrix} \tag{30}$$

That is a general expression for the sensitivity considering the entire domain piezoelectric. However, in the current problem the design domain consists of elastic material (non-piezoelectric) not including the piezoceramic. In this case, the terms:

$$\frac{\partial \mathbf{K}_{u\phi}}{\partial A_n} \text{ and } \frac{\partial \mathbf{K}_{\phi\phi}}{\partial A_n}$$

are equal to zero. Therefore, the previous equation simplifies to [39]:

$$\frac{\partial L_2(\mathbf{U}_1,\mathbf{\Phi}_1)}{\partial A_n} = -\mathbf{U}'_2 \frac{\partial \mathbf{K}_{uu}}{\partial A_n} \mathbf{U}_1 \tag{31}$$

and $\partial \mathbf{K}_{uu}/\partial A_n$ will be zero for the elements in the piezoceramic domain (since they were not assigned design variables). $\partial \mathbf{K}_{uu}/\partial A_n$ depends on $\partial \mathbf{c}^E/\partial A_n$ which is calculated using the data of the table that describes the variation of the property $\mathbf{c}^E$ relative to the design variables a_n and b_n. This table is built by calculating the effective properties of the unit cells through the homogenization method, considering discretized values of a_n and b_n, as discussed in Section 4.2. The data are interpolated by Bezier curves, and the derivatives are obtained.

The sensitivity of the mean compliance is obtained by considering $\mathbf{U}_2=\mathbf{U}_1=\mathbf{U}_3$ and $\mathbf{\Phi}_1=\mathbf{\Phi}_2=\mathbf{\Phi}_3$ [39]:

$$\frac{\partial L_3(\mathbf{U}_3,\mathbf{\Phi}_3)}{\partial A_n} = -\mathbf{U}'_3 \frac{\partial \mathbf{K}_{uu}}{\partial A_n} \mathbf{U}_3 \tag{32}$$

4.5 Examples

Examples will be presented to illustrate the design of flextensional actuators using the proposed method. The design domains used for the examples below are shown in Fig. 10a and b. They consist of a domain of piezoceramic that remains unchanged during the optimization and a domain S of some elastic material (steel, aluminum, polymer, etc.) where the optimization is conducted. Brass is considered in these examples.

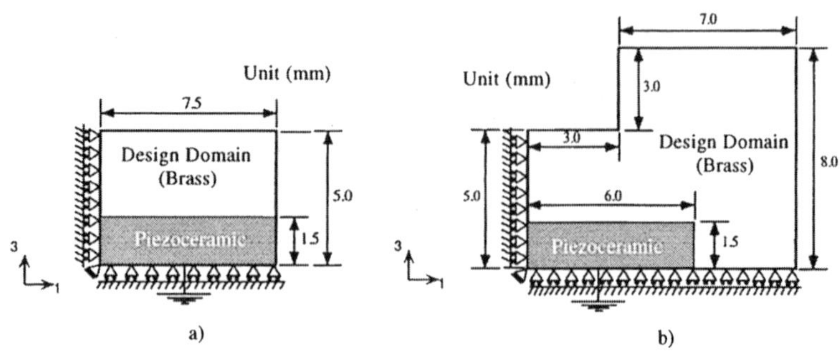

Figure 10. Design domains considered [48]

The domain of Fig. 10a has 800 elements (rectangle discretized by a 40 x 20 mesh) and the domain of Fig. 10b has 1136 elements. The mechanical and electrical boundary conditions for both domains are shown in the same figures. Only one-quarter of the domain is considered since the applications considered have two symmetry axes. For each example, the correspondent interpretation of the topology is also presented.

Table 1 presents the piezoelectric material properties used in the simulations. The brass material used has Young modulus equal to 10.6 10^{10} N/m^2 and Poisson's ratio equal to 0.3. Two-dimensional elements under plane-strain assumption are used in the finite element analysis. For most examples, the total volume constraint of the material is considered to be 30% of the volume of the whole domain without piezoceramic (domain S).

4.5.1 A Multilayer Actuator

An interesting example is the design of a multilayer actuator. This actuator is obtained by connecting many piezoelectric actuators in series, such as the well-known moonie actuator, for example. The mechanical connections between the single actuators define the performance of the multilayer actuator in terms of output displacement and generative force. Recently, based on the moonie actuator design, a new design of actuator called cymbal [18] was developed that presents a larger output displacement and generative force than the moonie design and also has a better performance in multilayer configuration. This can be explained because, in the moonie actuator, the flexibility in the coupling structure is provided by a compliant joint in the middle top of the coupling structure. When this transducer is assembled in multilayer configuration, the contact between them occurs in a small area (and not a point) at the compliant joint position. Therefore, this reduces the flexibility of the multilayer device, decreasing its performance in terms of output displacement. In the cymbal configuration, a flat region is provided for their connection and the flexibility is transfered to compliant joints at the ends of this region, providing a multilayer configuration with better performance. The cymbal design was developed using the finite element method and experimental verification, and seems to be close to an optimum solution. The objective of this example is to design a single actuator that generates a large output displacement and generative force in multilayer configuration. The design domain and load condition considered for this example are shown in Fig. 11.

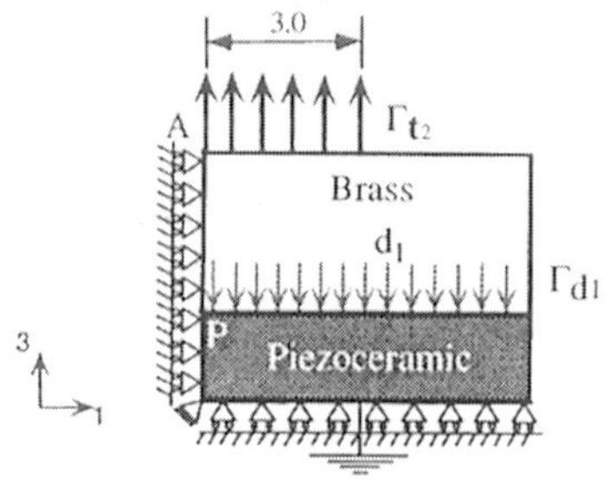

Figure 11. Design domain and load condition considered [48]

The optimization problem is defined as the maximization of the deflection at region $\Gamma \mathbf{t}_2$ in the direction of the specified dummy load when electrical charges d_1 are applied to the piezoceramic at electrode Γd_1 (see Fig. 11). The mean compliance at region $\Gamma \mathbf{t}_2$ is to be minimized since the connection between the single actuators in the multilayer configuration occurs in this region. The objective function described in equation (29) was used considering the coefficient w equal to 0.8, 0.9, and 1.0 and frequency $\omega = 0$. For w equal to 1.0 no minimization of mean compliance is considered in region $\Gamma \mathbf{t}_2$. The results obtained using the proposed method are shown in Figs. 12a-c (the figures must be reflected to both symmetry axes).

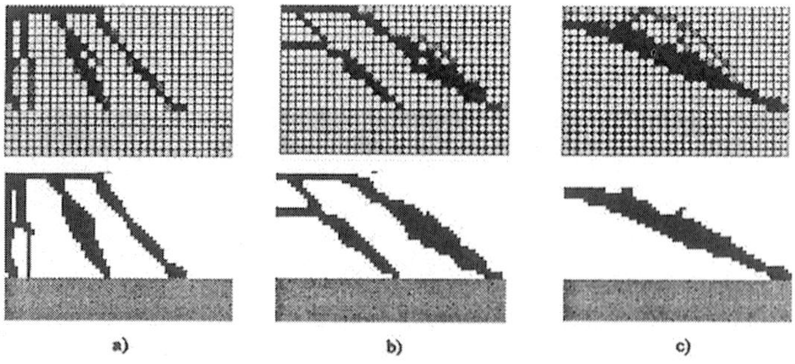

Figure 12. Topologies obtained for the single element of a multilayer actuator, and corresponding image interpretations: a) w is set equal to 0.8; b) w is set equal to 0.9; c) w is set equal to 1.0. [48]

The structure's stiffness in the vertical direction is reduced as the weight coefficient w increases, which means that the value of the mean compliance increases, as expected. Therefore, the generative force in the vertical direction is expected to decrease while the vertical magnification displacement between points A and P (see Fig. 11) is expected to increase with w. This is confirmed by the values of generative force and vertical displacement ratio between points A and P presented in Table 8, considering an applied voltage of 1000V/mm to the piezoceramic. Table 8 also presents the corresponding values of the first resonance frequency and the performance characteristic d_{33}. This quantity is defined as the ratio between the output displacement u_{out} in a defined point and direction (point A, for example) and the input voltage V_{inp} in the piezoceramic. It can also be defined as the ratio between the charge generated in the electrodes Q_{sc} considering short-circuit conditions, due to an applied force F_{ap} (in the same specified point and direction)[27], that is:

$$d_{33} = \frac{u_{out}}{V_{inp}} = \frac{Q_{sc}}{F_{ap}} \ (C/N) \tag{33}$$

Its value is also a measurement of the transducer response.

Table 1. Vertical displacement ratio (D.R.), generative force (G.F), performance d_{33}, and resonance frequency f_r for the actuators shown in the figures (applied voltage equal to 1000 V/mm).

Figure	D.R.	G.F. (N)	d_{33} (*n*C/N)	f_r (*k*Hz)
12a	1.48	0.491	0.491	66.69
12b	4.22	1.24	1.24	43.84
12c	7.73	2.44	2.44	30.38

Notice that the value of d_{33} increases and the value of resonance frequency f_r decreases with the compliance. These values were calculated through FEM analysis using FEM models based on the interpreted images obtained.

The coupling structures obtained remind the configuration of the cymbal transducer. Figures 13a and b show the image of the corresponding multilayer actuator configuration of Figs. 12b and c. Figure 13c shows the corresponding deformed configuration (obtained using FEM) for the single element of the actuator shown in Fig. 13b.

4.5.2 A Flextensional Gripper

This example illustrates the design of a flextensional gripper, that is, a gripper actuated by a piezoceramic (or stack of piezoceramics). The design domain and load conditions for this problem are shown in Fig. 14.

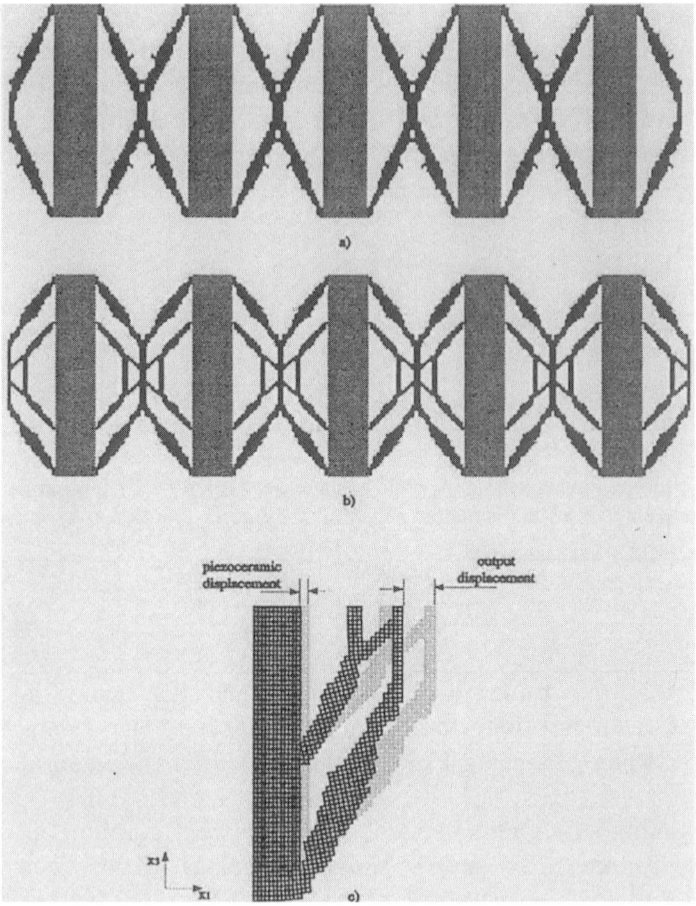

Figure 13. a) and b) show the multilayer configurations for the actuators shown in Figs. 12b and c; c) corresponding deformed configuration obtained using FEM for b). [48]

The function of the gripper is (1) to deform along the direction of the dummy force to grasp a workpiece at point A when the electrical charges are applied to electrode Γd_1 (see Fig. 14), and (2) to hold the workpiece while the external electrical charges are continuously applied. The minimization of mean compliance is considered at point A. In this example, the total volume constraint of the material is 25% of the volume of the whole domain without piezoceramic. Figures 15a, b, and c show the topologies obtained for the flextensional grippers considering the coefficient w equal to 0.5, 0.6, and 1.0 (each figure must be reflected to the vertical symmetry axis).

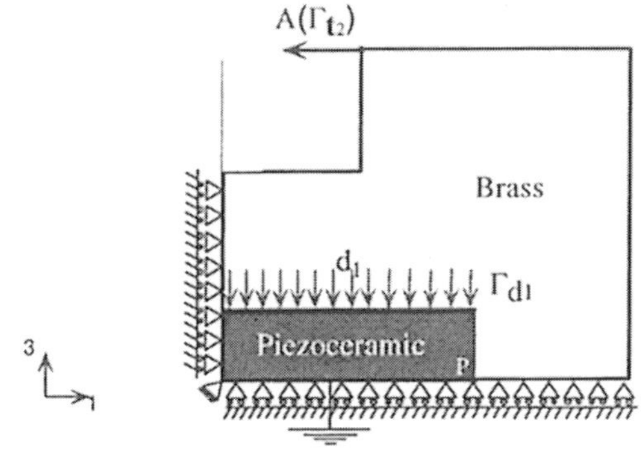

Figure 14. Design domain and load conditions considered [48]

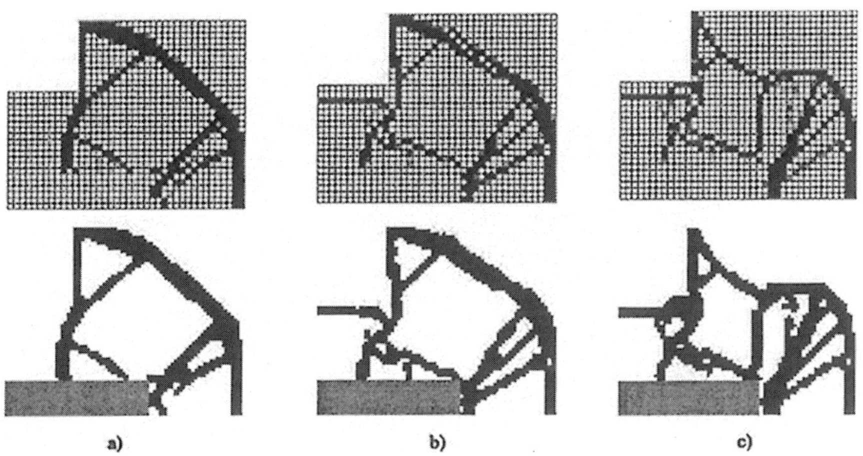

Figure 15. Topologies obtained for the flextensional grippers, and corresponding image interpretations: a) w is set equal to 0.5; b) w is set equal to 0.6; c) w is set equal to 1.0. [48]

The horizontal displacement ratios between points A and P (see Fig. 14) ranges from 0.63 in Fig. 15a to 1.63 in Fig. 15c, considering an applied voltage to the piezoceramic electrodes equal to 1000 V/mm. This shows the increase of the structure compliance between these two points with the weight coefficient w.

Figures 16a and b show the entire image and the corresponding deformed configuration (obtained using FEM) of the gripper shown in Figs. 15c.. Noticed that the coupling structure mechanism increases the piezoceramic lateral displacement.

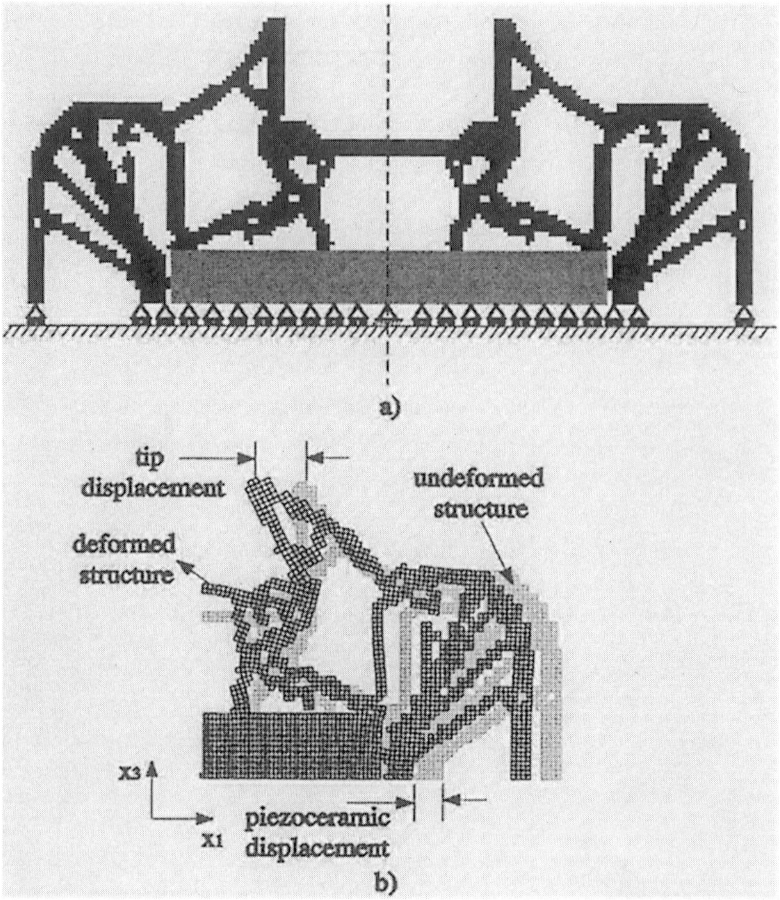

Figure 16. a) Image of the entire gripper of Fig. 15c; b)corresponding FEM deformed configuration [48]

Even though the examples presented herein are limited to two-dimensional (2D) models, the method introduced is general and can be applied to the design of three-dimensional (3D) actuators.

4.6 Manufactured Prototypes

Figure 17 shows some prototypes of flextensional actuators that were built by connecting an aluminum flexible structure to rectangular piezoceramics made of PZT5A. These ceramics are polarized in the z-direction and the electrode surfaces are normal to this direction.

The PZT5A piezoceramic was bonded to the aluminum flexible structure by using a thin film of epoxy resin in both cases. Special care was taken to electrically insulate the electrodes of piezoceramic from the flexible structure in some transducers, to avoid a short circuit, because of the closed shape of their end cap, which connects top and bottom electrodes.

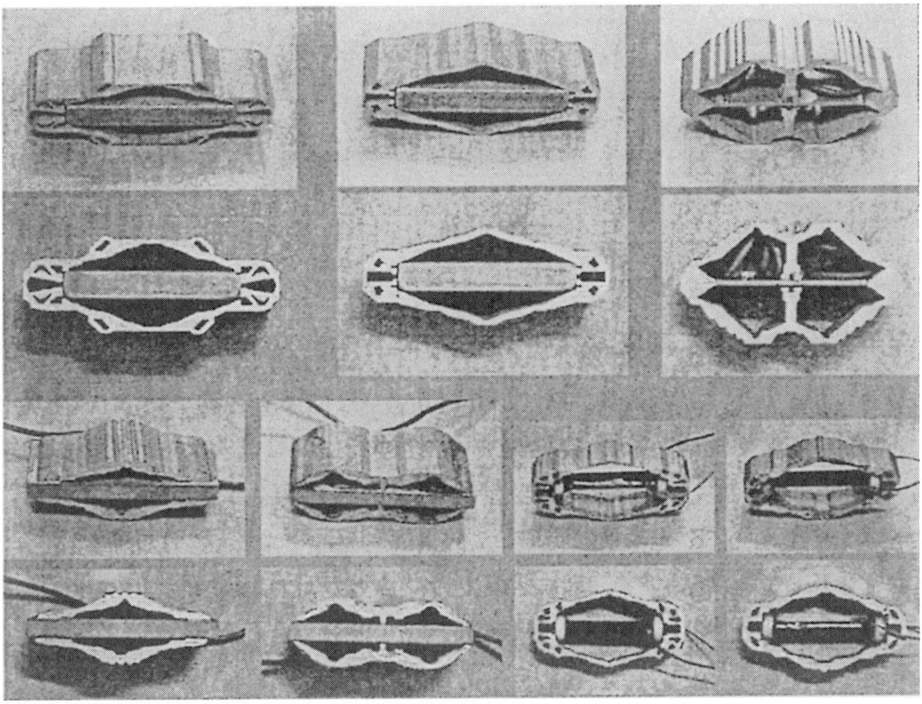

Figure 17. Prototypes of flextensional piezoelectric actuators: (a) fla1025 and (b) f2b0830. They have mirrors bonded on it for laser interferometry measurement purpose. (prototyping of designed flextensional actuators with G. Nader)

These actuators can also be manufactured at the microscale by using MEMS fabrication techniques. The main problem is the deposition of piezoceramic but there are techniques to deposit thin piezoelectric films.

5. CONCLUSION

The results presented give us an idea about the potential of applying optimization methods to design MEMS actuated by piezolectric materials. Optimization methods consist of a powerful tool to design these devices. However, this field is still in the beginning stages since many other optimization techniques can be applied and many other piezoelectric transducers can be designed. For example, the presented method for designing flextensional actuators has been recently extended for designing linear piezoelectric motors [45]. A linear piezoelectric motor is essentially a "multiflextensional" actuator, that is, it consists of a multiflexible compliant mechanism actuated by many piezoceramics attached to it. The design requires the flexible structure to generate different output displacements for different applied loads (or voltages). In this sense, we can think about designing micro-robot structures such as micromanipulators, micro-grippers, micro-pliers and other micro-tools actuated by piezoelectric materials.

In a recent work, Li *et al.* [46] have applied genetic algorithms to find the optimum position of the piezoceramic in the flextensional actuator domain for each topology optimization iteration. In the piezocomposite design field, optimization method was applied to design composite laminated structures with piezoelectric laminate obtained high improvements in its performance [47]. Therefore, new designs are expected to come.

In addition, the idea can be extended to other types of piezoelectric transducers [48] and transducers involving other types of phenomena such as shape memory alloys actuators or magnetostrictive transducers.

REFERENCES

1. Newnham, R. E. and G. R. Ruschau. "Smart Electroceramics". *Journal of the American Ceramic Society.* **74** No.3 (1991): 463-480.
2. Ried, R.P., Kim, E.S., Hong, D.M., and Muller, R.S., "Piezoelectric Microphone with On-Chip CMOS Circuits," *IEEE/ASME Journal of Microelectromechanical Systems*, vol. 2, pp. 111-120, September 1993.
3. Huang, Y.L., Zhang, H., Kim, E.S., Kim, S.G., and Jeon, Y.B., "Piezoelectrically Actuated Microcantilever for Actuated Mirror Array Application," Technical digest of *Solid-State Sensor and Actuator Workshop*, Hilton Head Island, SC, June 2 - 6, 1996, pp. 191-195.
4. Huang, D. and Kim, E.S., "Micromachined Acoustic-Wave Liquid Ejector," *IEEE/ASME Journal of Microelectromechanical Systems*, vol. 10, pp. 442-449, September 2001.
5. Vertsanova, O., Perlov, I., and Tsyganok, B., "Modelling of Processes of Photoacoustic Diagnostic with Piezoelectric Detection," *Proceedings of the 2000 Modeling and Simulation of Microsystems conference*, pp. 249-252.
6. Fernandez, J. F., A. Dogan, Q..M. Zhang, J.F. Tressler, and R.E. Newnham. "Hollow Piezoelectric Composites". *Sensors and Actuators A.* **51** (1996): 183-192.
7. McCammon, D. F. and W. Thompson Jr. "The Design of Tonpilz Piezoelectric Transducers Using Nonlinear Goal Programming". *The Journal of the Acoustical Society of America.* **68** No.3 (1980): 754-757.
8. Simson, É. A. and A. Taranukha. "Optimization of the Shape of a Quartz Resonator". *Acoustical Physics.* **39** No.5 (1993): 472-476.
9. Kim, J., V. V. Varadan, and V. K. Varadan. "Finite Element-Optimization Methods for the Active Control of Radiated Sound from a Plate Structure". *Smart Materials & Structures.* **4** (1995): 318-326.
10. Brissaud, M. "Characterization of Piezoceramics". *IEEE Transactions on Ultrasonics, Ferroelectrics and Frequency Control.* **38** No.6 (1991): 603-617.

11. Allik, H. and T.J.R. Hughes. "Finite Element Method for Piezoelectric Vibration". *International Journal of Numerical Methods in Engineering*. **2** (1970): 151-157.
12. Naillon, M., R.H. Coursant, F. Besnier. "Analysis of Piezoelectric Structures by a Finite Element Method", *Acta Electronica*, **25** 4 (1983): 341-362.
13. Lerch, R. "Simulation of Piezoelectric Devices by Two-and Three-Dimensional Finite Elements". *IEEE Transactions on Ultrasonics, Ferroelectrics and Frequency Control*. **37** No.2 (1990): 233-247.
14. Guo, N., P. Cawley, and D. Hitchings. "The Finite Element Analysis of the Vibration Characteristics of Piezoelectric Discs". *Journal of Sound and Vibration*. **159** No.1 (1992):115-138.
15. Kharouf, N. and P. R. Heyliger. "Axisymmetric Free Vibrations of Homogeneous and Laminated Piezoelectric Cylinders". *Journal of Sound and Vibration*. **174** No.4 (1994): 539-561.
16. Challande, P. "Optimizing Ultrasonic Transducers Based on Piezoelectric Composites Using a Finite-Element Method". *IEEE Transactions on Ultrasonics, Ferroelectrics and Frequency Control*. **37** No.2 (1990):135-140.
17. Dogan, A., S. Yoshikawa, K. Uchino, and R. E. Newnham."The Effect of Geometry on the Characteristics of the Moonie Transducer and Reliability Issue". *Proceedings of IEEE Ultrasonics Symposium*. (1994): 935-939.
18. Dogan, A., K. Uchino, and R.E. Newnham. "Composite Piezoelectric Transducer with Truncated Conical Endcaps "Cymbal"". *IEEE Transactions on Ultrasonics, Ferroelectrics and Frequency Control*. **44** No.3 (1997): 597-605.
19. Onitsuka, K., A. Dogan, J. F. Tressler, Q. Xu, S. Yoshikawa, and R. E. Newnham. "Metal-Ceramic Composite Transducer, The "Moonie"". *Journal of Intelligent Material Systems and Structures*. **6** (1995): 447-455.
20. ANSI/IEEE Std 176-1978. "*IEEE Standard on Piezoelectricity, Special Issue of Transactions on Sonics and Ultrasonics*". **SU-31** No.2 (1984).
21. Eer Nisse, E. P. "Variational Method for Electroelastic Vibration Analysis", *IEEE Transactions on Sonics and Ultrasonics*. **SU-14**, No.4: (1967): 153-160
22. Kunkel, H. A., Locke, S. and Pikeroen, B. "Finite Element Analysis of Vibrational Modes in Piezoelectric Ceramic Disks", *IEEE Transactions on Ultrasonics, Ferroelectrics and Frequency Control*. **37** No.4 (1990):316-328.
23. Modulef, http://www-rocq.inria.fr/modulef/english.html (1999)
24. Bathe, K. J. *Finite Element Procedures*. Englewood Cliffs, N.J.:Prentice-Hall, (1996).
25. Rolt, K. D. "History of the Flextensional Electroacoustic Transducer". *The Journal of the Acoustical Society of America*. **87** No.3 (1990): 1340-1349.
26. Decarpigny, J-N, B. Hamonic, and O. B. Wilson Jr. "The Design of Low-Frequency Underwater Acoustic Projectors: Present Status and Future Trends". *IEEE Journal of Oceanic Engineering*. **16** No.1 (1991): 107-122.
27. Xu, Q.C., S. Yoshikawa, J.R. Belsick, and R.E. Newnham. "Piezoelectric Composites with High Sensitivity and High Capacitance for Use at High Pressures". *IEEE Transactions on Ultrasonics, Ferroelectrics and Frequency Control*. **38** No.6 (1991): 634-639.
28. Xu, Q.C., A. Dogan, J.F. Tressler, S. Yoshikawa, and R. E. Newnham. "Ceramic-Metal Composite Actuator". *Proceedings of IEEE 1991 Ultrasonic Symposium*. (1991): 923-928.
29. Newnham, R.E., A. Dogan, Q.C. Xu, K. Onitsuka, J.F. Tressler, and S. Yoshikawa. "Flextensional "Moonie" Actuators". *Proceedings of IEEE Ultrasonics Symposium*. (1993): 509-513.
30. Ananthasuresh, G. K. and S. Kota. "Designing Compliant Mechanisms". *ASME Mechanical Engineering*. November (1995): 93-96.

31. Frecker, M.I., G.K. Ananthasuresh, S. Nishiwaki, N. Kikuchi, and S. Kota. "Topological Synthesis of Compliant Mechanisms Using Multi-criteria Optimization". *Journal of Mechanical Design, Transaction of the ASME.* **119** (1997): 238-245.
32. Ananthasuresh, G.K., S. Kota, and Y. Gianchandani. "Systematic Synthesis of Microcompliant Mechanisms - Preliminary Results". *Proceedings of the Third National Conference on Applied Mechanisms and Robotics.* Cincinnati Ohio (1993): Paper 82.
33. Ananthasuresh, G.K., S. Kota, and N. Kikuchi. "Strategies for Systematic Synthesis of Compliant MEMS". *Proceedings of the 1994 ASME Winter Annual Meeting.* Chicago Illinois DSC-55-2 (1994): 677-686.
34. Nishiwaki, S., M.I. Frecker, S. Min, and N. Kikuchi. "Optimization of Compliant Mechanisms Using the Homogenization Method". *International Journal for Numerical Methods in Engineering.* **42**, No.3 (1998): 535-559.
35. Sigmund, O. "On the Design of Compliant Mechanisms Using Topology Optimization". *Mechanics of Structures and Machines* **25** No.4 (1997): 495-526.
36. Silva, E.C.N., S. Nishiwaki, J.S.O. Fonseca, A.T. Crumm, G.A.Brady, F.R.M. Espinosa, J.W. Halloran, and N. Kikuchi "Topology Optimization Applied to the Design of Piezocomposite Materials and Piezoelectric Actuators", *Proceedings of Mathematics and Control in Smart Structures - 5th SPIE (Annual International Symposium on Smart Structures and Materials),* San Diego, California, (USA), (1998): 216-227.
37. Silva, E.C.N., S. Nishiwaki, and N. Kikuchi "Design of Piezoelectric Transducers Using Topology Optimization", *Proceedings of Fourth World Congress on Computational Mechanics (WCCM-IV),* Buenos Aires, Argentina, (CDROM), (1998).
38. Silva, E.C.N., S. Nishiwaki, and N. Kikuchi "Topology Optimization Design of Flextensional Actuators", *IEEE Transations on Ultrasonics, Ferroelectrics and Frequency Control* **47** (3) (2000): 657-671.
39. Canfield, S. and M. Frecker "Topology Optimization of Compliant Mechanical Amplifier for Piezoelectric Actuator". *Structural Multidisciplinary Optimization,* **20**, (2000): 269-279.
40. Kota, S., J. Hetrick, Z. Li, and L. Saggere "Tailoring Unconventional Actuators Using Compliant Transmissions: Design Methods and Applications". *IEEE/ASME Transactions on Mechatronics* **4** No.4 (1999): 396-408.
41. Barber, J. M. *Elasticity.* Kluwer Academic, Dordrecht, (1992).
42. Bendsøe, M. P. and N. Kikuchi "Generating Optimal Topologies in Structural Design Using a Homogenization Method". *Computer Methods in Applied Mechanics and Engineering.* **71** (1988): 197-224.
43. Pedersen, P. "On Optimal Orientation of Orthotropic Materials", *Structural Multidisciplinary Optimization* **1** No.2 (1989): 101-106.
44. Silva, E.C.N. "Design of Piezoelectric Motors Using Topology Optimization", *Proceedings of Modeling, Signal Processing, and Control in Smart Structures - 8th SPIE (Annual International Symposium on Smart Structures and Materials),* Newport Beach, California (USA) (2001): 513-524.
45. Li, Y., X. Xin, N. Kikuchi, and K. Saitou "Optimal Shape and Location of Piezoelectric Materials for Topology Optimization of Flextensional" *Proceedings of Sixth U.S. National Congress on Computational Mechanics (USNCCM),* Dearborn, Michigan (USA) (2001): 566.
46. Correia, V.M.F., C.M.M. Soares, and C.A.M. Soares "A Numerical Model for the Optimal Design of Composite Laminated Structures with Piezoelectric Laminae", *Proceedings of Mathematics and Control in Smart Structures - 6th SPIE (Annual International Symposium on Smart Structures and Materials),* Newport Beach, California (USA), (1999).

47. Silva, E.C.N. and N. Kikuchi "Design of Piezoelectric Transducers Using Topology Optimization", *Journal of Smart Materials and Structures* No.8 (1999): 350-364.
48. Silva, E.C.N., S. Nishiwaki, and N. Kikuchi, "Design of Piezocomposite Materials and Piezoelectric Transducers Using Topology Optimization - Part II", *Archives of Computational Methods in Engineering,* Vol. 6, No.3 (1999): 191-222.

Chapter 6

SYNTHESIS OF PIEZOCOMPOSITES

Emílio Carlos Nelli Silva
Department of Mechatronics and Mechanical Systems Engineering. Polytechnic School from University of São Paulo, Brazil.

Synopsis: Piezocomposite materials are largely applied to acoustical devices such as sonars and ultrasonic transducers. Their development has also been based on the use of simple analytical models, test of prototypes, and analysis using the finite element method (FEM). Thus, following the idea of applying synthesis methods to design smart materials and MEMS structures, this chapter presents systematic synthesis methods for piezocomposite materials by using the topology optimization combined with the homogenization technique. In this method, a piezocomposite material with improved electromechanical efficiency is obtained by designing its unit cell topology. Beginning with a brief introduction to piezocomposite materials, its performance measurement, the concept of the homogenization method, and its manufacturing technique, the chapter provides a self-contained description of the synthesis method for these materials. The examples presented show that the synthesis method is indeed a promising tool to design these smart materials.

1. PIEZOCOMPOSITE DESIGN

The term piezocomposite applies to any composite resulting from the combination of any piezoelectric material (polymer or ceramic) with other non-piezoelectric materials, including air-filled voids [1]. Usually these combinations result in new piezocomposites that offer substantial advantages over conventional piezoelectric materials. These advantages are high electromechanical coupling, which measures energy conversion in the piezocomposite and therefore its sensitivity, and low acoustic impedance, which helps to transmit acoustic waves to media such as human body tissue or water [1,2,3]. By changing the volume fraction of the constituents, the shape of the inclusions, or even the topology of the unit cell, we can obtain different effective properties. Therefore, in designing the piezocomposite we can tailor the properties to a specific application, which cannot be done with a single material. Of the currently available configurations of piezocomposite, the 1-3 piezocomposite, which consists of rods of piezoceramic PZT immersed in a polymer matrix as shown in Fig. 1a, has been the focus of most studies. Another typical configuration is the 2-2 piezocomposite, shown in Fig. 1b. Here, "1-3" and "2-2" designate the connectivity of the piezocomposite material; however, many other connectivities, such as 0-3 and 1-3, are also possible [4].

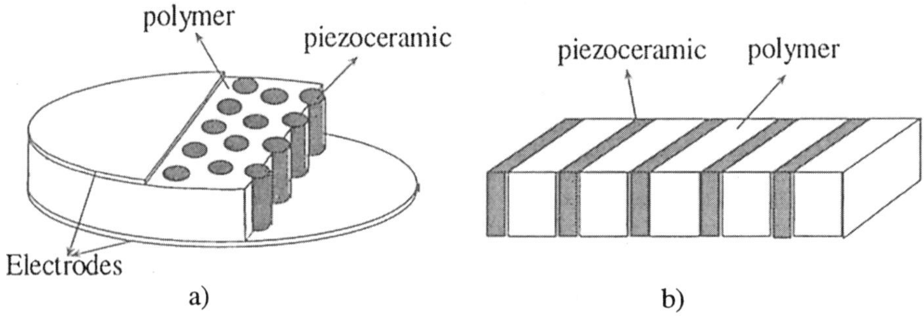

Figure 1. a) 1-3 piezocomposite; b) 2-2 piezocomposite. [36]

In piezocomposite applications, we assume that the excitation wavelengths are so large that the detailed structure of the unit cell does not matter, and the material may be considered as a new homogeneous medium with new "effective" material properties. Then, the excitation (acoustic, for example) will average out over the fine scale variations of the composite medium, in the same way as averaging occurs in the micron-sized grain structure in a conventional pure ceramic. In this case, the piezocomposite functions as a true composite material [1,5]. When wavelengths are not large enough relative to the size of the unit cell, the piezocomposite will present a dispersive behavior with scattering occurring inside the unit cells making its behavior extremely difficult to model. Homogeneous behavior can be assured by reducing the size of the unit cell (or microstructure) relative to the excitation wavelength. However, it is not always possible to guarantee that the size of the microstructure (or *unit cell*, the smallest structure that is periodic in the composite) is smaller than the wavelength.

Assuming the piezocomposite is a homogeneous medium, its behavior can be characterized by Eq. (5) in Chapter 5, by substituting all properties by the composite's effective properties (or homogenized properties) into these equations [5]. These effective properties can be obtained using the homogenization method presented in Section 2. Therefore, the constitutive equations of the composite material considering homogenized properties become:

$$\begin{cases} \langle \mathbf{T} \rangle = \mathbf{c}^{\mathrm{E}}{}_{\mathrm{H}} \langle \mathbf{S} \rangle - \mathbf{e}_{\mathrm{H}} \langle \mathbf{E} \rangle \\ \langle \mathbf{D} \rangle = \mathbf{e}^{t}{}_{\mathrm{H}} \langle \mathbf{S} \rangle + \varepsilon^{S}{}_{\mathrm{H}} \langle \mathbf{E} \rangle \end{cases} \tag{1}$$

or

$$\begin{cases} \langle \mathbf{S} \rangle = \mathbf{s}^{\mathrm{E}}{}_{\mathrm{H}} \langle \mathbf{T} \rangle + \mathbf{d}_{\mathrm{H}} \langle \mathbf{E} \rangle \\ \langle \mathbf{D} \rangle = \mathbf{d}^{t}{}_{\mathrm{H}} \langle \mathbf{T} \rangle + \varepsilon^{\mathrm{T}}{}_{\mathrm{H}} \langle \mathbf{E} \rangle \end{cases} \tag{2}$$

where:

$$\langle \ldots \rangle = \frac{1}{|\mathrm{V}|} \int_{V} . \mathrm{dV} \tag{3}$$

and the subscript "H" refers to the homogenized properties. $\mathbf{s}^{\mathrm{E}}{}_{\mathrm{H}}$ is the homogenized compliance tensor under short-circuit conditions, $\varepsilon^{\mathrm{T}}_{\mathrm{H}}$ is the homogenized clamped body dielectric tensor, and $\mathbf{d}_{\mathrm{H}}$ is the homogenized piezoelectric stress tensor.

In the following sections, the subscript "H" will be omitted for the homogenized properties for the sake of brevity. As a convention, the polarization axis of the piezoelectric material will be considered in the *3rd* (or *z*) direction.

1.1 Performance Characteristics of Piezocomposite Materials

The main applications of piezocomposites are the generation and detection of acoustic waves. It can be classified into: low frequency (hydrostatic operation mode, such as some hydrophones and naval sonars) and high frequency (ultrasonic transducers for imaging). At low-frequency applications, the operation of the device is quasistatic since the operational frequency of the device is generally smaller than the device's first resonance frequency.

In both cases, the quantities that describe the performance of the composite are obtained by considering only the constitutive properties (neglecting the effects of inertia) as described in [5, 6].

1.1.1 Low-Frequency Applications

Consider an orthotropic composite under hydrostatic pressure P as shown in Fig. 2.

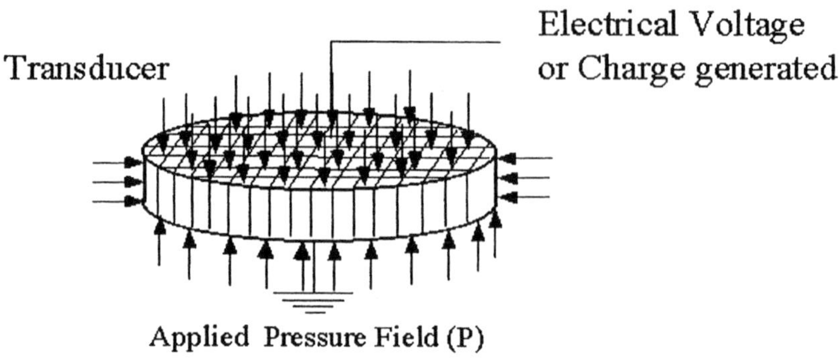

Figure 2. Piezocomposite transducer under hydrostatic pressure.

The composite's response can be measured by three different quantities [7]:

- Hydrostatic Coupling Coefficient (d_h):

$$d_h = \frac{\langle D_3 \rangle}{P} = d_{33} + d_{23} + d_{13} \tag{4}$$

- Figure of Merit ($d_h g_h$):

$$g_h = \frac{\langle E_3 \rangle}{P} = \frac{d_h}{\varepsilon^T{}_{33}} \Rightarrow d_h g_h = \frac{d^2{}_h}{\varepsilon^T{}_{33}} \tag{5}$$

- Hydrostatic Electromechanical Coupling Factor (k_h):

$$k_h = \sqrt{\frac{d^2{}_h}{\varepsilon^T{}_{33} s^E{}_h}} \tag{6}$$

where

$$s_h = (\langle \varepsilon_1 \rangle + \langle \varepsilon_2 \rangle + \langle \varepsilon_3 \rangle)/P$$

is the dilatational compliance. For an orthotropic material:

$$s^E_h = s^E_{11} + s^E_{22} + s^E_{33} + 2s^E_{12} + 2s^E_{13} + 2s^E_{23}$$

and the coefficients s^E_{kl} are those defined in equation (2).

These quantities can be written in terms of the properties described in equation (1) by using equation (7) in Chapter 5. The resulting analytical expressions are very complicated and will not be presented here. They can be easily calculated by using a computer.

The definitions presented above consider an orthotropic piezocomposite material. If a transversely isotropic composite in the *12* (or *xy*) plane is considered ($s^E_{13}=s^E_{23}$, $s^E_{11}=s^E_{22}$, $d_{13}=d_{23}$), then expressions of d_h, $d_h g_h$, and k_h are simplified to:

$$d_h = \frac{\langle D_3 \rangle}{P} = d_{33} + 2d_{13} \qquad d_h g_h = \frac{d^2_h}{\varepsilon^T_{33}} \qquad k_h = \sqrt{\frac{d^2_h}{\varepsilon^T_{33} s^E_h}} \tag{7}$$

where

$$s^E_h = 2s^E_{11} + 2s^E_{12} + 4s^E_{13} + s^E_{33}$$

In terms of the properties described in equation (1), these quantities can be written as in [7, 8]:

$$d_h = \mathbf{i}\mathbf{C}^{-1}\mathbf{g} \quad d_h g_h = \frac{(\mathbf{i}\mathbf{C}^{-1}\mathbf{g})^2}{\varepsilon_{33} + \mathbf{g}^t\mathbf{C}^{-1}\mathbf{g}} \quad k^2_h = \frac{(\mathbf{i}\mathbf{C}^{-1}\mathbf{g})^2}{(\varepsilon_{33} + \mathbf{g}^t\mathbf{C}^{-1}\mathbf{g})\mathbf{i}\mathbf{C}^{-1}\mathbf{i}^t} \tag{8}$$

where:

$$\mathbf{C} = \begin{bmatrix} \kappa & c_{13} \\ c_{13} & c_{33} \end{bmatrix} \quad \mathbf{g} = \begin{Bmatrix} e_{11} \\ e_{33} \end{Bmatrix} \quad \mathbf{i} = \{1 \quad 1\} \tag{9}$$

and $\kappa = (c_{11} + c_{12})/2$ is the transverse bulk modulus. All the properties considered in the definitions are homogenized properties.

1.1.2 High-Frequency Applications

In ultrasonic applications, thin plates of the piezocomposite are excited near their thickness-mode resonance. In this case, the quantity that describes the performance of the ultrasonic transducer is the Electromechanical Coupling Factor (k_t) [6] given by:

$$k_t = \sqrt{\frac{e^2_{33}}{c^D_{33}\varepsilon^S_{33}}} \tag{10}$$

where the properties are defined in equation (1) and $c^D_{33}=c^E_{33}+(e_{33})^2/\varepsilon^S_{33}$.

In addition to k_t, two other quantities are of interest for ultrasonic transducers in ultrasonic imaging: acoustic impedance ($Z=(c^D_{33}\rho)^{1/2}$) and longitudinal velocity ($v_t=(c^D_{33}/\rho)^{1/2}$), where ρ is the composite density). To increase the performance of the transducer, we must try to maximize k_t and minimize the acoustic impedance Z. A low acoustic impedance guarantees good coupling between the transducer and the external medium (usually water or a water-like medium). In addition, to model the composite as a homogeneous medium, the wavelength (γ) must be much larger than the dimension of the piezocomposite microstructure (as discussed above). Therefore, considering the relation $v_t=\gamma f$, where f is the specified operational frequency of the transducer, the longitudinal velocity v_t must be kept high; otherwise it causes a reduction of γ [6].

2. HOMOGENIZATION METHOD

In the piezocomposite design, the calculation of effective properties is necessary to obtain its performance. This can be achieved by applying the homogenization method that plays an important role in the design method. Therefore, a brief introduction about the homogenization method is given in this section.

Homogenization method allows the calculation of effective properties of a complex periodic material (composite) based only on the topology information of its unit cell (or microstructure), that is, type of materials and its distribution in the unit cell [9,10]. Among the main assumptions of the homogenization method is the fact that the composite material must be periodic and the scale of unit cell must be very small compared to the size of entire composite part and larger than the material grain size. There are other methods that allow us to obtain effective properties of a composite material, however, the main advantage of the homogenization technique is that it needs only the information about the unit cell which can have any complex shape.

The idea of homogenization is very useful as it allows us to save the effort when modeling parts made of composite materials. As an example consider Fig. 3. The perforated beam described in the top part, apparently seems to be very difficult to model since at first sight we would have to build an FEM model of the beam by considering all holes. However, it can be understood as a continuous beam (no holes) made of a material with properties equal to the effective properties of a "composite" material whose unit cell consists of a square with a circular hole inside, that is, a homogenized material. Therefore, if we have the effective properties of this composite material, the beam can be modelled as a homgeneous medium by building a simple FEM model with corresponding homogenized properties.

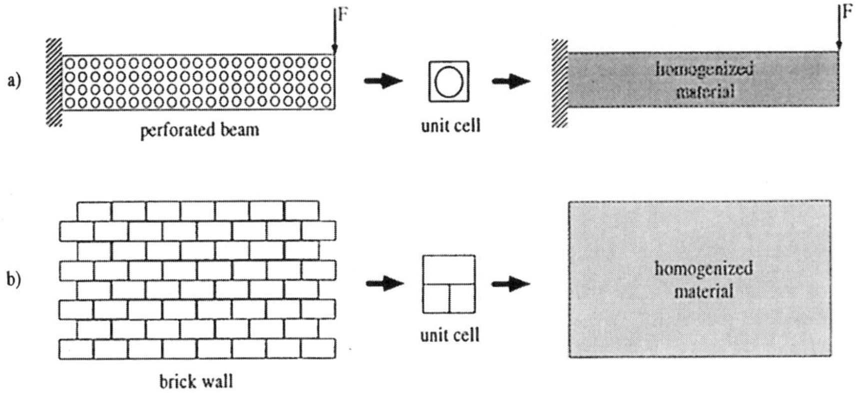

Figure 3. Homogenization concept.

The second example in Fig. 3 is a wall made of bricks. This wall can be understood as a composite material whose unit cell is described in the figure. The same concept can be applied to model the wall. It is important to mention that the size of phenomenon we are interested in analyzing will determine if homogenization concept can be applied or not. For example, suppose that we want to model a bullet hitting the wall. If the bullet size is much larger than unit cell size, homogenization can be applied and the entire wall can be modeled as a homogeneous material with corresponding effective properties. However, if the bullet size is of the same order of wall unit cell size (i.e., the brick), then homogenization cannot be applied and a detailed FEM model of the wall must be built taking into account the unit cell details.

Homogenization equations are obtained by first expanding displacement **u** in zeroth order and first order terms. The zeroth order terms represent the "average" values of displacement over the piezocomposite domain scale (**x**) while first order terms represent the variation of displacement in the unit cell domain scale (**y**), that is:

$$\mathbf{u}^{\varepsilon} = \mathbf{u}_0(\mathbf{x}) + \varepsilon \mathbf{u}_1(\mathbf{x}, \mathbf{y}) \tag{11}$$

where ε is a small number. This expression is substituted in the energy formulation for the medium and by using variational calculus, the so called *homogenization* equations related only to the **y** scale are extracted. Essentially, the meaning of homogenization equations consists of applying different load cases to the unit cell to calculate its response to these load cases. Based on these responses, the composite effective properties are obtained. Since the unit cell may have a complex shape, these equations are solved by using FEM.

The basic homogenization equations applied to calculate the effective properties of elastic materials are presented below.

$$\frac{1}{|Y|} \int_Y \left[c_{ijkl}(\mathbf{x}, \mathbf{y}) \left(\delta_{im}\delta_{jn} + \frac{\partial \chi_i^{(mn)}}{\partial y_j} \right) \right] S_{kl}(\mathbf{v})dY = 0 \quad \forall \mathbf{v} \in H_{per}(Y, R^3) \tag{12}$$

$$H_{per}(Y) = \{v \in H_1(Y) \mid v \text{ takes equal values on opposite sides of } Y\}$$
$$H_{per}(Y, R^3) = \{\mathbf{v} = (v_i) \mid v_i \in H_{per}(Y), i = 1, 2, 3\}$$

This homogenization formulation has no limitations regarding volume fraction or shape of the composite constituents, and is based upon assumptions of periodicity of the microstructure and the separation of the microstructure scale from the component scale through an asymptotic expansion.

Consider now a composite material under dynamic excitation (electrical or mechanical). If the operational wavelength is larger than the unit cell dimensions it seems natural that homogenization equations can be applied. This situation is called a static case. If operational wavelength is smaller than the unit cell dimensions, then unit cell scale will affect the calculation of effective properties, that is, the effective properties will have a dispersive behavior as explained in Section 1.1. Essentially, what happens is that if the wavelength is smaller than unit cell dimensions there will be wave reflections inside of the unit cell, and this effect must be taken into account in the homogenization equations. This situation is called a dynamic case and homogenization equations must be developed again originating the so called homogenization equations for dynamic case.

For a piezoelectric medium, the homogenization theory for piezoelectricity considering the static case (where the operational wavelength is much larger than the unit cell dimensions) was developed by Telega [11]. Galka *et al.* [12] presented the homogenization equations and effective properties for thermopiezoelectricity considering the static case. For dynamic applications (wavelength is of the same size as, or smaller than, the unit cell dimensions), Turbé and Maugin [13] developed a homogenization formulation to obtain the dynamical effective properties of the piezoelectric medium. In the limit of the static (and low-frequency) case, they recovered the expressions derived by Telega [11]. Finally, Otero *et al.* [14] developed general homogenized equations and effective properties for (heterogeneous and periodic) piezoelectric medium by considering terms of infinite order in the homogenization asymptotic expansion.

The homogenization equations for piezoelectric medium considering the static case are [11]:

$$
\begin{aligned}
&\frac{1}{|Y|}\int_Y \left[c^E_{ijkl}(\mathbf{x},\mathbf{y})\left(\delta_{im}\delta_{jn} + \frac{\partial \chi_i^{(mn)}}{\partial y_j}\right) + e_{ikl}(\mathbf{x},\mathbf{y})\frac{\partial \psi^{(mn)}}{\partial y_i}\right] S_{kl}(\mathbf{v})\,dY = 0 \\
&\quad \forall \mathbf{v} \in H_{per}(Y, R^3) \\
&\frac{1}{|Y|}\int_Y \left[e_{kij}(\mathbf{x},\mathbf{y})\left(\delta_{im}\delta_{jn} + \frac{\partial \chi_i^{(mn)}}{\partial y_j}\right) - \varepsilon^S_{ik}(\mathbf{x},\mathbf{y})\frac{\partial \psi^{(mn)}}{\partial y_i}\right] \frac{\partial \varphi}{\partial y_k}\,dY = 0 \\
&\quad \forall \varphi \in H_{per}(Y)
\end{aligned}
\tag{13}
$$

and:

$$
\begin{aligned}
&\frac{1}{|Y|}\int_Y\left[c^E_{klij}(\mathbf{x},\mathbf{y})\frac{\partial\Phi_k^{(m)}}{\partial y_l}+e_{kij}(\mathbf{x},\mathbf{y})\left(\delta_{mk}+\frac{\partial R^{(m)}}{\partial y_k}\right)\right]S_{ij}(\mathbf{v})dY=0\\
&\quad\forall\mathbf{v}\in H_{per}(Y,R^3)\\
&\frac{1}{|Y|}\int_Y\left[e_{kij}(\mathbf{x},\mathbf{y})\frac{\partial\Phi_i^{(m)}}{\partial y_j}-\varepsilon^S_{ik}(\mathbf{x},\mathbf{y})\left(\delta_{mi}+\frac{\partial R^{(m)}}{\partial y_i}\right)\right]\frac{\partial\varphi}{\partial y_k}dY=0\\
&\quad\forall\varphi\in H_{per}(Y)
\end{aligned}
\tag{14}
$$

These equations are equivalent to equation (12) for elastic medium. They were obtained by expanding piezocomposite displacement **u** and electrical potential ϕ in zeroth order and first order terms. The zeroth order terms represent the "average" values of these quantities over the piezocomposite domain scale (**x**) while first order terms represent the variation of these quantities in the unit cell domain scale (**y**), that is:

$$
\begin{aligned}
\mathbf{u}^{\varepsilon}&=\mathbf{u}_0(\mathbf{x})+\varepsilon\mathbf{u}_1(\mathbf{x},\mathbf{y})\\
\phi^{\varepsilon}&=\phi_0(\mathbf{x})+\varepsilon\phi_1(\mathbf{x},\mathbf{y})
\end{aligned}
\tag{15}
$$

The characteristic functions χ_i,ψ,Φ_i,R represent the displacement and electrical response of the unit cell to applied load cases (see Fig. 5).

By using FEM formulation to solve equations (13) and (14), the following matrix system is obtained [15]:

$$
\begin{aligned}
&\begin{bmatrix}\mathbf{K}_{uu} & \mathbf{K}_{u\phi}\\ \mathbf{K}^t_{u\phi} & -\mathbf{K}_{\phi\phi}\end{bmatrix}\begin{Bmatrix}\hat{\chi}^{(mn)}\\ \hat{\psi}^{(mn)}\end{Bmatrix}=\begin{Bmatrix}\mathbf{F}^{(mn)}\\ \mathbf{Q}^{(mn)}\end{Bmatrix}\\
&\begin{bmatrix}\mathbf{K}_{uu} & \mathbf{K}_{u\phi}\\ \mathbf{K}^t_{u\phi} & -\mathbf{K}_{\phi\phi}\end{bmatrix}\begin{Bmatrix}\hat{\Phi}^{(m)}\\ \hat{R}^{(m)}\end{Bmatrix}=\begin{Bmatrix}\mathbf{F}^{(m)}\\ \mathbf{Q}^{(m)}\end{Bmatrix}
\end{aligned}
\tag{16}
$$

where $\hat{\chi},\hat{\psi},\hat{\Phi},\hat{R}$ are the corresponding nodal values of the characteristics functions χ_i,ψ,Φ_i,R respectively, and

$$
\begin{aligned}
&F^{e(mn)}_{iI}=-\int_{\Omega^e}c^E_{ijmn}\frac{\partial N_I}{\partial y_j}d\Omega^e;\quad Q^{e(mn)}_I=-\int_{\Omega^e}e_{kmn}\frac{\partial N_I}{\partial y_k}d\Omega^e;\\
&F^{e(m)}_{iI}=-\int_{\Omega^e}e_{mij}\frac{\partial N_I}{\partial y_j}d\Omega^e;\quad Q^{e(m)}_I=-\int_{\Omega^e}\varepsilon^S_{mj}\frac{\partial N_I}{\partial y_j}d\Omega^e;
\end{aligned}
\tag{17}
$$

The other terms were defined in Chapter 5. Analyzing Eqs. (16) we conclude that for the 3D problem, there are nine load cases to be solved independently. Six of them come from Eqs. (13), where the indices *m* and *n* can be 1, 2, or 3, resulting in the following *mn* combinations: *11*, *22*, *33*, *12* or *21*, *23* or *32*, and *13* or *31*. The remaining three load cases come from Eqs. (14) where the index *m* can be *1*, *2*, and *3*. For example, in the 2D problem, there are five load cases to be solved independently (see Fig. 5). Three of them come from equations (13), where the indices *m* and *n* can be *1* or *3*, resulting in the combinations *11*, *33*, and *13* or *31*, for *mn*. The other two load cases come from equations (14) where the index *m* can be *1* and *3*. All load cases must be solved by enforcing periodic boundary conditions in the unit cell for the displacements and electrical potentials. The displacements and electrical potential of some point of the cell must be prescribed to overcome the non-unique solution of the problem; otherwise the problem will be ill posed [16]. The choice of the point of the prescribed values does not affect the homogenized coefficients since only derivatives of the characteristic functions are used for their computation. The numerical solution of matrix system (35) has already been discussed in Chapter 5.

The applied load cases for the bi-dimensional problem considering elastic material (Eq. (12)) and piezoelectric material (Eqs. (13) and (14)) are described in Figs. 4 and 5, respectively.

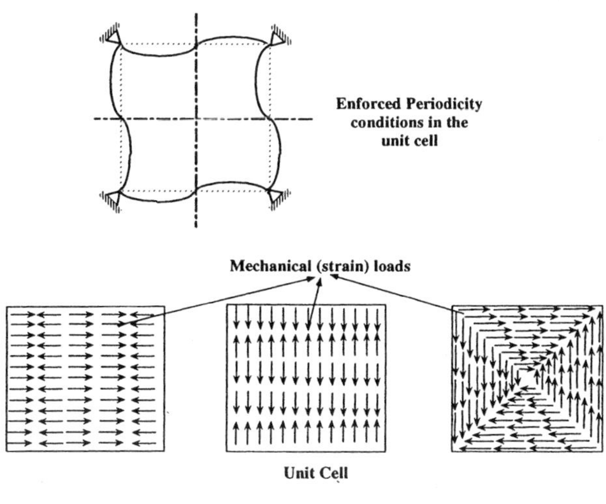

Figure 4. Load cases for homogenization of elastic materials.

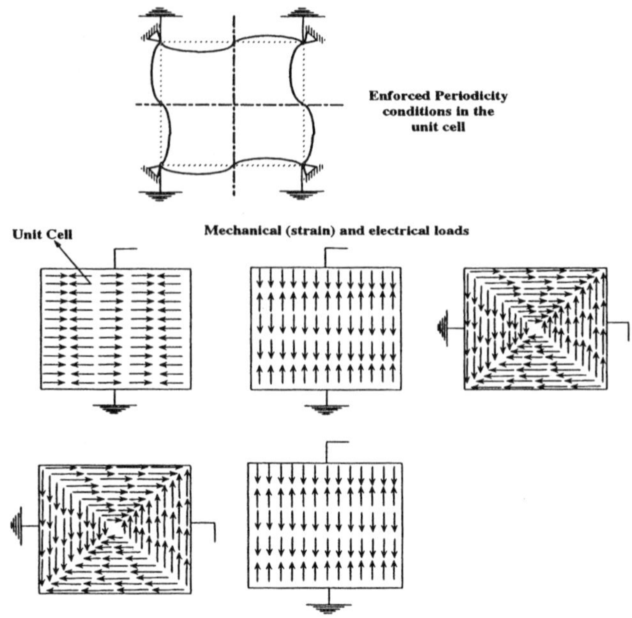

Figure 5. Load cases for homogenization of piezoelectric materials.

After solving for the characteristic displacements and electrical potentials, the effective properties are computed by using Eq. (16):

$$c^{H}_{pqrs}(\mathbf{x}) = \frac{1}{|Y|}\int_{Y}\left[c^{E}_{pqrs}(\mathbf{x},\mathbf{y}) + c^{E}_{pqkl}(\mathbf{x},\mathbf{y})\frac{\partial \chi_{k}^{(rs)}}{\partial y_{l}} + e_{kpq}(\mathbf{x},\mathbf{y})\frac{\partial \psi^{(rs)}}{\partial y_{k}}\right]dY$$

$$e^{H}_{prs}(\mathbf{x}) = \frac{1}{|Y|}\int_{Y}\left[e_{prs}(\mathbf{x},\mathbf{y}) + e_{pij}(\mathbf{x},\mathbf{y})\frac{\partial \chi_{i}^{(rs)}}{\partial y_{j}} - \varepsilon^{S}_{pk}(\mathbf{x},\mathbf{y})\frac{\partial \psi^{(rs)}}{\partial y_{k}}\right]dY$$

$$\varepsilon^{H}_{pq}(\mathbf{x}) = \frac{1}{|Y|}\int_{Y}\left[\varepsilon^{S}_{pq}(\mathbf{x},\mathbf{y}) + \varepsilon^{S}_{pj}(\mathbf{x},\mathbf{y})\frac{\partial R^{(q)}}{\partial y_{j}} - e_{pij}(\mathbf{x},\mathbf{y})\frac{\partial \Phi_{i}^{(q)}}{\partial y_{j}}\right]dY \tag{18}$$

To illustrate the homogenization concept for piezoelectricity the results of calculated effective properties for some piezocomposites are presented. The first example considers a bi-dimensional model of a 2-2 piezocomposite unit cell as shown in Fig. 6 which is made of PZT5/Spurr (see Table 1).

By using homogenization method described above, its effective properties were calculated and thus its performance characteristics were obtained. The graphics shown in Figs. 7a, 7b, 8a, and 8b show the performances characteristics d_h, $d_h g_h$, k_h, and k_t, respectively, as a function of the volume fraction for this piezocomposite design. From these graphics, maximum values of the performance can be obtained and are presented on Table 2. Values of these performances considering the unit cell fully made of PZT5 material are also presented. As we can see, the performance values obtained for the piezocomposite are better than the performance of the pure piezoceramic. The optimized piezocomposite designs presented below will be compared with the results presented in Table 2.

Table 1. Material properties of PZT5 and Spurr polymer

Mat.	c^E_{11}	c^E_{12}	c^E_{13}	c^E_{33}	c^E_{1212}	c^E_{1313}	e_{13}	e_{33}	e_{15}	$\varepsilon^S_{11}/\varepsilon_0$	$\varepsilon^S_{33}/\varepsilon_0$
				10^{10}	N/m^2			C/m^2			
PZT5	12.1	7.54	7.52	11.1	2.30	2.10	-	15.8	12.3	1650	1700
Spurr	0.53	0.31									4
Styc.	1.23	0.52									4

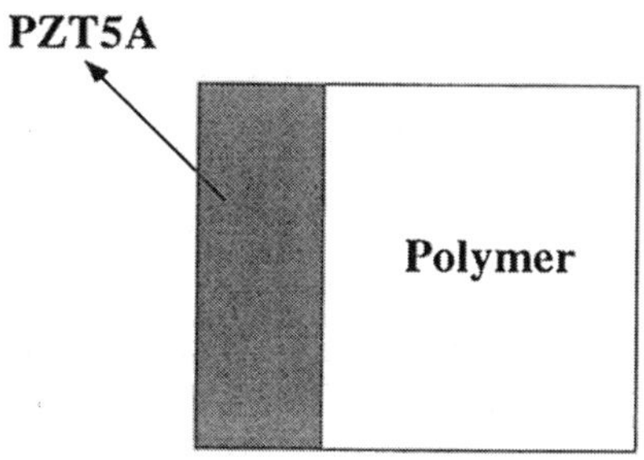

Figure 6. Model of a 2-2 piezocomposite unit cell. [36]

Table 2. Performance characteristics of the 2-2 piezocomposite unit cell considered

Unit cell	$\lvert d_h \rvert$ (*p*C/N)	$d_h g_h$ (*p*m²/N)	k_h	k_t
PZT5(full)	68.2	0.222	0.145	0.361
PZT5/Spurr	935	3.2	0.145	0.49

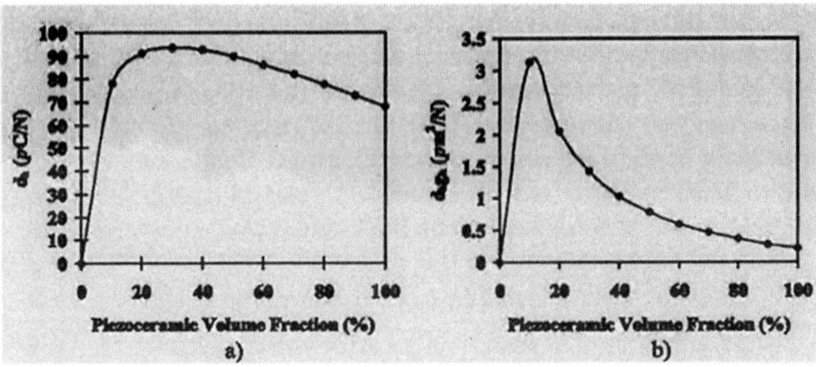

Figure 7. Performances d_h and d_hg_h as a function of the PZT5 volume fraction for the 2-2 piezocomposite unit cell shown in Fig. 6. [36]

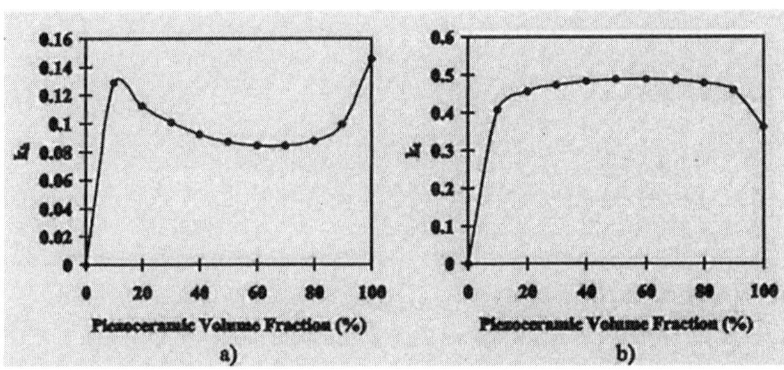

Figure 8. Performances k_h and k_t as a function of the PZT5 volume fraction for the 2-2 piezocomposite unit cell shown in Fig. 6. [36]

The second example considers three-dimensional models of typical designs of 1-3 piezocomposite unit cells as shown in Figs.9a-c. These unit cells consist of piezoceramic inclusions of different shapes (circular and square) immersed in a matrix polymer. By using homogenization method, a study of the performance characteristics (d_h, d_hg_h, and k_h) for these piezocomposites is presented [16].

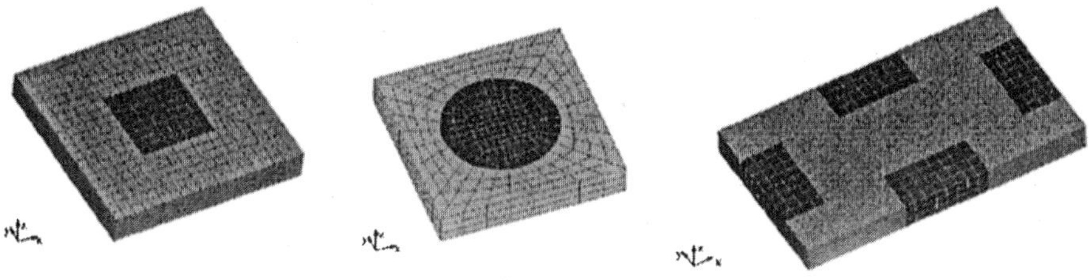

a) b) c)

Figure 9. Unit cells made of: a) square piezoceramic inclusion immersed in a polymer in a square array; b) circular piezoceramic inclusion immersed in a polymer in a square array; c) square piezoceramic inclusions immersed in a polymer in a hexagonal (staggered) array.

For the square shape, square and hexagonal (staggered) arrays are considered, as shown in Figs. 9a and 9c, respectively. For the circular shape, only the square array was considered. The performance characteristics are obtained as a function of the volume fraction, as shown in Fig. 10 for d_h. Similar curves are obtained for $d_h g_h$ and k_h and are compatible with results obtained in the literature by using other methods [7].

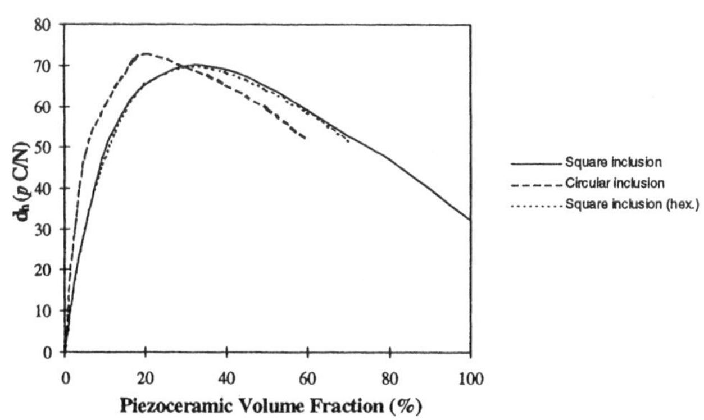

Figure 10. Comparison among the hydrostatic coupling coefficient (d_h) obtained by using homogenization for different types of PZT5A/Stycast unit cells, as a function of the piezoceramic volume fraction. [36]

Table 3 presents the maximum value of the performance characteristics for each type of unit cell, and for which volume fraction they occur. Noticed that the maximum values for d_h, $d_h g_h$, and k_h, have the same order for these unit cell types. From these values, we can have an idea of the maximum performance achievable when only the volume fraction, shape, and configuration are changed. The optimized 3D piezocomposite designs presented below will be compared with the results presented in Table 5.3.

Table 3. Performance characteristics of 1-3 piezocomposites shown in the figures

Unit Cell	Max. d_h (*p*C/N)	V.F.(%)	Max. $d_h g_h$ (*p*m^2/N)	V.F.(%)	Max. k_h	V.F.(%)
PZT5A(full)	32.	100.	0.068	100.	0.078	100.
Fig. 9a	70.	30.	2.16	10.	0.14	15.
Fig. 9b	72.9	20.	2.17	5.	0.14	10.
Fig. 9c	69.8	30.	2.19	10.	0.14	15.

3. PIEZOCOMPOSITE DESIGN PROBLEM

The main goal in the transducer design is to increase the response of the transducer that can be achieved, for example, by increasing the electromechanical energy conversion in the material. Among the most important factors on which the energy conversion depends are the properties of the piezoelectric material. In the optimization of a piezocomposite the objective is to obtain an improvement in its performance characteristics (see Section 1.1), usually by changing the volume fractions of constituent materials, its properties, shape of inclusions, and the mechanical properties of the polymer matrix in the composite unit cell.

In studies related to piezocomposite optimization, a model for calculating the effective properties of the piezocomposite (either analytically or numerically) is first developed. Using this model, curves are obtained to describe the performance characteristics as a function of the volume fraction of the ceramic inclusion in the polymer matrix. Based on these curves, the volume fraction that maximizes the performance characteristics can be determined. Usually, piezocomposites with simple geometry such as 0-3, 3-3, 2-2, and 1-3 piezocomposites are considered. Most part of works related to piezocomposite follow this procedure [17,18,19].

However, Smith's work [20] initiated our understanding of the mechanism behavior of the piezocomposite, demonstrating its high dependence on the Poisson's ratio of the polymer matrix. Essentially, what happens is that the PZT ceramic has a high d_{33} coefficient and a high but negative d_{13} coefficient. Therefore, the ceramic presents a high response when subjected to uniaxial stress (*3* direction), but a very low response when subjected to hydrostatic stress (lateral and vertical stresses), since in this case, the response will be equal to the d_{33} coefficient subtracted from the d_{13}. As a consequence, a transducer (e.g., a hydrophone) fully made of PZT will present a very low response in low-frequency applications when the transducer is subjected to hydrostatic stresses. In the case of the composite, by changing the Poisson's ratio of the polymer matrix, we can change the way the lateral stress is transmitted to the ceramic rods avoiding that effect, and thus improve the performance. A similar effect also occurs [6] in high-frequency applications (thickness-mode operation).

In a pioneering study, Smith [20] discussed the use of polymer materials with negative Poisson's ratio to increase the electromechanical energy conversion of the piezocomposite. In this case, the lateral stresses are transformed by the polymer matrix to help to compress the ceramic. Avellaneda and Swart [7], using a more sophisticated theoretical approach, discussed the limits of the optimal Poisson's ratio that increases the performance, confirming the improvement by using negative Poisson's ratio materials proposed by Smith [20]. They also studied the influence of porosity in the polymer matrix, which resulted in different effective properties and also improved performance. However, they limited the analysis to transversely isotropic materials.

Based on their study, Gibiansky and Torquato [8,21] applied an optimization approach to find the optimum properties of the polymer matrix material that maximizes the performance characteristics of the piezocomposite. No constraint in the characteristics of the material was considered and as a result, anisotropic materials were obtained. These anisotropic properties can be implemented by building composite materials with unit cells made of an optimum arrangement of pores. These polymer unit cells were designed later by Sigmund *et al.* [22], combining Gibiansky and Torquato's work with topology optimization and the homogenization approach. Stereolithography techniques have been used for manufacturing these unit cells.

In a parallel study, Hossack and Auld [23] tried to improve composite performance by suggesting some configurations of 1-3 piezocomposites with holes in the polymer matrix. Essentially, the holes provide some change in the Poisson's ratio behavior of the polymer matrix, which, as described in the above theoretical studies, improve performance. However, there were no criteria to define the position of the holes. Hossack and Auld [18] noticedd a much larger improvement for hydrostatic applications rather than high-frequency applications. Finally, Sanchez and Espinosa [19] using an analytical model, analyzed the performance improvement in thickness mode operation (high-frequency) in a 2-2 piezocomposite modified by grooves made in the polymer part. They also achieved a modest improvement.

In fact, these previous studies suggested that, to increase the performance of these materials, we must seek a new set of properties in the space of achievable properties of piezoelectric materials. This can be done by designing the unit cell of the piezocomposite. This is called *material design*. The objective of material design is to generate composite materials with prescribed or improved properties not found in common materials. This can be achieved by modifying the microstructure of the composite materials (see

Chapter 7 for additional examples). Currently, in composites such as fiber- or sphere-reinforced and laminated materials, the change in the properties is obtained by modifying the location, orientation, material constituents, or volume fraction of the fiber, sphere, or laminar inclusion. This allows some control of the composite properties. This approach can also be observed in the development of the piezocomposite materials [24,25]. However, heuristic methods have been used to define these modifications. By using topology optimization, a general approach for designing these materials can be provided.

In the process of designing materials with prescribed and improved properties, a natural question arises: What are the achievable properties in a material design? To answer this question for elastic materials, Milton and Cherkaev [26], based on the fact that the constitutive elasticity matrix must be positive definite, proved the existence of materials for all thermodynamically admissible set by layering and combining an infinitely rigid material with voids (infinitely compliant). However, the extreme of this admissible set (for example, isotropic material with Poisson's ratio equal to -1) cannot be reached in practice since an infinitely rigid material does not exist. Considering materials with finite properties, other bounds were derived in the past. We can cite for example, the work of Hashin and Shtrikman [27] which obtained the bounds for an isotropic mixture of isotropic materials using energy analysis methods, and the work of Lipton [28] (among others) which defined the bounds for orthotropic mixtures of isotropic materials.

In the case of piezoelectric materials, let's consider Eq. (1) written in the form:

$$\begin{Bmatrix} \langle \mathbf{S} \rangle \\ \langle \mathbf{D} \rangle \end{Bmatrix} = \begin{bmatrix} \mathbf{s}^{E}{}_{H} & \mathbf{d}_{H} \\ \mathbf{d}^{t}{}_{H} & \varepsilon_{H} \end{bmatrix} \begin{Bmatrix} \langle \mathbf{T} \rangle \\ \langle \mathbf{E} \rangle \end{Bmatrix} \tag{19}$$

In this case, the matrix can be interpreted as a supertensor that contains elastic, piezoelectric, and dielectric properties. By changing the topology of the unit cell, we can obtain new effective properties in this space [16]. Smith [29] discusses attainable properties, considering the positive definiteness of this tensor. No bounds were obtained for piezoelectric materials; however, bounds for elastic and conductivity properties of mixtures of two materials (not necessarily isotropic) were developed by Gibiansky and Torquato [8]. These bounds could be expanded for piezoelectric materials.

Topology optimization can be combined with the homogenization method, allowing us to design unit cells of complex shape. Therefore, a richer class of material properties can be achieved, and new kinds of composite materials can be generated rather than the usual fiber-reinforced or laminate types. These new materials may have unusual properties such as negative Poisson's behavior [30,31,32] or zero thermal expansion [33].

In next section, a method is described for the design of piezocomposite microstructures with high performance characteristics using the homogenization method and topology optimization techniques [15,16]. The topology of the microstructure is obtained by searching for the optimal distribution of two phases (material and void phases) in the composite unit cell (fixed domain). The homogenization method for piezoelectricity described in Section 2 is applied, using the finite element method, to calculate the effective properties of the piezocomposite.

3.1 Formulation of Optimization Problem

The optimization problem consists of finding a distribution of material that individually maximizes a performance characteristic such as the hydrostatic coupling coefficient (d_h), the hydrostatic figure of merit ($d_h g_h$), and the electromechanical coupling factor (k_h) defined in Section 1.1 in terms of the homogenized properties of the piezocomposite.

The material model adopted for topology optimization formulation is the density method approach which assumes that the properties in a given element are simply some fraction x_n times the material properties of the basic material. This is an "artificial" material model for intermediate densities, but since we obtain an entirely solid material or void in each element, this is a valid approach. However, during the optimization process intermediate densities are allowed and consist of artificial (non-existing) materials. Even though, the design procedure is greatly simplified using this model, the optimization problem may not be sufficiently relaxed [32].

The basic material has stiffness tensor c^0_{ijkl}, piezoelectric tensor e^0_{ijk}, and dielectric tensor ε^0_{ij}. If the basic material considered in the analysis is the polymer (isotropic), the piezoelectric coefficients are zero. Therefore, the local tensor properties in each element n can be expressed in terms of one design variable x_n times the basic material property:

$$c^n_{ijkl} = x_n c^0_{ijkl}; \qquad e^n_{ijk} = x_n e^0_{ijk}; \qquad \varepsilon^n_{ij} = x_n \varepsilon^0_{ij} \tag{20}$$

where x_n represents the amount of basic material in that element (local density) that ranges from x_{low} to 1. For $x_n = x_{low}$ the element is a "void" and for $x_n = 1$ the element assumes the properties of the solid material.

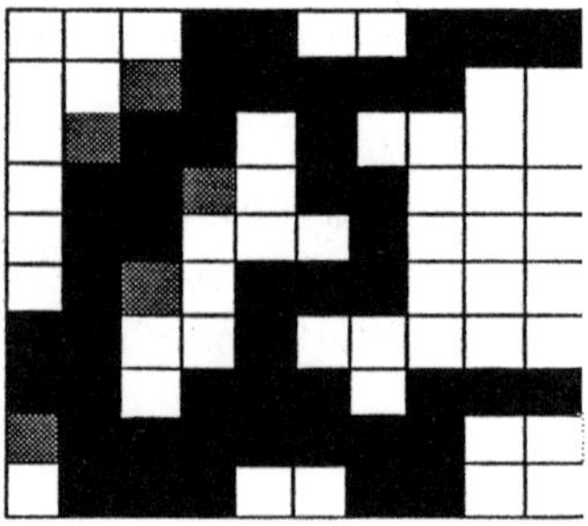

Figure 11. Material distribution in the unit cell domain

Considering that the design domain was discretized in N finite elements, the material type changes from element to element during the optimization. The design problem consists of finding the fraction x_n amount of material in each element as described in Fig. 11 so that the objective function is maximized or minimized. Therefore, the optimization problem can be stated as:

$$\underset{\mathbf{x}_{dv}}{\textit{Maximize}}: F(\mathbf{x}_{dv}), \text{ where } \mathbf{x}_{dv}=[x_1, x_2,\ldots,x_n,\ldots,x_{NDV}]$$

$$\begin{aligned} \textit{Subject to}: \; & c^E_{ijkl} \geq c_{low}, \text{ i,j,k,l are specified values} \\ & W > W_{low} \\ & 0 < x_{low} \leq x_n \leq 1 \\ & \text{symmetry conditions} \end{aligned}$$

where x_n is the design variable that represents the fraction (or "density") of material in each finite element of the design domain, as explained above. There are NDV design variables, NDV being equal to the total number of finite elements N less the number of elements that were explicitly excluded from the optimization. $F(\mathbf{x}_{dv})$ is the performance characteristic to be maximized. The maximization of these performance quantities are achieved by reducing the stiffness of the composite material in some direction. Therefore, to avoid a low stiffness in the final design, a lower-bound constraint on the effective stiffness is specified. Since the microstructure is orthotropic, we must choose a stiffness coefficient for the constraint (c_{low}). A "cost function" constraint [16] was also added to control the amount of gray scale in the final design. A lower-bound x_{low} is also specified for design variables x_n to avoid numerical problems (singularity of the stiffness matrix in the finite element formulation). Therefore, the bounds for the design variable are $0<x_{low} \leq x_n \leq 1$. Finally, we address the constraint related to the symmetry conditions. Considering the design domain a rectangular (2D problem) or a cubic (3D problem) unit cell, we can define many kinds of

symmetries relative to the symmetry axes of the rectangular domain (2D problem) or the three middle section planes of the cubic domain (3D problem), respectively. Symmetry reduces the computational cost, which is very high in 3D problems.

A flow chart of the optimization algorithm describing the steps involved is shown in Fig. 12.

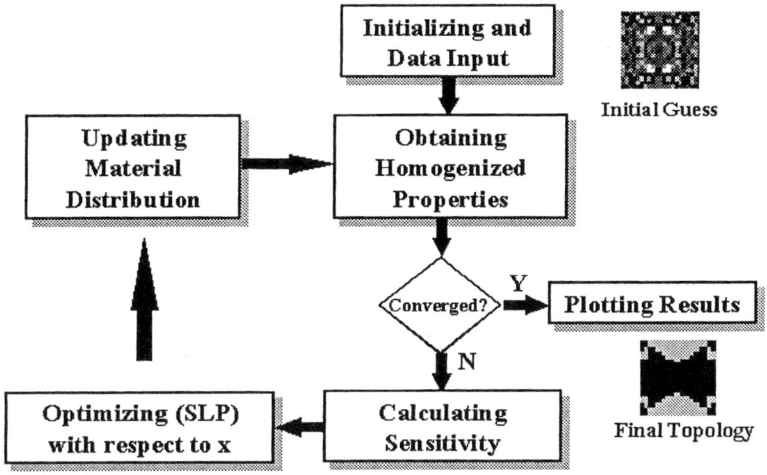

Figure 12. Flow chart of optimization procedure.

The initial domain is discretized by finite elements and the design variables are the values of x_n in each finite element. The iteration must start with an initial guess of the distribution of x_n. This initial distribution cannot be uniform, otherwise equal values of gradients are obtained in relation to all design variables, and therefore the optimization method does not have a preferential direction for starting the search.

The effective properties of the composite are obtained using the homogenization method, allowing calculation of the performance characteristic that is being optimized and the constraints. The next step is the optimization.

There are plenty of optimization algorithms available for solving nonlinear optimization problems with constraints. Most part of these algorithms are based on the so called mathematical programming techniques and sequential approximation methods. For a review and description of these algorithms the reader should refer to [34].

Among algorithms based on sequential approximation methods, sequential linear programming (SLP) algorithm has been successfully applied to solve topology optimization problems, in particular piezocomposite and piezoeletric actuators. In this algorithm, the nonlinear optimization problem is approximated by a sequence of linear optimization subproblems obtained by locally approximating objective function and constraints at the current design point by using first order terms of a Taylor's series expansion. At every iteration, the linear subproblem is solved by using linear programming method to find the optimum for that approximation. The result from the last approximation subproblem is updated as a current design point for the next approximation subproblem and this procedure is repeated along iterations. In each iteration, moving limits are defined for the design variables. Typically, during one iteration, the design variables will be allowed to change by 5-15% of their original values. The method continues until the final optimum for the nonlinear problem is reached. Therefore, SLP consists of the sequential solution of approximate linear subproblems. Figure 13 describes this process. SLP is recommended when several different objective functions and constraints are considered, and there are a large number of design variables [34,35].

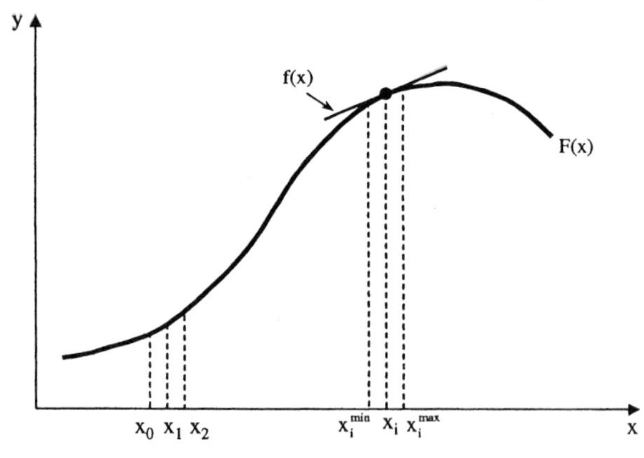

Figure 13. Steps of SLP algorithm.

All methods based on mathematical programming techniques can obtain only a local optimum, that is, they cannot guarantee that the global optimum will be found.

The linearization of the problem (Taylor series) in each iteration requires the sensitivities (gradients) of the objective function and constraints relative to x_n. Sensitivities for the piezocomposite design problem were derived in detail in [16] and essentially they measure how much the values of effective properties change when the amount of material is changed in each finite element of unit cell domain. After optimization, a new set of design variables x_n is obtained and updated in the unit cell. A new homogenization is done and the convergence is verified by comparing the value of the objective function with the value in the previous iterations. The iteration proceeds until the convergence of objective function is achieved.

The problem is inherently non-convex. For each starting point we may get different unit cells, or at least the many definitions of periodicity for a same unit cell. Different mesh sizes may imply different resulting topologies.

4. EXAMPLES

Two kinds of initial domains are considered [15]. In the first model, there are two materials in the unit cell: polymer and piezoelectric materials. The polymer is the design domain and the piezoelectric domain remains unchanged during this analysis. This configuration avoids the difficulties in manufacturing piezoceramics with complex topologies. The goal is to obtain a unit cell that presents complex topology only in the polymer domain. In the second model the entire domain consists of piezoceramic (no polymer at all).

The initial configuration considered for the unit cell in this first model is shown in Fig. 6. The initial "volume fraction" (vol %) of ceramic was set to 20%. This volume fraction was chosen concerning the robustness of the piezocomposite prototype that was manufactured. A second optimization loop can be considered to obtain the optimum volume fraction of piezoceramic for each polymer topology obtained. However, a constraint of 20% in the volume fraction must be imposed to guarantee the robustness of the final design. The properties used are presented in Table 1.

Figures 14 through 17 show optimized microstructure topologies obtained. The performance characteristics for these topologies are listed in Table 4. Symmetry conditions in *1* (or *x*) and *3* (or *z*) axes were considered.

Table 4. Performance characteristics of the optimized unit cells

Figure	Design	c^E_{11} (10^9N/m^2)	$\lvert d_h\rvert$ (*p*C/N)	$d_h g_h$ (*p*m^2/N)	k_h	k_t
14	Max $\lvert d_h\rvert$	0.8	253.	15.3	0.11	0.52
15	Max $d_h g_h$	0.8	249.	15.0	0.11	0.52
16	Max. k_h	3.0	217.	11.3	0.18	0.51
17	Max. k_t	0.8	229.	12.6	0.10	0.52
18	Max. $\lvert d_h\rvert$	$c^E_{33}>10^{10}$	259.4	19.	0.4	0.52

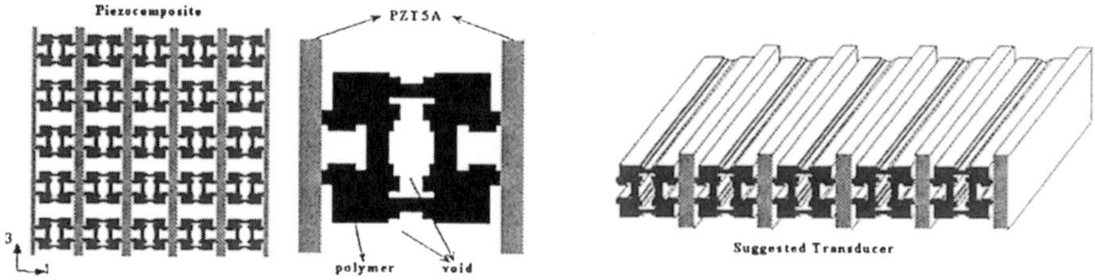

Figure 14. Microstructure that maximizes $|d_h|$. Corresponding suggested transducer design

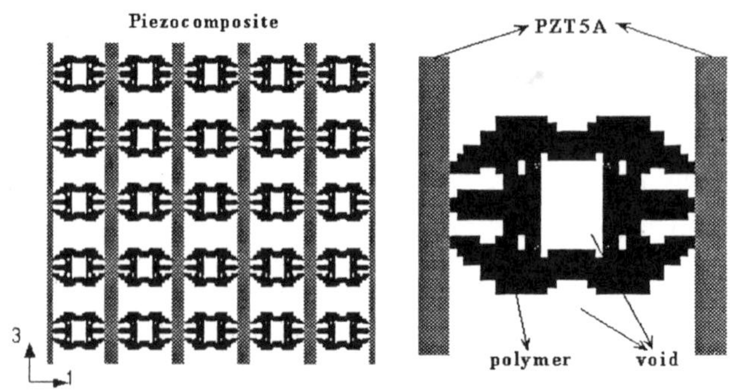

Figure 15. Microstructure that maximizes $d_h g_h$ [36]

Figure 14 shows a topology that maximizes $|d_h|$. Figure 15 shows a microstructure that has an optimized value for $d_h g_h$. Figure 16 shows a unit cell that maximizes k_h, and Fig. 17 shows one that maximizes k_t. The stiffness constraint in each case is presented in Table 4.

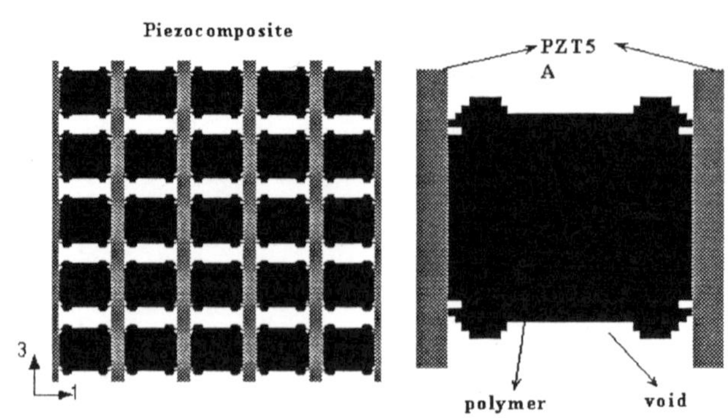

Figure 16. Microstructure that maximizes k_h [36]

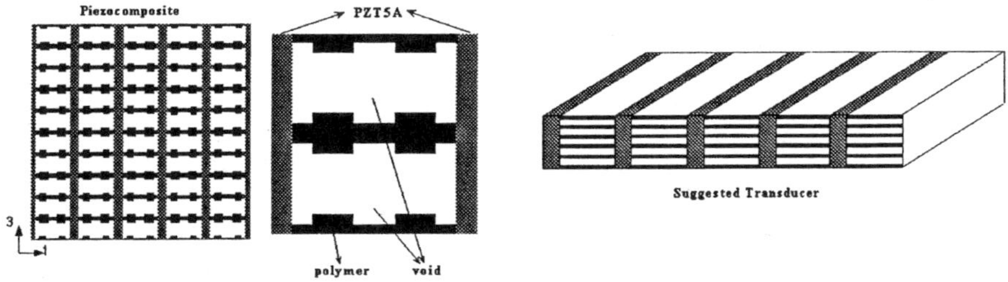

Figure 17. Microstructure that maximizes k_t. Corresponding suggested transducer design

The result shown in Fig. 16 is a layered structure between the piezoelectric rods. The explanation for this is that, to maximize k_h, the optimization procedure tries to maximize d_h, and minimize the dilatational compliance and the dielectric constant ε^S_{33} (see equation 6). This is well achieved with the layered structure, since it decouples the piezoelectric effect due to the horizontal forces (coefficient d_{13}) in relation to the piezoelectric effect due to the vertical forces (coefficient d_{33}), increasing the hydrostatic coupling coefficient d_h. In the usual piezoceramic material the sign of d_{13} coefficient is opposite relative to the sign of d_{33}, reducing the value of d_h. Besides, the effective dielectric constant is also minimized due to the increase of amount of void in the unit cell. However, there is a trade-off between the minimization of the effective dielectric property and the minimization of the dilatational compliance (or maximization of dilatational stiffness) limiting the maximization of k_h to a certain level. The maximization is also limited by the stiffness constraint imposed in the *1* direction (c^E_{11}). The result shown in Fig. 17 that maximizes k_t (thickness mode applications) is also a layered structure. The explanation is similar: the optimization procedure tries to minimize the dielectric coefficient ε^S_{33} and the stiffness coefficient c^D_{33} (see equation 10). The first one is minimized by increasing the amount of void in the structure, and the created layered structure disconnects the polymer in the *3* direction, reducing the stiffness coefficient in this direction. In the result from Fig. 14, in addition to the layered structure, the polymer microstructure design presents a negative poisson's ratio behavior.

Comparing the values in Table 4 with the values in Table 2, the following improvements can be verified in relation to the maximum values achieved by changing only the volume fraction of the 2-2 piezocomposite unit cell in Fig. 6: 2.7 times in $|d_h|$, 4.8 times in $d_h g_h$, 1.2 times in k_h, and 1.1 times in k_t. In relation to the pure piezoelectric material, there was an improvement of 3.7 times in $|d_h|$, 69 times in $d_h g_h$, 1.2 times in k_h, and 1.4 times in k_t. These are very good improvements in relation to the usual design of the 2-2 piezocomposite. In addition, the impedance (*Z*) is lower because density is lower and wave propagation velocity (v_t) in the *3* direction remains almost the same since the stiffness in this direction is not considerably changed.

In a second model the entire design domain is made of piezoelectric material. An example of optimized unit cell topology that maximizes $|d_h|$ is shown in Fig. 18 and its performance characteristics are presented in Table 4. Symmetry conditions in *1* (or *x*) and *3* (or *z*) axes were considered.

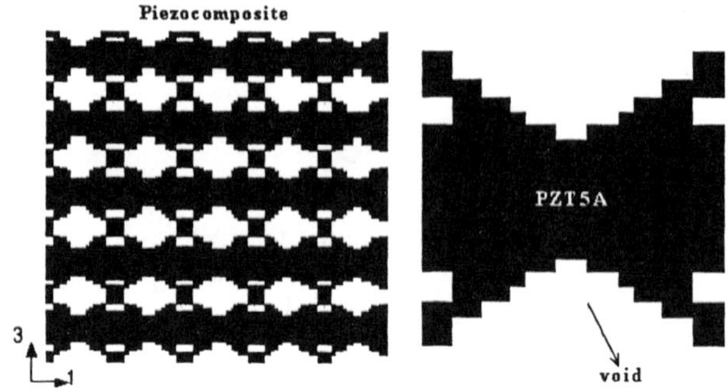

Fig. 18. Microstructure that maximizes $|d_h|$. The entire initial domain is made of piezoelectric material [36]

Comparing its performance values with values in Table 2, the following improvement can be verified in relation to the maximum values achieved by changing only the volume fraction of the 2-2 piezocomposite unit cell in Fig. 6: 2.8 times in $|d_h|$, 5.9 times in $d_h g_h$, 2.8 times in k_h, and 1.1 times in k_t. In relation to the pure piezoelectric material, there was an improvement by a factor of 3.8 in $|d_h|$, 85.5 in $d_h g_h$, 2.8 in k_h, and 1.44 in k_t. By specifying lower stiffness constraints, further improvement can be achieved. It is noticed that in general the design with only piezoceramic provides larger improvement than the design with piezoceramic and polymer. Therefore, these results show that the performance of piezoelectric materials can be improved by distributing voids in an optimized way.

Considering 3D microstructures the optimized results obtained are shown in Figs. 19-21. The initial domain consists of a cubic block made of piezoceramic. In all the examples, symmetry in relation to the three mediator planes was considered.

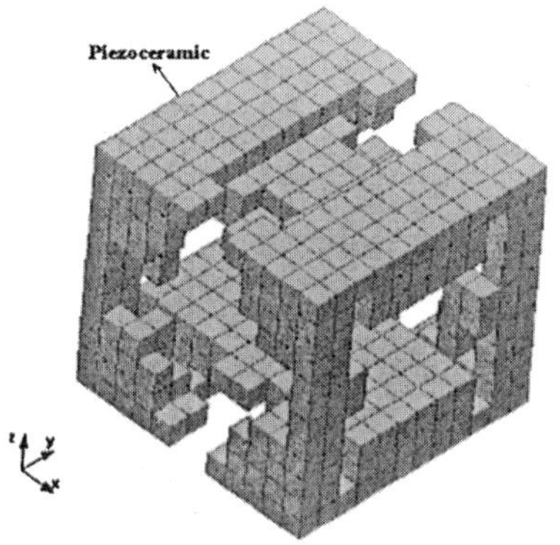

$$c_H = \begin{bmatrix} 1.323 & 0.223 & 0.016 & 0. & 0. & 0. \\ 0.223 & 1.254 & 0.047 & 0. & 0. & 0. \\ 0.016 & 0.047 & 0.420 & 0. & 0. & 0. \\ 0. & 0. & 0. & 0.223 & 0. & 0. \\ 0. & 0. & 0. & 0. & 0.064 & 0. \\ 0. & 0. & 0. & 0. & 0. & 0.031 \end{bmatrix} \times 10^{10}\ \mathrm{N/m^2}$$

$$e_H = \begin{bmatrix} 0. & 0. & -0.02 \\ 0. & 0. & -0.1 \\ 0. & 0. & 1.57 \\ 0.04 & 0. & 0. \\ 0. & 0.14 & 0. \end{bmatrix} \mathrm{C/m^2}$$

$$\varepsilon_H = \begin{bmatrix} 2.67 & 0. & 0. \\ 0. & 2.33 & 0. \\ 0. & 0. & 0.68 \end{bmatrix} \times 10^{-9}\ \mathrm{F/m}$$

Figure 19. Microstructure that maximizes $|d_h|$. The effective elastic (c^H), piezoelectric (e^H), and dielectric ($\mathcal{E}^H$) properties of the unit cell are presented [36]

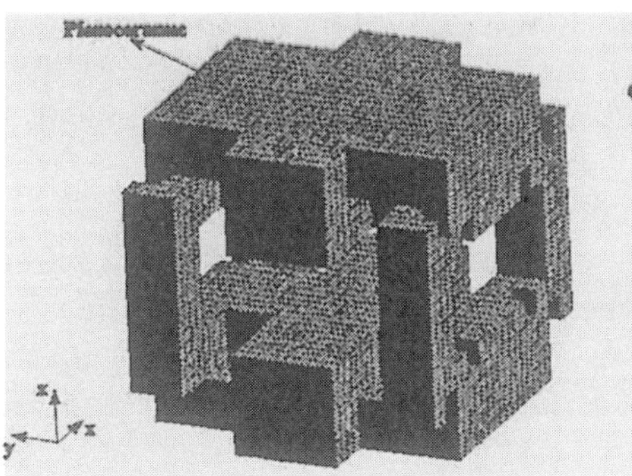

$$c_H = \begin{bmatrix} 2.963 & 0.963 & 0.318 & 0. & 0. & 0. \\ 0.963 & 2.828 & 0.2 & 0. & 0. & 0. \\ 0.316 & 0.2 & 0.5 & 0. & 0. & 0. \\ 0. & 0. & 0. & 0.484 & 0. & 0. \\ 0. & 0. & 0. & 0. & 0.229 & 0. \\ 0. & 0. & 0. & 0. & 0. & 0.071 \end{bmatrix} \times 10^{10}\ \mathrm{N/m^2}$$

$$e_H = \begin{bmatrix} 0. & 0. & 0.47 \\ 0. & 0. & -0.06 \\ 0. & 0. & 1.6 \\ 1.1 & 0. & 0. \\ 0. & 0.37 & 0. \end{bmatrix} \mathrm{C/m^2}$$

$$\varepsilon_H = \begin{bmatrix} 4.41 & 0. & 0. \\ 0. & 4.43 & 0. \\ 0. & 0. & 1.03 \end{bmatrix} \times 10^{-9}\ \mathrm{F/m}$$

Figure 20. Microstructure that maximizes $|d_h|$. The effective elastic (c^H), piezoelectric (e^H), and dielectric ($\mathcal{E}^H$) properties of the unit cell are presented [36]

Figures 19 and 20 show two different unit cells that maximize d_h, obtained by specifying different values of stiffness constraint in the z direction.

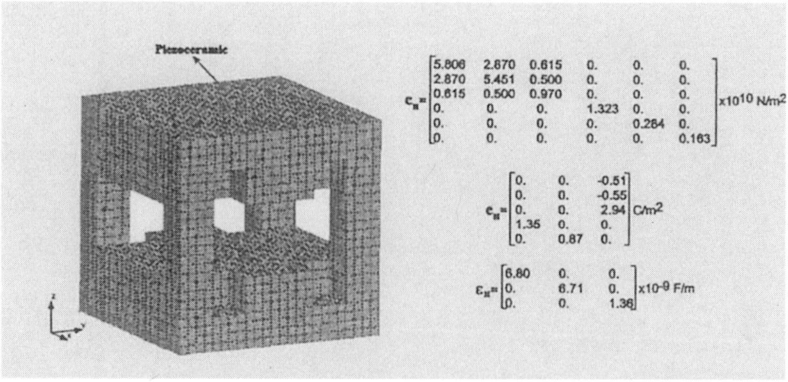

Figure 21. Microstructure that maximizes $d_h g_h$. The effective elastic (c^H), piezoelectric (e^H), and dielectric (ε^H) properties of the unit cell are presented [36]

Figure 21 shows unit cell topology that maximizes $d_h g_h$. Table 5 presents performance characteristics for the unit cells obtained. The improvement is confirmed by comparing these results with the values in Table 3. A stiffness constraint in the coefficient c^E_{3333} was specified. The improvement of $d_h g_h$ in Fig. 21 is lower due to the large stiffness constraint specified. Improvement in d_h also improves $d_h g_h$, which suggests that a multiobjective function including these two performances can be optimized. The homogenized elastic, piezoelectric, and dielectric properties of each unit cell are presented next to the figures.

Table 5. Performance characteristics of the 3D optimized unit cells

Figure	Design	c^E_{33} (10^9N/m^2)	$\lvert d_h \rvert$ (*p*C/N)	$d_h g_h$ (*p*m^2/N)	k_h
19	Max. $\lvert d_h \rvert$	4.2	350.0	97.0	0.52
20	Max. $\lvert d_h \rvert$	5.0	307.0	60.0	0.52
21	Max. $d_h g_h$	9.7	281.0	33.0	0.55

4.1 Piezocomposite Manufacturing

In this section, some manufacturing techniques that can be used for building the optimized composites are described. These techniques have been studied in the Department of Materials Science and Engineering of The University of Michigan - Ann Arbor.

4.1.1 Microfabrication by Coextrusion Technique

This technique allows us to manufacture, at the microscale, a two-dimensional array of unit cells with any complex topology made of ceramic, through a process based on successive thermo-plastic coextrusions [36]. The array of unit cells is actually the composite material. Essentially, the method can be described in the following way, as shown in Fig. 22:

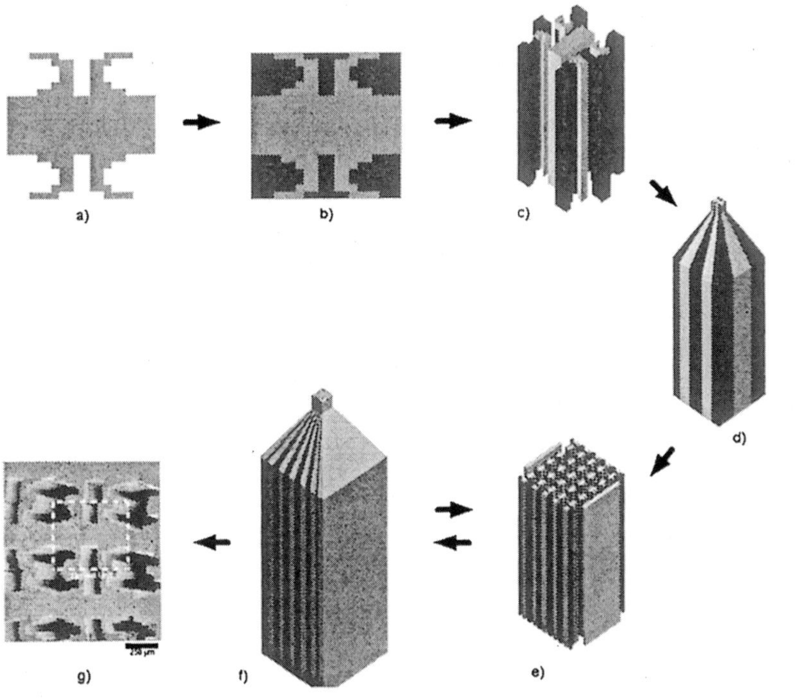

Figure 22. Piezocomposite material manufacturing sequence: a) theoretical unit cell image; b) unit cell image considering the fugitive material filling the holes; c) unit cell large-scale model to be extruded; d) first thermo-plastic extrusion; e) assemblage of extruded unit cell models; f) second extrusion; g) final SEM image of the prototype built. [36]

- First, a large model of the unit cell (dimensions approximately 25mm x 25mm) is built considering the details of the complex topology, and the topology's holes are filled with the "fugitive" material. This fugitive material will be eliminated during the pyrolosis of the polymer, generating the holes;

- This unit cell model is submitted to a thermo-plastic extrusion that reduces its size, usually by a factor of 6;

- These extruded unit cells are put together, and the entire set is extruded again, reducing its size by a factor of 6 (total reduction equal to 17).
- This extruded array of unit cells can be put together with other extruded arrays, and the entire set can be extruded again. This process is repeated until the desired microscale dimension for each unit cell of the composite is reached.

The final scale that can be achieved will depend on how complex is the topology of the unit cell. Usually, a square unit cell with dimensions equal to 80 μm x 80 μm with 5 μm features can be easily achieved.

At the end, a piece of piezocomposite (array of unit cells) with dimensions approximately 4mm x 4mm with a large height can be obtained; however, parts larger than this can be made. For more details refer to Crumm and Halloran [37]. The fact that the final unit cell presents a large aspect ratio is an advantage in relation to other ceramic microfabrication techniques, and also, to lithography methods.

Using this technique, a prototype based on the piezocomposite shown in Fig. 18 was manufactured. The SEM image of the piezocomposite is shown in Fig. 23.

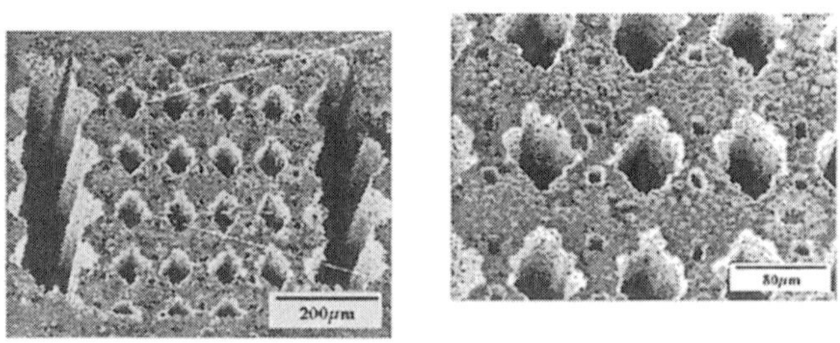

a)

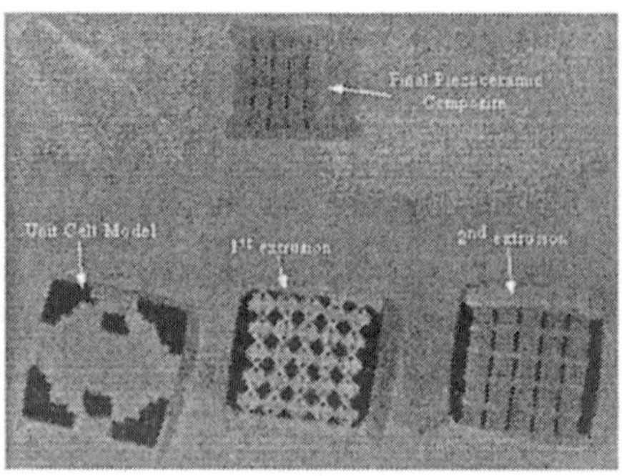

b)

Figure 23. a) SEM image of prototype built based on piezocomposite shown in Fig. 18; b) Extrusion sequence and final piezocomposite part. [36]

4.1.2 Stereolithography Technique

Stereolithography consists of a laser beam controlled by a computer that hardens a liquid polymer in a tank layer by layer building the complex part (layered manufacturing technique). Essentially, once we have the solid image of the part in the computer (obtained using a CAD for example), we can build it [39]. It can also build ceramic parts.

The first prototype presented is described in Fig. 24b. It consists of one layer of the piezocomposite shown in Fig. 14. The transducer is made of commercial piezoceramic bars together with a polymer part with complex shape. The complex polymer part is built using stereolithography. The polymer part is glued to the piezoceramic.

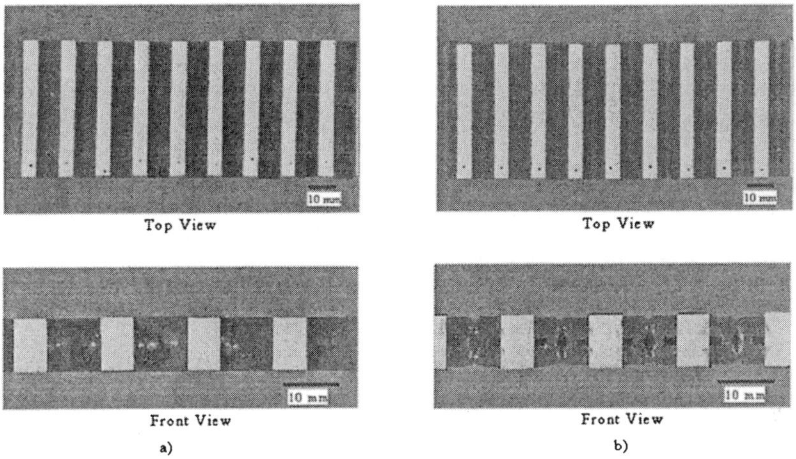

Figure 24. a) Photo of the reference transducer; b) photo of the optimized transducer. [36]

To compare the improvement, a reference transducer which consists of a common 2-2 piezocomposite was also built. Figure 24 shows the optimized and reference transducer.

The manufacturing of prototypes of 3D piezocomposite unit cells described in Figs. 19-21 is still a challenge. However, stereolithography methods seem to be the way. Figure 25 shows some polymer prototypes of these 3D piezocomposites designs (Figs. 19-21) just for display purposes. Each one consists of a 5x5x5 array of unit cells with dimensions 8mm x 8mm x 8mm, 6mm x 6mm x 6mm, and 4mm x 4mm x 4mm, respectively [38].

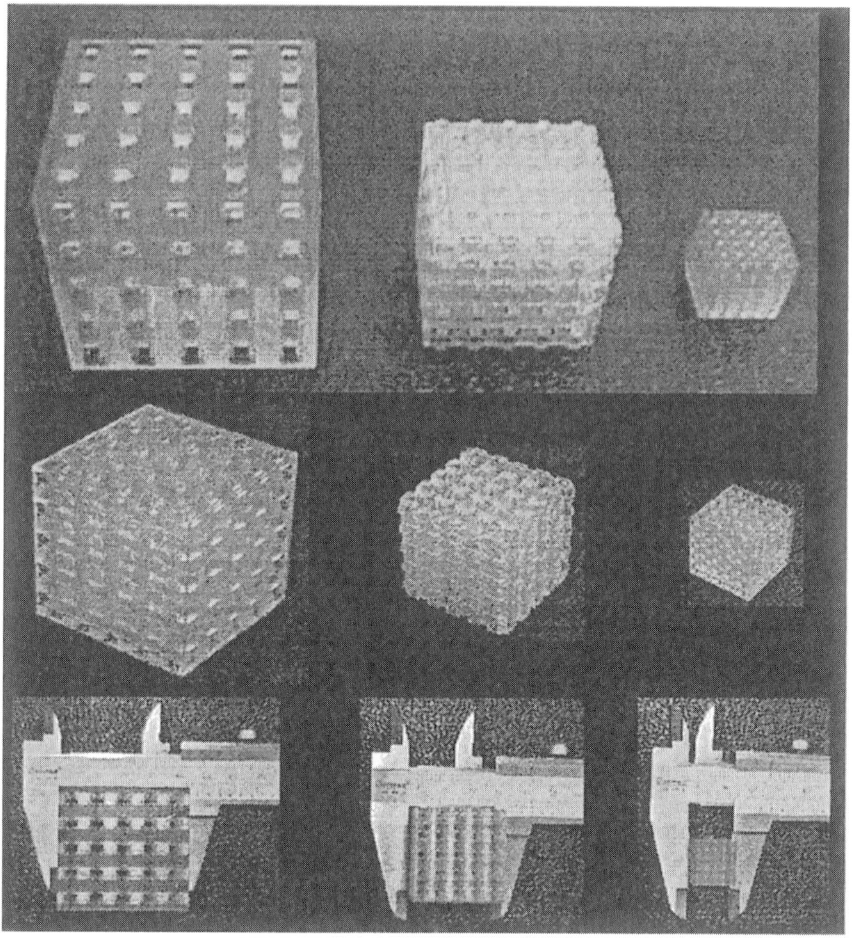

Figure 25. Polymer prototypes built based on piezocomposite designs shown in Figs. 19-21 [38]

4.2 Experimental Results

Considering initially the prototype shown in Fig. 24b, the optimized transducer was tested for ultrasonic applications (high frequency) and hydrostatic applications (low frequency) and the performance values were compared with the performances measured for a common 2-2 piezocomposite (reference) built with the same materials (see Fig. 24a). These experiments were conducted in the "Instituto de Acústica" in Madrid, Spain. Results are presented in Table 6 for the optimized and reference transducer.

From Table 6 we can see that, as expected in the simulations, the experiments confirm a high improvement of the performances for hydrostatic applications and a modest improvement for high-frequency applications.

Table 7 presents the equivalent homogenization results obtained by simulation using PZT5A/Stycast (see Table 5.1 for properties). Comparing the results from Table 7 with the results from Table 6, we can see that the simulation and experimental results agree very well for the optimized transducer.

Now the piezocomposite prototype shown in Fig. 23 built using the microfabrication technique will be considered. This piezocomposite is made only of piezoceramic with microholes distributed in an optimized way. Table 7 shows the performances measured. Only hydrostatic measurements were performed.

Table 6. Measured and simulated performance characteristics from prototypes made of polymer and piezoceramic

Transducer	$\|d_h\|$ (pC/N)	$d_h g_h$ (pm^2/N)	k_t
Reference (simulation)	102.1	2.29	0.58
Optimized (simulation)	229.1	10.56	0.655
Optimized (measured)	246.	10.4	0.70

Table 7. Measured performance characteristics from prototype built using the microfabrication technique (only piezoceramic).

Prototype	$\|d_h\|$ (pC/N)	$d_h g_h$.(pm^2/N)
Solid PZT	185.	0.967
Optimized	315.	20.4

The measured values for the optimized piezocomposite block were compared with the measured values of a solid piezoceramic block made with the same piezoceramic material used to build the optimized piezocomposite block. The values presented are averaged values obtained from a small number of samples.

The experimental results presented on Table 7 confirm the large performance improvement predicted. Simulation results are not presented since the properties of the PZT material used are unknown and they must be measured.

References [3,40] present the improvement of the performances d_h and $d_h g_h$ obtained for different piezocomposite designs developed in the literature. The maximum value of d_h presented is equal to 265 pC/N. It was achieved by using a design variation of the 1-3 piezocomposite called 1-3 diced copper encapsulated composite. The values obtained with the two-dimensional design using topology optimization (the manufactured prototypes, for example) (see Table 4) are of the same order, or higher than the above value. In addition, the three-dimensional designs presented can reach even higher d_h values (see Table 5). Regarding the quantity $d_h g_h$, the maximum value reported is equal to 30,000 fPa^{-1}. It was achieved by using another design variation of the 1-3 piezocomposite called 1-3-1 piezocomposite. Some values obtained with the two-dimensional design (see Table 4) are of the same order. However, the three-dimensional designs presented can reach higher $d_h g_h$ values (see Table 5).

5. CONCLUSIONS

The results presented give us an idea about the potential of applying optimization methods to design smart materials. However, synthesis methods for designing these materials are still at the beginning stage. Applications of piezocomposite material for SAW devices can be considered. In a recent work, topology optimization method was applied to design band-gap materials [41]. In addition, an emerging field that can take advantage of applying synthesis methods is the FGM (Functionally Graded Materials) smart materials [42,43]. The techniques applied to manufacture these materials allow us to obtain a material with graded properties along its domain. This idea is close related to the principle used in the topology optimization formulation where the optimum solution actually consists of graded properties ("gray scale"). Nowadays, scientists spend time developing techniques, such as filters, to come up with 0-1 results in topology optimizations due to the limitations of manufacturing techniques, however with FGM technology this will be no longer necessary in a near future.

A common question that arises regarding the topology optimization results is "Are they really better than current designs?" That is a tricky question to answer. The problem is that most of the current designs were obtained by empirical methods and not by systematic mathematical methods. Therefore, it is very difficult to make a fair comparison since it may happen that the current design (empirical) corresponds to a point in the design space that was not reached by the optimization method (which usually can only find local optima). As a consequence, the current design may have, in some cases, a slightly better performance than the optimization solution. However, this does not mean that the optimization design method is not useful. The main advantage of applying optimization methods is to systematize the design procedure for these transducers, that is, by changing the initial design specifications even novice designers can obtain a new design with a good performance in a less time without the need for previous experience with the design problem. The main problem of empirical solutions is that it may take years to develop them and usually only few people with large experience in the field are involved in this process. By developing systematic design methods, it is easier to make transducer technology more accessible and to spread it widely in the engineering community.

REFERENCES

1. Smith, W. A. "The Role of Piezocomposites in Ultrasonic Transducers". *Proceedings of IEEE Ultrasonic Symposium.* (1989):755-765.
2. Smith, W. A. "The Application of 1-3 Piezocomposites in Acoustic Transducers". *Proceedings of 7th International Symposium on Applications of Ferroelectrics.* (1990): 145-152.
3. Safari, A. "Development of Piezoelectric Composites for Transducers". *Journal de Physique III.* **4** No.7 (1994): 1129-1149.
4. Newnham, R. E., D. P. Skinner, and L. E. Cross. "Connectivity and Piezoelectric--Pyroelectric Composites". *Materials Research Bulletin.* **13** (1978): 525-536.
5. Smith, W. A. "Modeling 1-3 Composite Piezoelectrics: Hydrostatic Response". *IEEE Transactions on Ultrasonics, Ferroelectrics, and Frequency Control.* **40** No.1 (1993): 41-49.
6. Smith, W. A. and B. A. Auld. "Modeling 1-3 Composite Piezoelectrics: Thickness-Mode Oscillations". *IEEE Transactions on Ultrasonics, Ferroelectrics, and Frequency Control.* **38** No.1 (1991): 40-47.
7. Avellaneda, M. and P. J. Swart. "The Role of Matrix Porosity and Poisson's Ratio in the Design of High-Sensitivity Piezocomposite Transducers". *Adaptive Structures and Composite Materials: Analysis and Application - ASME Aerospace Division* AD **45** (1994): 59-66.
8. Gibiansky, L. V. and S. Torquato. "On the Use of Homogenization Theory to Optimally Design Piezocomposites for Hydrophone Applications". *Journal of the Mechanics and Physics of Solids.* **45** No.5 (1997): 689-708.
9. Sanchez-Palencia, E. "*Non-Homogeneous Media and Vibration Theory*". Lectures Notes in Physics 127, Springer, Berlin, (1980).
10. Guedes, J. M. and N. Kikuchi. "Preprocessing and Postprocessing for Materials Based on the Homogenization Method with Adaptive Finite Element Methods". *Computer Methods in Applied Mechanics and Engineering.* **83** (1990): 143-198.
11. Telega, J.J. "Piezoelectricity and Homogenization: Application to Biomechanics". *Continuum Models and Discrete Systems.* **2** in: G.A. Maugin, ed., Longman, London (1990): 220-230.
12. Galka, A., J.J. Telega, and R. Wojnar. "Homogenization and Thermopiezoelectricity". *Mechanics Research Communications.* **19** No.4 (1992): 315-324.
13. Turbé, N. and G. A. Maugin. "On the Linear Piezoelectricity of Composite Materials". *Mathematical Methods in the Applied Sciences* **14** No.6 (1991): 403-412.

14. Otero, J.A., J.B. Castillero, and R.R. Ramos. "Homogenization of Heterogeneous Piezoelectric Medium". *Mechanics Research Communications.* **24** No.1 (1997): 75-84.
15. Silva, E.C.N., J.S.O. Fonseca, and N. Kikuchi. "Optimal Design of Piezoelectric Microstructures". *Computational Mechanics.* **19** No.5 (1997): 397-410.
16. Silva, E.C.N., J.S.O. Fonseca, and N. Kikuchi, "Optimal Design of Periodic Piezocomposites", *Computer Methods in Applied Mechanics and Engineering*, **159**, (1998):49-77.
17. Haun, M. J. and R. E. Newnham. "An Experimental and Theoretical Study of 1-3 and 1-3-0 Piezoelectric PZT-polymer Composites for Hydrophone Applications". *Ferroelectrics.* **68** (1986): 123-139.
18. Hossack. J. A. and G. Hayward. "Finite-Element Analysis of 1-3 Composite Transducers". *IEEE Transactions on Ultrasonics, Ferroelectrics, and Frequency Control.* **38** No.6 (1991): 618-629.
19. Sanchez, S. and F. R. M. de Espinosa. "Modeling (2-2) Piezocomposites Partially Sliced in the Polymer Phase". *IEEE Transactions on Ultrasonics, Ferroelectrics, and Frequency Control.* **44** No.2 (1997): 287-296.
20. Smith, W. A. "Optimizing Electromechanical Coupling in Piezocomposites Using Polymers with Negative Poisson's Ratio", *Proceedings of IEEE Ultrasonic Symposium.* **1** (1991): 661-666.
21. Gibiansky, L. V. and S. Torquato. "Optimal Design of 1-3 Composite Piezoelectrics". *Structural Multidisciplinary Optimization.* **13** No.1 (1997): 23-38.
22. Sigmund, O., S. Torquato, and I. A. Aksay. "On the Design of 1-3 Piezocomposites Using Topology Optimization". *Journal of Material Research* **13** No.4 (1998): 1038-1048.
23. Hossack, J.A. and B.A. Auld. "Improved Efficiency Piezoelectric Ceramic/Polymer CompositeTransducers". *Proceedings of Ultrasonics Symposium.* (1992): 523-526.
24. Hayward, G. and J. Bennett. "Assessing the Influence of Pillar Aspect Ratio on the Behavior of 1-3 Connectivity Composite Transducers". *IEEE Transactions on Ultrasonics, Ferroelectrics, and Frequency Control.* **43** No.1 (1996): 98-108.
25. Bennett, J. and G. Hayward. "Design of 1--3 Piezocomposite Hydrophones Using Finite Element Analysis". *IEEE Transactions on Ultrasonics, Ferroelectrics and Frequency Control.* **44** No.3 (1997): 565-574.
26. Milton, G. W. and A.V. Cherkaev. "Which Elasticity Tensors are Realizable?". *Journal of Engineering Materials and Technology.* **117** (1995): 483-493.
27. Hashin, Z. and S. Shtrikman. "A Variational Approach of the Theory of Elastic Behavior of Multiphase materials". *Journal of the Mechanics and Physics of Solids.* **11** (1963): 127-140.
28. Lipton, R. "Optimal Bounds on Effective Elastic Tensors for Orthotropic Composites". *Proceedings of the Royal Society of London Serie A.* **444** No.1921 (1994): 399-410.
29. Smith, W. A. "Limits to the Enhancement of Piezoelectric Transducers Achievable by Materials Engineering". *Proceedings of IEEE Ultrasonics Symposium.* **1** (1992): 697-702.
30. Fonseca, J.S.O. "*Design of Microstructures of Periodic Composite Material*". Ph.D. Dissertation, N. Kikuchi, advisor, The University of Michigan (1997).
31. Sigmund, O. "Materials With Prescribed Constitutive Parameters: and Inverse Homogenization Problem". *International Journal of Solids and Structures.* **31** No.17 (1994): 2313-2329.
32. Sigmund, O. "Tailoring Materials with Prescribed Elastic Properties". *Mechanics of Materials.* **20** (1995): 351-368.
33. Sigmund, O. and S. Torquato. "Design of Materials with Extreme Thermal Expansion Using a Three-Phase Topology Optimization Method". *Journal of Mechanics and Physics of Solids* **45** No.6 (1997): 1037-1067.

34. Haftka, R. T. and Z. Gürdal *"Elements of Structural Optimization"*, Solid Mechanics and its Applications, Kluwer Academic Publishers, Dordrecht, The Netherlands, (1995).
35. Vanderplaatz, G.N. *"Numerical Optimization Techniques for Engineering Design: with Applications"*, McGraw-Hill, NY (1984).
36. Silva, E.C.N., S. Nishiwaki, J.S.O. Fonseca,, F.M. Espinosa, A.T. Crumm, G.A. Brady, J.W. Halloran, and N. Kikuchi, "Design of Piezocomposite Materials and Piezoelectric Transducers Using Topology Optimization - Part I", *Archives of Computational Methods in Engineering* **6** No.2 (1999): 117-182.
37. Crumm, A. T. and J.W. Halloran, "Fabrication of Micro-Configured Multicomponent Ceramic", *Journal of American Ceramic Society* **81** No.4 (1998):1053-1057.
38. Reilly, C. and J.W. Halloran, personal communication, (2001).
39. Jacobs, P. F. *"Stereolithography and Other RP&M Technologies from Rapid Prototyping to Rapid Tooling"*. Society of Manufacturing Engineers in Cooperation with the Rapid Prototyping Association of SME. ASME Press, New York, (1996).
40. Safari, A., G. S. Gong, J. Giniewicz, and R. E. Newnham. "Composite Piezoelectric Sensors", *Proceedings of the Twenty-First University Conference on Ceramic Science.* Plenum New York NY (1986): 445-454.
41. Sigmund, O. and Jensen, J.S. "Topology Optimization of Elastic Band Gap Structures and Waveguides", *in* H. A. Mang, F.G. Rammerstorfer and J. Eberhardsteiner (eds) *Proceedings of the Fifth World Congress on Computational Mechanics*, Vienna University of Technology, Austria, (2002).
42. Ruys, A.J., Popov, E.B., Sun, D., Russell, J.J., Murray, C.C.J., "Functionally Graded Electrical/Thermal Ceramic Systems", *Journal of the European Ceramic Society*, No.21 (2001): 2025–2029.
43. He, X.Q., Ng, T.Y., Sivashanker, S., Liew, K.M., "Active Control of FGM Plates with Integrated Piezoelectric Sensors and Actuators", *International Journal of Solids and Structures*, 38 (2001): 1641 – 1655.

Chapter 7

SYNTHESIS OF PERIODIC MICRO MECHANISMS

Extremal material design by topology optimization

Ole Sigmund
Department of Mechanical Engineering, Solid Mechanics, Nils Koppels Alle 404, Technical University of Denmark, DK-2800, Lyngby, Denmark
sigmund@mek.dtu.dk

Abstract Extremal materials may exhibit mechanism like behaviours. Examples are negative Poisson's ratio materials that expand transversely when pulled horizontally. Such materials are often composed of periodic arrays of simple microscopic mechanisms and may therefore be called Periodic Micro Mechanisms (PMMs). This chapter describes how PMMs can be systematically synthesized using the topology optimization method. A number of examples including design of negative Poisson's ratio, negative thermal expansion coefficient and band gap PMMs are given.

Keywords: Topology optimization, MEMS, material design, compliant mechanisms, homogenization.

Introduction

Periodic Micro Mechanisms (PMMs)[1] are a special kind of *materials*. Usually, the bulk form of materials is shaped in different ways to create structures and mechanisms that support and/or trasmit loads and motion. However, at the microstructural level, materials themselves can exhibit mechanism-like behaviors. Examples of PMMs are shearing materials (Fig. 7.1), negative Poisson's ratio materials (expand vertically when pulled horizontally), negative

[1]A recent paper [1] names the same kind of periodic mechanisms *Machine-Augmented Composites* (MACs), however, the author prefers the term *mechanisms* which covers passive as well as active materials as opposed to the term *machines* that normally is used for active structures.

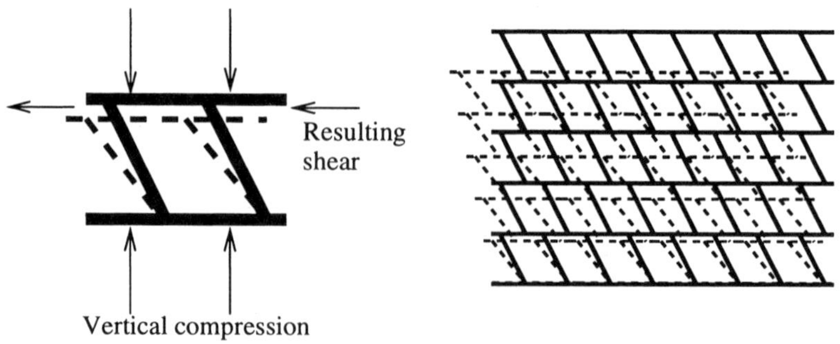

Figure 7.1. A shearing mechanism (left) and a periodic shearing mechanism (right). A vertical compression results in shearing of the periodic mechanism. Dashed lines indicate the deformed mechanisms.

thermal expansion materials, piezo-composites, band gap materials (materials that prohibit wave-propagation in certain frequency intervals), smart materials and many others. The special behaviours of PMMs originate from the topology of the single mechanisms (base cells) from which they are built.

Figure 7.1 shows a very simple shearing PMM which was synthesized based on intuition. There is no guaranty, however, that the mechanism is optimal, e.g. it could be that there exists a mechanism that converts the compression to a larger shearing or that converts the compression to shearing with less loss of energy. Also, we would like to synthesize periodic mechanisms with more complex responses where the intuitive approach is infeasible. Therefore, we need a systematic method for the design of PMMs. A systematic design methodology that has proven its worth in a myriad of structural, thermal and other physical problems is *topology optimization* (original citation [2] and a recent book [3]).

The topology optimization method was originally developed for the design of mechanical structures but was extended to material design (PMMs) by the author and others (low-density, truss-like materials [4–6], higher density, continuum-type materials [4, 7–16], conduction problems [17], three phase elastic and thermoelastic problems [18, 19, 12, 20], piezoelectric transducers [21–23], strength based design [24], viscoelastic materials [25], and phononic and photonic band gap materials [26–29]).

The basic idea of the topology optimization method is to divide the structure (base cell) into a large number of discrete elements (finite elements) and assign a design variable (area, material density or type) to each element (see Fig. 7.2). The base cell may be discretized using truss, frame or continuum finite elements. The optimization problem then consists in determining the optimal distribution of a limited amount of material such that an objective function is optimized and some constraints are fulfilled.

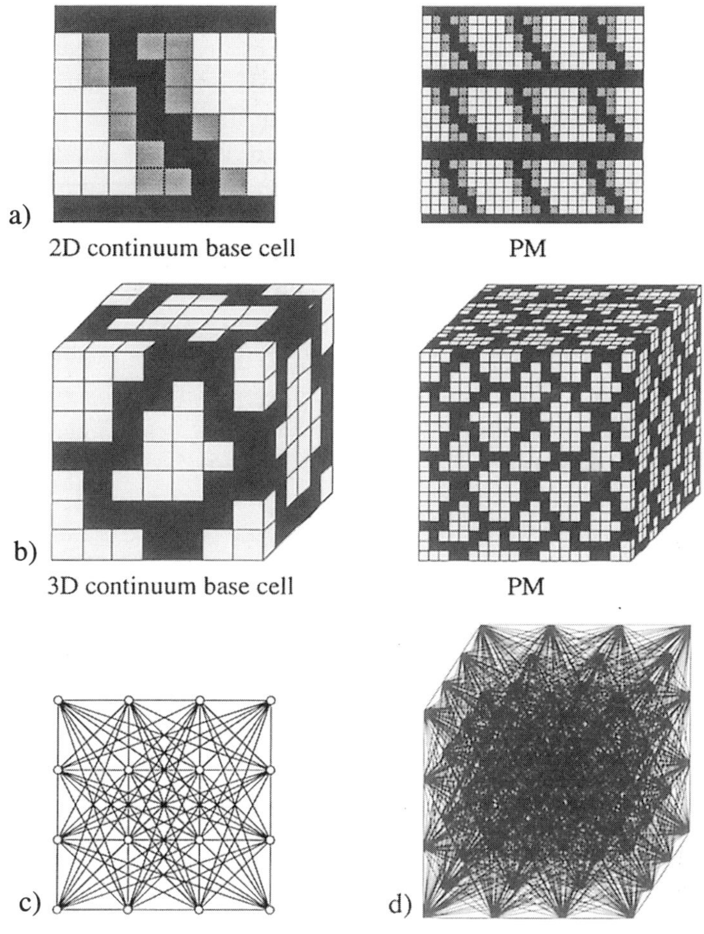

Figure 7.2. Base cells, or design domains for PMM synthesis. a) Two dimensional continuum design domain and 3 by 3 PMM, b) three dimensional continuum design domain and 3 by 3 by 3 PMM. For the continuum case the material density or type in each element are design variables. c) and d) Base cells for for 2D and 3D truss PMM synthesis. For the truss case, the areas of the individual bars are design variables.

Design of PMMs may be seen as mechanism design problems where the mechanisms have periodic boundary conditions[2]. Finding the effective behaviour of such periodic mechanisms can be done by the *homogenization method*. Although there exists an extensive mathematical literature on the theory, the basic idea is simple. The effective properties of the PMM can be found

[2]In fact, some of the first works on topology optimization of MEMS and compliant mechanisms [7, 30] were inspired by the authors work on material and PMM design [5–7]

by subjecting the discretized base cell (a single mechanism) to test strain fields and then the effective properties are calculated from the finite element response. In two dimensions, the test strain fields correspond to a horizontal, a vertical and a shear load. In three dimensions there are six test fields.

PMMs have a wide range range of applications. In aerospace and automotive industry vehicle weight can be reduced and safety or driving comfort can be increased by using materials with tailored properties like extremal stiffness, crash absorption or vibration and sound damping properties. In biomechanics, tailored PMMs may substitute biological tissue [31] and in sensors and actuators, one can improve sensitivity or actuation capability by optimizing the microstructures. Materials with negative or neutral thermal expansion coefficients are important in precision structures like optical circuits and instruments. Many of the extremal materials like negative Poisson's ratio materials, negative Young's modulus materials [32] and others may not have direct applications at present but will surely find use in the future.

Two dimensional PMMs and extremal materials may be manufactured using the same techniques that are used for manufacturing of MEMS [33]. Three dimensional PMMs may be manufactured by rapid prototyping processes, extrusion techniques and others. Simple microstructures may at present be manufactured in large batches whereas complex (multi material, three dimensional or extremely small) microstructures only can be processed as small samples. However, with the rapid growth of micro manufacturing methods these limitations may soon be overcome.

This chapter is built up as follows. Section 1 gives a brief introduction to homogenization theory and defines the finite element problems that need to be solved in order to find the effective behaviour of a PMM and the sensitivities with respect to design variations. Section 2 discusses the definition of a general optimization formulation that can be used for a number of PMM synthesis problems. Section 3 discusses the numerical implementation of the topology optimization method. Section 4 shows a number of example applications and finally section 6 draws the conclusions and discusses the open problems and future applications.

1. Numerical homogenization, FE-modelling and sensitivity analysis

The behaviour of a linear elastic PMM can be described by the constitutive law $\sigma_{ij} = E^H_{ijkl}\varepsilon_{kl}$. The effective (homogenized) constitutive tensor E^H_{ijkl} describes the relation between applied strains ε_{kl} and resulting stresses σ_{ij}.

Written in matrix form the constitutive law is

$$\left\{ \begin{array}{c} \sigma_{11} \\ \sigma_{22} \\ \sqrt{2}\sigma_{12} \end{array} \right\} = \left[\begin{array}{ccc} E^H_{1111} & E^H_{1122} & \sqrt{2}E^H_{1112} \\ E^H_{1122} & E^H_{2222} & \sqrt{2}E^H_{2212} \\ E^H_{1112} & E^H_{2212} & 2E^H_{1212} \end{array} \right] \left\{ \begin{array}{c} \varepsilon_{11} \\ \varepsilon_{22} \\ \sqrt{2}\varepsilon_{12} \end{array} \right\} . \qquad (1)$$

For example, the stresses for an applied vertical strain field $\{\varepsilon\} = \{0, 1, 0\}$ will be $\{\sigma_{11}, \sigma_{22}, \sqrt{12}\sigma_{12}\} = \{E^H_{1122}, E^H_{2222}, \sqrt{2}E^H_{2212}\}$.

The effective constitutive tensor may be found by the homogenization method [34]. In the following, the necessary equations will be given in continuum form and in Finite Element (FE) notation.

The homogenized stiffness tensor is by integration over the base cell Y found as

$$E^H_{ijkl} = \frac{1}{|\mathrm{Y}|} \int_{\mathrm{Y}} \left[E_{ijkl} - E_{ijpq} \frac{\partial \chi^{kl}_p}{\partial y_q} \right] dy \; , \qquad (2)$$

where the Y-periodic test fields χ^{kl}_p are found as the solutions to the equilibrium equations

$$\int_{\mathrm{Y}} E_{ijpq} \frac{\partial \chi^{kl}_p}{\partial y_q} \frac{\partial \phi_i}{\partial y_j} dy = \int_{\mathrm{Y}} E_{ijkl} \frac{\partial \phi_i}{\partial y_j} dy \quad \text{for all Y} - \text{periodic } \phi \, . \qquad (3)$$

In a somewhat simpler notation, (2) may be written as

$$E^H_{ijkl} = \frac{1}{|\mathrm{Y}|} \int_{\mathrm{Y}} E_{pqrs} (\varepsilon^{0(ij)}_{pq} - \varepsilon^{*(ij)}_{pq})(\varepsilon^{0(kl)}_{rs} - \varepsilon^{*(kl)}_{rs}) \; dy \; , \qquad (4)$$

where $\varepsilon^{*(kl)}_{pq} = \frac{1}{2}(\frac{\partial \chi^{kl}_p}{\partial y_q} + \frac{\partial \chi^{kl}_q}{\partial y_p})$ and ε^0_{ij} corresponds to the three (2D) or six (3D) unit test strain fields. Furthermore, (3) may be rewritten as

$$\int_{\mathrm{Y}} E_{ijpq} \frac{\partial \chi^{kl}_p}{\partial y_q} \frac{\partial \phi_i}{\partial y_j} dy = \int_{\mathrm{Y}} E_{ijpq} \varepsilon^{0(kl)}_{pq} \frac{\partial \phi_i}{\partial y_j} dy \quad \text{for all Y} - \text{periodic } \phi \, . \quad (5)$$

In practice, the equilibrium equations (3) are solved as a finite element problem with three or six load cases

$$\boldsymbol{K}_{per} \boldsymbol{\chi}^{kl} = \boldsymbol{f}^{kl} \; , \qquad (6)$$

where the global stiffness matrix is calculated as the usual assembly of element stiffness matrices ($\boldsymbol{K} = \sum_e \boldsymbol{K}_e = \sum_e \int_{\mathrm{Y}_e} \boldsymbol{B}^T_e \; \boldsymbol{E}_e \; \boldsymbol{B}_e \; dy$) and $\boldsymbol{K}_{per}$ is the global stiffness matrix corrected for periodic boundary conditions such that the resulting displacements $\boldsymbol{\chi}^{kl}$ are Y-periodic. This can be obtained either by a penalty approach or by assigning equal node numbers to opposing boundary nodes. The force vector is found from

$$\boldsymbol{f}^{kl} = \sum_e \int_{\mathrm{Y}_e} \boldsymbol{B}^T_e \boldsymbol{E}_e \; \boldsymbol{\varepsilon}^{0(kl)} dy \; = \boldsymbol{K} \boldsymbol{\chi}^{0(kl)}, \qquad (7)$$

where $\chi^{0(ij)}$ is the displacement field corresponding to the uniform test strain field $\varepsilon_{pq}^{0(ij)}$, $\boldsymbol{B}_e$ is the finite element strain-displacement matrix and $\boldsymbol{E}_e$ is the element constitutive tensor in matrix form. After (6) has been solved, the effective properties may be found as

$$E_{ijkl}^{H} = \frac{1}{|\mathrm{Y}|}\sum_{e}(\boldsymbol{\chi}^{0(ij)} - \boldsymbol{\chi}^{ij})^{T}\int_{\mathrm{Y}_e}\boldsymbol{B}_e^{T}\boldsymbol{E}_e\boldsymbol{B}_e dy\,(\boldsymbol{\chi}^{0(kl)} - \boldsymbol{\chi}^{kl}) = \frac{1}{|\mathrm{Y}|}\sum_{e}(\boldsymbol{\chi}^{0(ij)} - \boldsymbol{\chi}^{ij})^{T}\,\boldsymbol{K}_e\,(\boldsymbol{\chi}^{0(kl)} - \boldsymbol{\chi}^{kl}). \tag{8}$$

For the continuum base cells, the variables in the design problem are the element relative densities ρ_e. The ultimate goal of the design problem will be to end up with a discretized base cell where each element is either solid or void, i.e. $\rho_e = 1$ or $\rho_e = 0$. However, this integer type problem has so far been impossible to solve for finer discretizations and therefore the discrete problem is usually converted into a continuous problem where the densities can take any value between 0 and 1. This allows the optimization problem to be solved by efficient gradient based mathematical programming algorithms. A commonly used approach is to let the element (local) constitutive matrix $\boldsymbol{E}_e$ be a function of the continuous design variable $\rho_e \in]0,\ 1]$, i.e.

$$\boldsymbol{E}_e = \boldsymbol{E}(\rho_e) = \rho_e^p \boldsymbol{E}^0, \tag{9}$$

where $\boldsymbol{E}^0$ is the constitutive matrix for solid material and p is a penalization factor. The penalization factor ($p > 1$) results in low stiffness for intermediate densities and therefore makes intermediate densities un-economical compared to solid or void. This so-called power-law or SIMP (Solid Isotropic Material with Penalization) interpolation scheme that allows for the design of one-material porous PMMs may be extended to two material problems

$$\boldsymbol{E}(\rho_e) = \rho_e^p \boldsymbol{E}^1 + (1-\rho_e)^p \boldsymbol{E}^2, \tag{10}$$

and to three phase (two materials and void) design by adding a second variable pr. element

$$\boldsymbol{E}(\rho_e^1, \rho_e^2) = (\rho_e^2)^p \left[(\rho_e^1)^p \boldsymbol{E}^1 + (1-\rho_e^1)^p \boldsymbol{E}^2\right], \tag{11}$$

where $\boldsymbol{E}^1$ and $\boldsymbol{E}^2$ are the constitutive matrices for material phase 1 and 2, respectively.

For truss discretized base cells, the design variables ρ_e are simply the bar areas and there is no need for special interpolation schemes.

For all interpolation and discretization cases, the element stiffness matrices become functions of the design variables, i.e. $\boldsymbol{K_e} = \boldsymbol{K_e}(\rho_e)$ or $\boldsymbol{K_e} = \boldsymbol{K_e}(\rho_e^1, \rho_e^2)$.

The sensitivity of the effective properties with respect to a change in design variable ρ_e can be found as [3]

$$\frac{\partial E^H_{ijkl}}{\partial \rho_e} = \frac{1}{|\mathrm{Y}|}(\boldsymbol{\chi}^{0(ij)} - \boldsymbol{\chi}^{ij})^T \frac{\partial \boldsymbol{K}_e}{\partial \rho_e}(\boldsymbol{\chi}^{0(kl)} - \boldsymbol{\chi}^{kl}). \tag{12}$$

Here, $\boldsymbol{K}_e/\rho_e$ can be calculated analytically and there is thus no need for additional FE-analyses in order to perform the sensitivity analysis.

The FE-problem (6) can be solved by standard direct linear solvers taking the periodic boundary conditions into account. Alternatively, the problem may advantageously be solved using an Element-By-Element Preconditioned Conjugate Gradient (EBE-PCG) approach [19].

The homogenization formulation sketched above may relatively easily be applied to other physical problems. For example the thermal expansion coefficient may be found by solving an additional FE-problem

$$\boldsymbol{K}_{per}\boldsymbol{\chi}^{\alpha} = \boldsymbol{f}^{\alpha}, \tag{13}$$

where the thermal test field $\boldsymbol{\chi}^{\alpha}$ is again Y-periodic and the thermal finite element load vector is defined as

$$\boldsymbol{f}^{\alpha} = \sum_e \int_{Y_e} \boldsymbol{B}_e^T \boldsymbol{E}_e(\rho_e^1, \rho_e^2)\, \boldsymbol{\alpha}(\rho_e^2) dY. \tag{14}$$

Here, the element thermal expansion coefficient $\alpha(\rho_e^2)$ is interpolated by the expression [35, 19]

$$\alpha(\rho_e^2) = \frac{\alpha^1\, \kappa^1\, (\kappa^2 - \kappa(\rho_e^2)) - \alpha^2\, \kappa^2\, (\kappa^1 - \kappa(\rho_e^2))}{\kappa(\rho_e^2)\, (\kappa^1 - \kappa^2)}, \tag{15}$$

where κ^1 and κ^2 are the bulk moduli of phase 1 and 2, respectively. Note that the interpolation for the thermal expansion coefficient depends on the bulk modulus of the two-phase composite only.

Having solved (13), the effective thermal expansion coefficient tensor may be found as

$$\alpha_{ij}^H = \left(E_{ijkl}^H\right)^{-1} \frac{1}{|\mathrm{Y}|} \sum_e (\boldsymbol{\chi}^{0(\alpha)} - \boldsymbol{\chi}^{\alpha})^T\, \boldsymbol{K}_e\, (\boldsymbol{\chi}^{0(kl)} - \boldsymbol{\chi}^{kl}), \tag{16}$$

and the sensitivity is found similarly to (12).

Similar homogenization and sensitivity expressions may be derived for piezoelectricity [21, 22] and for many other problems, e.g. viscoelasticity [25] and wave-propagation problems [26, 27, 29, 28]. The latter case requires special non-periodic boundary conditions as will be discussed in Section 5.

2. Formulation of the optimization problem.

The goal of the optimization problem is to come up with mechanisms, that when repeated, provide some interesting behaviour of the resulting PMM. An example could be the shearing mechanism from Fig. 7.1. Here, the goal is to maximize the shear stress for a given horizontal strain, i.e. according to the remarks in the beginning of section 1, we would like to maximize E^H_{2212}. This maximization may result in a very flimsy material with very low stiffness in the vertical direction. Therefore, we may want to impose a minimum bound on the vertical Young's modulus, i.e. $E^H_2 = E^H_{2222} - \left(E^H_{1122}\right)^2 / E^H_{1111} \geq E^*_2$. An optimization problem that can be used to solve problems like this can be written as

$$\begin{aligned} &\min_{\boldsymbol{\rho}} c(E^H_{ijkl}(\boldsymbol{\rho})) \\ &\qquad \frac{1}{|\mathrm{Y}|}\sum_{e=1}^{N} v_e \rho_e \leq \vartheta, \\ &\qquad g_i(E^H_{ijkl}(\boldsymbol{\rho})) \leq g^*_i, \quad i = 1, \ldots, M \, , \\ &\qquad 0 < \rho_{min} \leq \rho_e \leq 1, \quad e = 1, \ldots, N \, , \end{aligned} \tag{17}$$

where ϑ is an upper bound on the volume fraction and the objective function $c(E^H_{ijkl})$ and constraints $g_i(E^H_{ijkl})$ are functions of the homogenized tensor.

The constraints in (17) may take different forms. As discussed above, it may for example be necessary to introduce a constraint in the minimum Young's modulus value. Also, it may be desirable to impose a constraint that ensures symmetries in the resulting material properties. Orthotropy (i.e. $E^H_{1112} = E^H_{2212} = 0$) may be obtained by imposing one or more lines of symmetry in the base cell. Square symmetry may for example be obtained by imposing one line of symmetry and adding the constraint $g_2 = (E^H_{1111} - E^H_{2222})^2/(E^H_{1111} + E^H_{2222})^2$ with $g^*_2 = \epsilon^2$, where ϵ is a small tolerance number.

For three phase problems and thermal problems, the objective function may be expanded to

$$\begin{aligned} &\max_{\boldsymbol{\rho}^1, \boldsymbol{\rho}^2} c(E^H_{ijkl}(\boldsymbol{\rho}^1, \boldsymbol{\rho}^2), \; \alpha^H_{ij}(\boldsymbol{\rho}^1, \boldsymbol{\rho}^2)) \\ &\text{s.t.} : \frac{1}{|\mathrm{Y}|}\sum_{e} v_e \rho^1_e \rho^2_e \leq \vartheta_1, \quad 0 < \rho_{min} \leq \rho^1_e \leq 1, \quad e = 1, \ldots, N \, , \\ &\qquad \frac{1}{|\mathrm{Y}|}\sum_{e} v_e \rho^1_e (1 - \rho^2_e) \leq \vartheta_2, \quad 0 \leq \rho^2_e \leq 1, \quad e = 1, \ldots, N \, , \\ &\qquad g_i(E^H_{ijkl}(\boldsymbol{\rho}^1, \boldsymbol{\rho}^2)) \leq g^*_i, \quad i = 1, \ldots, M \, , \end{aligned} \tag{18}$$

where the objective function now may depend on either or both of the effective constitutive tensor and the effective thermal expansion tensor. Both these tensors again depend on the two design variable vectors ρ^1 and ρ^2.

3. Numerical implementation

The optimization problems (17) and (18) are non-linear and cannot be solved in one step. Instead we solve a sequence of linearized sub-problems, where the linearizations are based on the sensitivities found in Section 1. The resulting optimization problems are large scale. Often the number of design variables will be several thousands and up to a million for 3d problems. At the same time the number of constraints is low - typically countable on one hand. Such problems can be solved by Optimality Criteria methods (for at most one constraint), by Sequential Linear Programming or by custom made codes like the Method of Moving Asymptotes (MMA) by [36]. The latter code uses first order sensitivity information combined with an adaptive convexification strategy based on asymptotes that are moved according to the convergence behaviour of previous steps. The MMA is presently the most popular code for advanced topology optimization problems. Quadratic programming methods are usually not used for topology optimization problems due to the huge cost and overhead involved in calculating the Hessian.

The solutions to (17) and (18) are mesh-dependent and non-unique. To circumvent the first problem and to some extent be able to control the latter, different schemes may be used to ensure well-posed problems [37, 38]. Here we suggest to use a scheme that modifies the sensitivity information such that element sensitivities become dependent on sensitivities within a fixed radius of the element [4, 30]. This means that we substitute the original sensitivities (12) with the filtered sensitivities

$$\widehat{\frac{\partial E^H_{pqrs}}{\partial \rho_k}} = \frac{1}{\rho_k \sum_{i=1}^{N} \hat{H}_i} \sum_{i=1}^{N} \hat{H}_i \, \rho_i \, \frac{\partial E^H_{pqrs}}{\partial \rho_i} \,, \tag{19}$$

where the *mesh-independent* convolution operator (weight factor) $\hat{H}_i$ is written as

$$\hat{H}_i = r_{\min} - \text{dist}(k,i), \quad \{i \in N \mid \text{dist}(k,i) \leq r_{\min}\}\,, \; k = 1,\ldots,N \,. \tag{20}$$

In this expression, the operator $\text{dist}(k,i)$ is defined as the distance between the center of the element k and the center of an element i. The convolution operator $\hat{H}_i$ is zero outside the filter area. The convolution operator for element i is seen to decay linearly with the distance from element k. It is worthwhile noting that the sensitivity (19) converges to the original sensitivity when the filter radius

$r_{\min}$ approaches zero and that all sensitivities will be equal (resulting in an even distribution of material) when $r_{\min}$ approaches infinity.

Although the computational implementation of the topology optimization problem may seem quite involved, a simple problem like stiffness optimization of mechanical structures may be programmed in just 99 lines of Matlab code including FE-analysis, sensitivity analysis, OC optimizer, mesh-independency filter and graphics [39]. The code can be downloaded from the author's web-page:
`http://www.topopt.dtu.dk`. With some knowledge of FE-modelling it is easy to modify this code for the extra load cases and periodic boundary conditions associated with the PMM optimization problem.

4. Examples

4.1 Shearing materials

As the first example, we want to optimize the shearing mechanism from Fig. 7.1. As discussed previously, maximizing the shear stress for a given vertical strain corresponds to maximizing E^H_{2212}. Allowing 50% of an isotropic base material with Young's modulus 1 and Poisson's ratio 0.3 to be distributed in the 2D base cell which is discretized with 40 by 40 4-node elements, we obtain the optimized PMM seen in Fig. 7.3a ($E^H_{2212} = 0.140$). It is seen that we get a laminated structure consisting of alternating layers of solid and void. Obviously, this is a very efficient mechanism in converting a vertical compression into a shear stress, however, the mechanism cannot carry any loads in neither the horizontal nor the vertical directions. Therefore, we add constraint on the effective Young's moduli, i.e. $E^H_1 \geq 0.1$ and $E^H_2 \geq 0.1$. This results in the optimized PMM seen in Fig. 7.3b ($E^H_{2212} = 0.127$). It is now seen that thin stringers between the layers ensure horizontal and vertical stiffness. Figure 7.3c ($E^H_{2212} = 0.108$) and Fig. 7.3d ($E^H_{2212} = 0.002$) show the optimized PMMs for Young's modulus constraints of 0.2 and 0.3, respectively. In the latter case it is seen that the mechanism type behaviour almost has disappeared because of the required high stiffness.

The results of this optimization problem correspond very well to the intuitive solution from Fig. 7.1. However, for more complex problems as shown in the following examples, it is difficult to come up with the solutions by intuition.

4.2 Negative Poisson's ratio materials

A material that expands vertically when pulled horizontally is called a negative Poisson's ratio material. A negative Poisson's ratio PMM which was invented by intuition is the inverted honeycomb [40]. However, this PMM only has negative Poisson's ratio in certain directions. In order to synthesize an

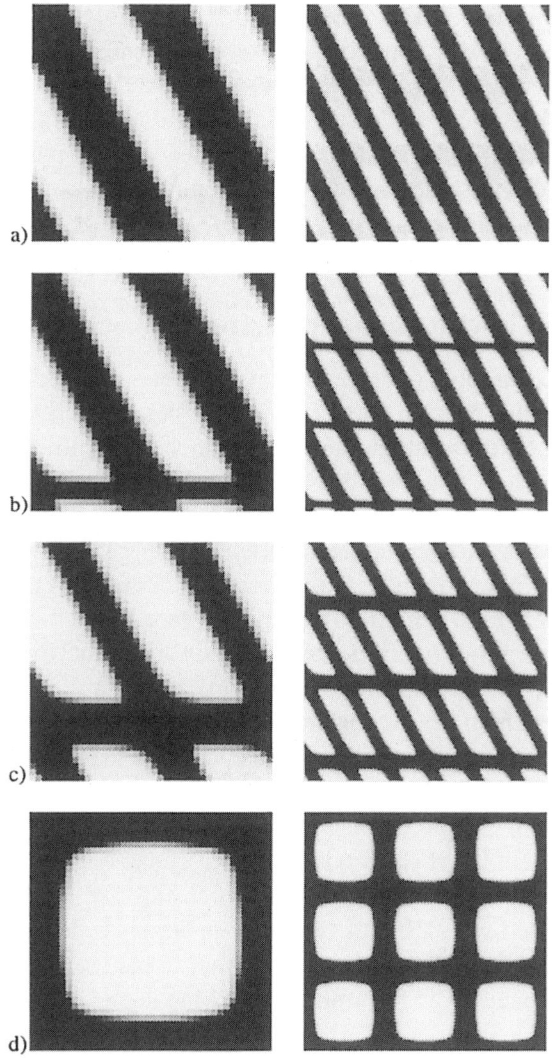

Figure 7.3. Left: optimized base cells and right: PMMs resulting from maximization of shear stress subject to a vertical strain. a) free Young's moduli ($E^H_{2212} = 0.140$), b) Young's moduli higher than 0.1 ($E^H_{2212} = 0.127$), c) Young's moduli higher than 0.2 ($E^H_{2212} = 0.108$) and d) Young's moduli higher than 0.3 ($E^H_{2212} = 0.002$).

isotropic negative Poisson's ratio PMM we minimize the Poisson's ratio subject to constraints on isotropy and minimum bulk modulus [33]. The resulting base cell topologies for truss and continuum-type base cells in two and three dimensions are seen in Fig. 7.4. In the figure the 2d continuum base cell was discretized by 60 by 60 elements, the 3d continuum base cell by 40 by 40 by

40 (64000) cubic elements and the 2d and 3d truss base cells where discretized by 120 and 2016 truss elements, respectively (as seen in Fig. 7.2).

A range of other negative Poisson's ratio materials have been suggested and discussed [41–44, 4–6].

Negative Poisson's ratio PMMs may be used as raw-plugs (easy to insert but difficult to pull out since they expand when pulled), hydrophones (see next section), toys (c.f. Hoberman Spheres a.o.) and may have other potentially rewarding applications [42].

4.3 Extremal thermal expansion coefficient

The previous examples considered the synthesis of purely passive PMMs, i.e. they required an external force or displacement to activate. A semi-active PMM could be thermally actuated, i.e. it produces some intricate displacement when heated.

For a two-phase PMM made from solid and void, the effective thermal expansion coefficient will always be the same as for the solid phase. For two-phase mixtures of two non-void material phases, the effective thermal expansion coefficient for any square or cubic symmetric mixture of the materials will always take values between the maximum and the minimum thermal expansion coefficients of the two constituent phases. However, for three-phase composites, it becomes more interesting. According to theory [45–47] it is possible to synthesize three-phase materials with effective thermal expansion coefficients that exceed those of the individual phases.

We start by an example resembling the shearing mechanism from before. This time we maximize the effective thermal shear coefficient α_{12}^H subject to constraints on the Young's moduli ($E_1^H > 0.1$ and $E_2^H > 0.1$). We allow the base cell to be filled with 25% of each of two constituent materials with equal Young's moduli ($E^1 = E^2 = 1$) and thermal expansion coefficients $\alpha^1 = 1$ and $\alpha^2 = 10$, respectively. The optimized PMM is shown in Fig. 7.5.

Another example is the minimization of the *isotropic* thermal expansion coefficient. Above mentioned theoretical bounds predict that it should be possible to synthesize negative thermal expansion PMMs from mixtures of two positive expansion phases and a void phase. The extreme thermal expansion is obtained at the cost of a decrease in the effective bulk modulus. The theoretical bounds on the range of attainable combinations of thermal coefficients and bulk modulus for a particular case are shown in Fig. 7.6a.

Figure 7.6a shows a graph of the thermal expansion coefficients as a function of the bulk moduli for some numerically obtained PMMs compared with the theoretical bounds. The design problem consists in extremizing the thermoelastic properties of a PMM consisting of 25% of a material with thermal expansion coefficient 10 (normalized value) and 25% of a material with thermal

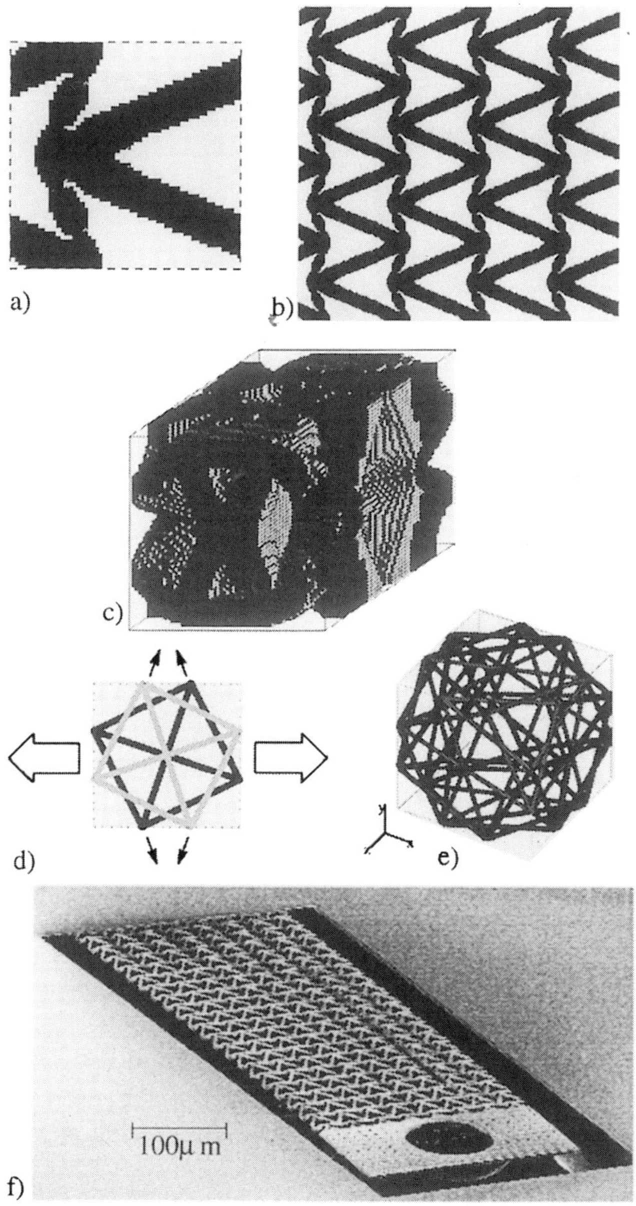

Figure 7.4. Different optimized negative Poisson's ratio PMMs. a) 2d continuum base cell and b) corresponding PMM. c) 3d continuum base cell. d) and e) 2d and 3d truss modelled base cells. f) Negative Poisson's ratio testbeam manufactured at MIC, DTU [33] by MEMS processing techniques.

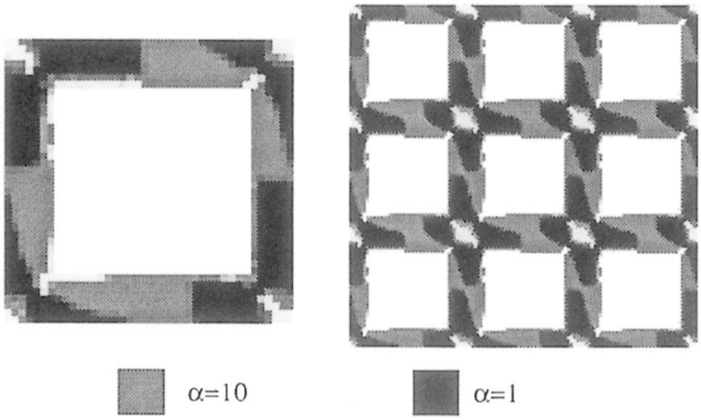

Figure 7.5. Optimized PMM for maximum thermal shear coefficient ($\alpha_{12}^H = 113$).

expansion coefficient 1. A resulting PMM with a negative thermal expansion coefficient of -4.02 is shown in Fig. 7.6b. This shows that it is actually possible to design PMMs with *negative* thermal expansion coefficients from constituent phases with *positive* thermal expansion coefficients.

In Fig. 7.6a one notes that the numerically obtained effective values are far away from the old bounds [45, 46] but very close to the newer bounds [47]. In fact, the substantial improvement of the new bounds compared to the old bounds was inspired by the numerical results by [19]. Using the translation method, [47] managed to make the theoretical bounds match the numerical results!

4.4 Piezoelectric transducers

Another three-phase PMM synthesis example is the design of hydrophones based on piezoelectric sensing. A piezoelectric material responds with an electric output when strained. For a typical ceramical piezoelectric rod, the electric field depends on the elongation of the rod. Simply said, this means that the electric output for a horizontal compression will have the negative sign of a that for a longitudinal compression. For hydrophones which should detect changes in hydrostatic pressures, this is a problem. Compression in all directions simultaneously will almost cancel the electrical output. In order to circumvent this problem it has been suggested to embed piezoelectric rods in a PMM with tailored properties (Fig. 7.7a). If the piezo rods are embedded in negative Poisson's ratio PMM, the transverse pressure will be converted into compression of the rods instead of extension, in turn causing a much larger output signal.

The synthesis procedure described in this chapter may be used to identify the PMM that maximizes the electric output of hydrophones. The problem corresponds to a three-phase material design problem of distributing a piezoelectric, a polymer and a void phase in a periodic base cell. However, for various reasons,

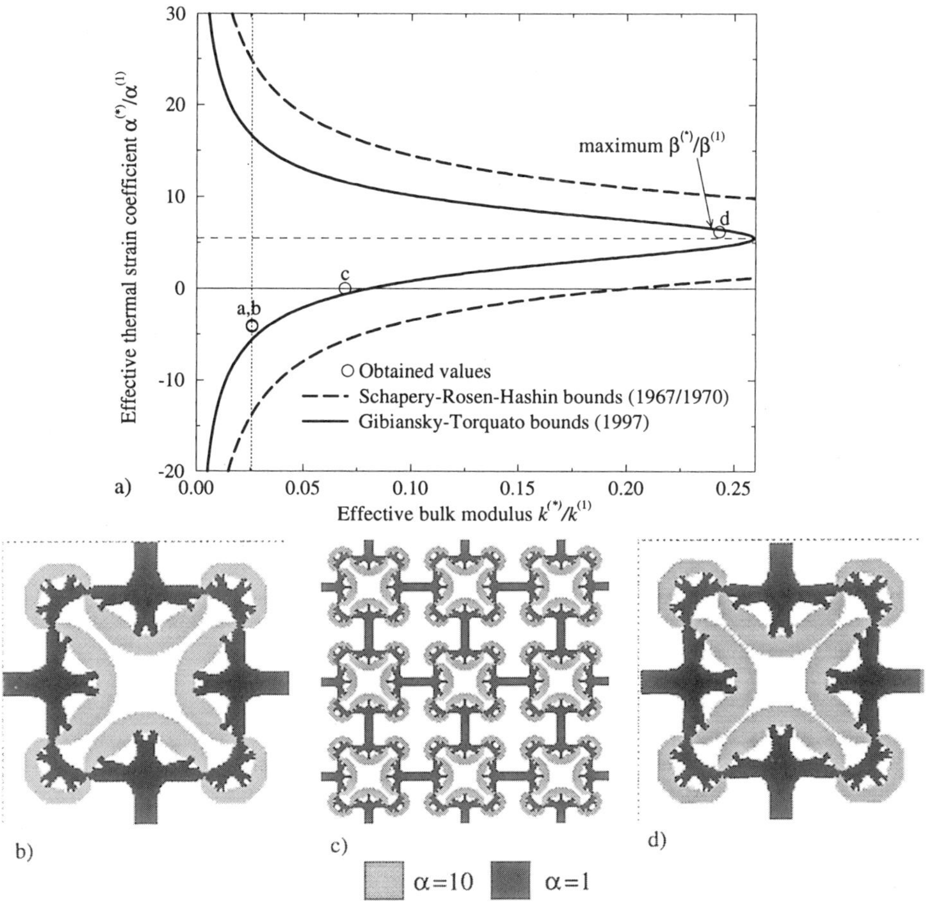

Figure 7.6. Design of PMMs with extreme thermoelastic properties using topology optimization. a) Thermal expansion coefficient-bulk modulus graph for specific thermoelastic composite including theoretical bounds and numerically obtained properties, b), c) and d) PMMs with negative thermal expansion coefficient (single base cell, 3 by 3 array and heated single cell). When heated, the cell contracts resulting in an effective negative thermal expansion coefficient (from [19]).

the length-scale of the PMM will often be much smaller than for the embedded rods. Therefore one may model the problem partly by effective medium theory and partly by numerical homogenization. This means that the effective properties of the PMM may be found by numerical homogenization as described in the previous sections whereas the effective properties of the mixture of the PMM and the piezoelectric rods may be found analytically [48]. The effective piezoelectric properties may thus be determined directly as functions of the ef-

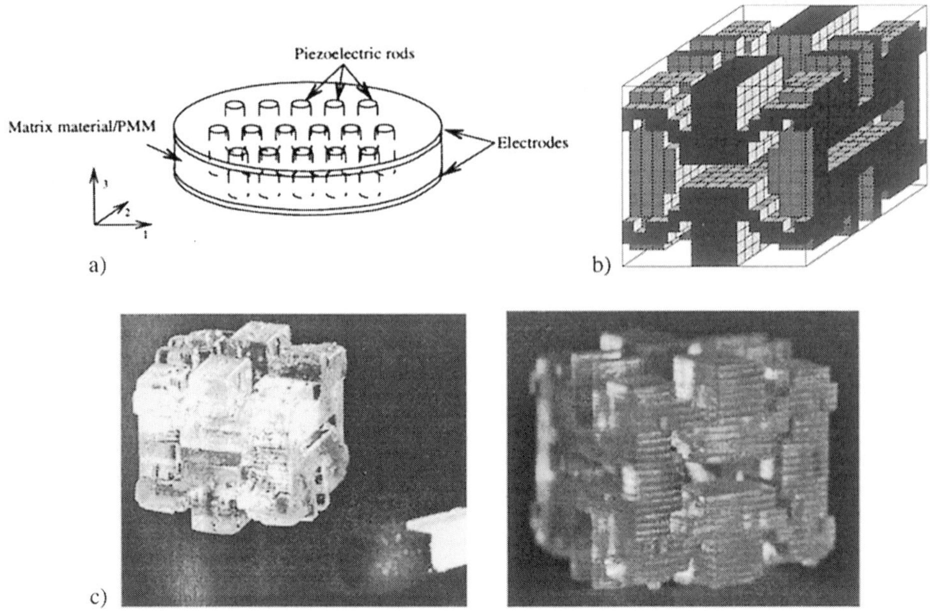

Figure 7.7. Design of hydrophones using topology optimization. a) Schematic of a 1-3 piezoelectric hydrophone, b) one base cell of hydrophone PMM optimized for piezoelectric charge coefficient and c) one base cell manufactured using stereolithography (from [49]).

fective matrix properties E^H_{ijkl}. The optimization problem may then be solved by the PMM design formulation given in (17).

Figure 7.7b shows one base cell of the PMM of a hydrophone optimized for maximum piezoelectric charge coefficient [49]. One observes that this PMM is transversely isotropic with a negative Poisson's ratio. The improvement compared to a solid polymer PMM is a factor of 11. Prototype PMMs have been manufactured in polymer using a stereolithography process (Fig. 7.7c).

5. Wave propagation

The mechanism behaviours of the PMMs considered so far could all be explained by the isolated behaviour of a single base cell. In other words: the mechanism worked for single base cells, a few base cells as well as for a huge periodic structure[3]. In this section we will discuss a special kind of PMMs that only works due to the coupling effects between a larger number of base cells.

[3]Note, however, that the effective properties are calculated for infinitely periodic PMMs so for finite PMMs the performance may be degraded.

The phenomenon is called band gaps and will be explained by the following example.

We consider wave propagation through composite media. The waves may be elastic, acoustic or electromagnetic, and the phenomenon to be exploited is that for some frequency bands it is possible to construct PMMs or materials that hinder wave propagation. This is called a band gap.

The phenomenon of band gaps in structures subject to periodic loads is illustrated in Fig. 7.8a and b. Here a two dimensional square domain is subjected to a periodic loading at the left edge and has absorbing boundary conditions along all four edges. The frequency of excitation of the structure in Fig. 7.8a is lower than for Fig. 7.8b. It is seen that waves propagate through the structures from left to right and that the absorbing boundary conditions damp the waves at the top and bottom. These are perfectly normal situations and could model surface waves on water, acoustic waves through air, out-of plane waves in an elastic structure, polarized electromagnetic waves, etc.

Now, if we introduce a periodic distribution of inclusions with different material properties than in the original structures, the situation changes. For the structure subjected to the lower excitation frequency (Fig. 7.8c), there is still propagation although the waves have different shapes. However, for the structure subjected to a higher excitation frequency (Fig. 7.8d) there seems to be no propagation at all. This illustrates the band gap phenomenon. A band gap material is defined as a PMM that does not allow wave propagation for certain frequency ranges. For elastic and acoustic waves the PMMs are called *phononic* band gap materials, for electromagnetic wave propagation the PMMs are called *photonic* band gap materials and the same principle on the atomic scale lies behind semiconductors. The length-scale of the PMMs in band gap materials is typically close to the wavelengths of the forbidden frequencies.

Band gap materials may be used for many purposes. An example is waveguides. The idea is illustrated in Figs. 7.8e and f. If we introduce a defect in the periodic structures from Figs. 7.8c and d, waves with frequencies outside the band gap may still propagate through the whole structure (Fig. 7.8e) but waves with frequencies within the band gap may now only propagate through the defect, resulting in a wave guide as seen in Fig. 7.8f. It is observed that it is actually possible to guide waves around a corner. This is especially interesting for light waves since it may allow for the manufacturing and design of so-called photonics based microchips which have much higher clock frequencies than conventional microchips based on electrical conduction.

Apart from semiconductors and wave guides, band gap materials may be used to generate frequency filters with control of pass or stop bands, as beam splitters, as sound or vibration protection devices, as perfect mirrors and in many other applications. The following discussion is based on two recent papers [28, 29] and is based on wave propagation in elastic media. However,

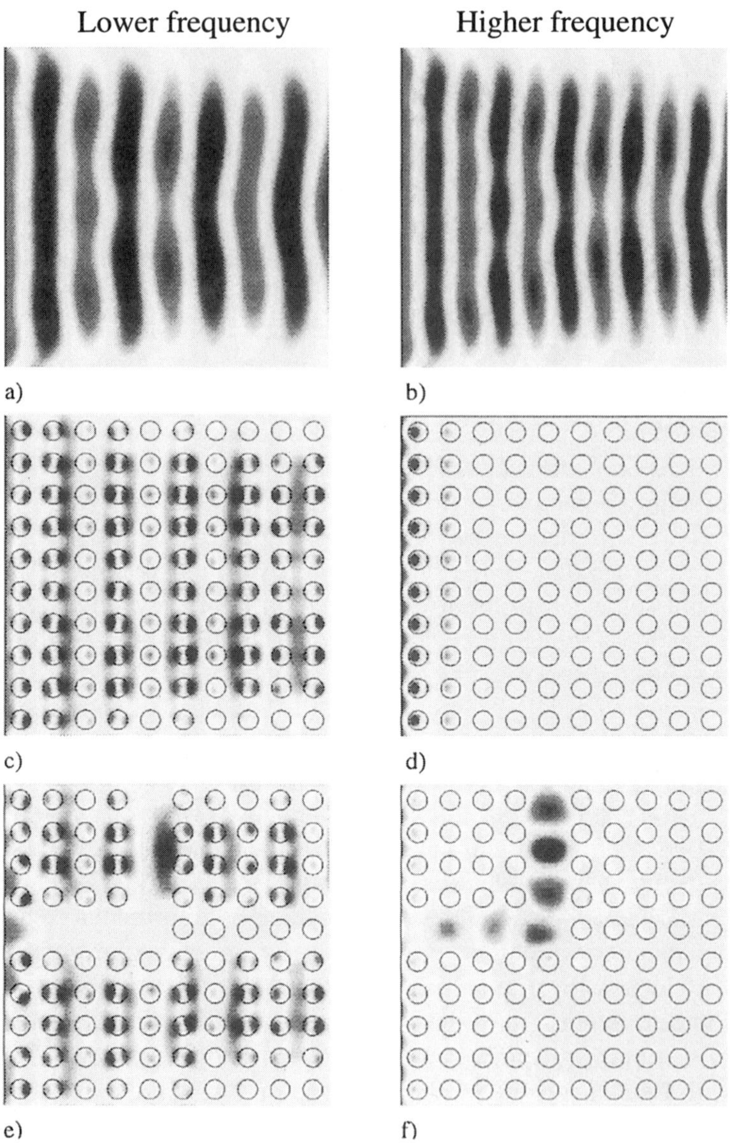

Figure 7.8. Scalar wave propagation in 2D domains with absorbing boundary conditions and forced vibrations at the left edge. a) Wave propagation through homogeneous structure, b) wave propagation with higher frequency through homogeneous structure, c) wave propagation through structure with periodic inclusions, d) (no) wave propagation with higher frequency through periodic structure, e) wave propagation through periodic structure with defect and f) wave guiding at higher frequency through a periodic structure with defect.

very similar equations and expressions are obtained for planar electromagnetic wave-propagation.

5.1 Modelling of wave propagation

Elastic wave propagation in an undamped elastic medium may be calculated by the FE-equation

$$\boldsymbol{K}\boldsymbol{u} - \boldsymbol{M}\ddot{\boldsymbol{u}} = \boldsymbol{0}\,, \tag{21}$$

where, as before, $\boldsymbol{K}$ is the global stiffness matrix and $\boldsymbol{M}$ is the global mass matrix, $\boldsymbol{u}$ is the global displacement vector and $\ddot{\boldsymbol{u}}$ is the global velocity vector.

For a large periodic structure we would like to find the "homogenized" wave-propagation properties. However, this is not as simple as for the static PMMs treated in Section 1 since the waves propagating through the PMM may be of the same length-scale as the microstructure whereas for the usual homogenization approach we had to assume that the microstructures were very small compared to the global deformation field. For the wave propagation case we would like to analyse the wave propagation properties for any frequency/wave-length and any direction of incoming waves. This problem can be solved using the so-called Floquet-Bloch wave theory [50, 51].

We analyse a single base cell as before and assume that the waves are harmonic. Then we can write the pointwise displacements as

$$u(\boldsymbol{y},\boldsymbol{k}) = v(\boldsymbol{y})e^{i\boldsymbol{k}^T\boldsymbol{y}}e^{i\omega t}\,, \tag{22}$$

where v is a Y-periodic displacement field, $\boldsymbol{y}$ is the spatial coordinate and $\boldsymbol{k}$ is the wave vector. For $\boldsymbol{k} = \boldsymbol{0}$, the solution mode $u(y)$ will be Y-periodic as for the χ^{kl} solution field in (6). For $\boldsymbol{k} = \boldsymbol{\pi}$, the solution mode will be 2Y-periodic, i.e. the wave will extend over two base cells in each direction. For other $\boldsymbol{k}$, the solution modes can take any kind of periodicity in all directions.

Inserting (22) in (21) and performing some algebra we end up with the eigenvalue problem

$$\left(\tilde{\boldsymbol{K}}(\boldsymbol{k}) - \omega^2\boldsymbol{M}\right)\boldsymbol{v} = \boldsymbol{0}\,, \tag{23}$$

where $\tilde{\boldsymbol{K}}$ is a complex but hermitian global “stiffnes” matrix corrected for the complicated boundary conditions. The FE eigenvalue equation (23) should in principle be solved for any wave vector $\boldsymbol{k}$. However, due to symmetry we may restrict the investigated values of the wave vector to the first Brillouin zone $\boldsymbol{k} \in [-\pi,\ \pi]^d$ (d is the dimension) [52]. This corresponds to analysing the structural response to incident waves of any possible wave length and direction. If we furthermore assume that the base cell is square symmetric (i.e. is quadratic and symmetric around horizontal, vertical and diagonal lines), we may restrict the range of wave vectors to the triangular region indicated in Fig. 7.9, left. It is generally accepted (but to the authors knowledge, not proved) that one only has to search the borders of the triangular region to obtain a description of the band gap structure of a periodic material. This means that the wave equation

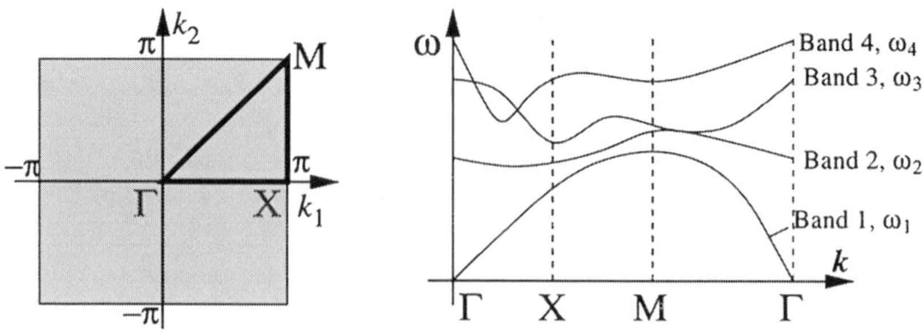

Figure 7.9. Left: The irreducible Brillouin zone indicating the wave vectors to be searched for the general 2D case (grey area). For square symmetry, the wave equation only has to be calculated for $\boldsymbol{k}$-vector values along the curve $\Gamma - X - M - \Gamma$. Right: Sketch of band structure indicating lowest four eigenvalues for wave vectors along the line $\Gamma - X - M - \Gamma$ in the irreducible Brillouin zone.

(23) only has to be solved for a number of wave vectors along the lines $\Gamma \to X$, $X \to M$ and $M \to \Gamma$ (see Fig. 7.9, left). Since $\tilde{\boldsymbol{K}}$ is a Hermitian matrix and $\boldsymbol{M}$ is real, symmetric and positive definite, the eigenvalues of (23) will all be real and positive.

If one solves for the first few eigenvalues of (23) for a number of $\boldsymbol{k}$ along $\Gamma - X - M - \Gamma$, the results can be plotted in a band diagram as sketched in Fig. 7.9, right. From the curves one may read the propagation modes for given frequencies.

Real band diagrams for out-of-plane polarized waves are shown in Fig. 7.10a and b for pure epoxy and duraluminum, respectively (data from [53]). It is seen that for these homogeneous materials there exist eigenmodes for any frequency, i.e. there are no band gaps. Figure 7.10c shows the band structure of duraluminum cylinders (radius equal to 30% of cell size) in an epoxy matrix. It is seen that there are ranges of frequencies with no corresponding eigenmodes. This means that no modes will propagate for these frequencies. There is a large band gap between the first and the second band (from 31 kHz to 43 kHz corresponding to a relative gap size of $\Delta f / f_0 = 0.32$) and a small gap ($\Delta f / f_0 = 0.11$) between the second and the third band. This means that no elastic waves with frequencies within the band gaps may propagate through the structure. The band gap zones are indicated with hatched regions in the diagram.

An obvious goal for the optimization of band gap materials is to maximize the relative band gap size. In this way the range of prohibited frequencies will be wider and more signals may be sent through a waveguide based on defects in the band gap material.

The design problem is a two material problem. We want to distribute two non-void phases in the design domain (base cell). For reasons that will become

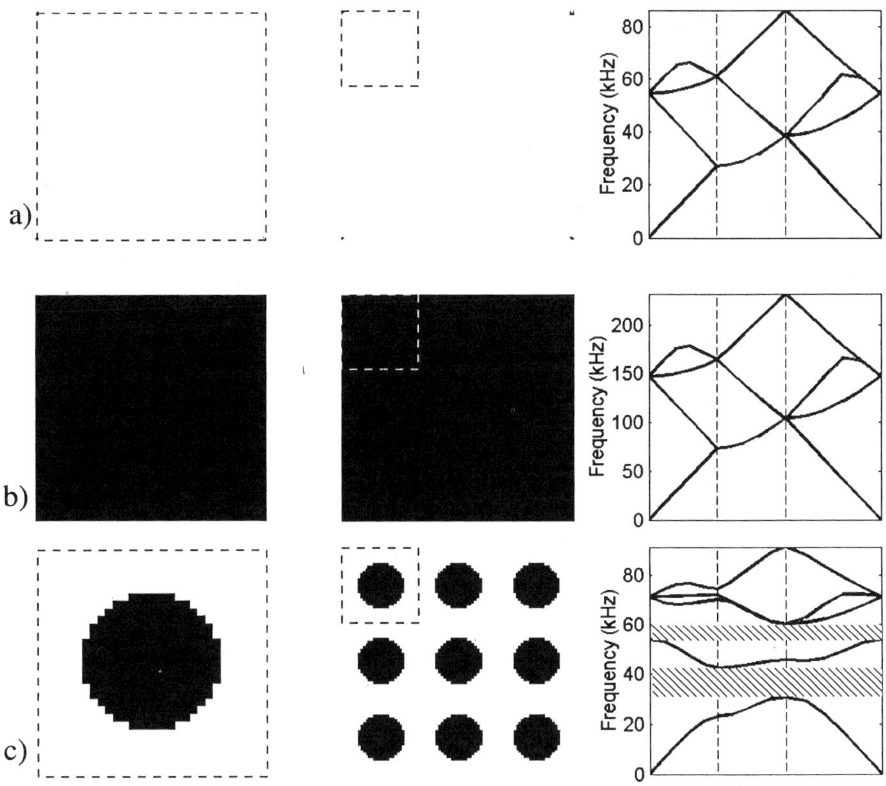

Figure 7.10. Left: single cell, middle: 3 by 3 arrays of cells and right: band structure of a) pure epoxy, b) pure duraluminum and c) duralumin cylinders (radius 30% of cell size) in epoxy. Hatched areas denote band gaps. The horizontal axes denote values of the wave vector $\boldsymbol{k}$ on the boundary of the irreducible Brillouin zone. The band diagrams are based on the solution of 15 eigenvalue problems with varying $\boldsymbol{k}$.

clear later, we here choose a linear material interpolation between the phases, i.e. $p = 1$ in (10).

The objective is to maximize the relative band gap size between to bands j and $j+1$, i.e. maximizing the lowest value of the overlying bands and minimizing the maximum value of the underlying bands. This can be written as a (double) max-min problem

$$\max_{\boldsymbol{\rho}} \left\{ c(\boldsymbol{\rho}) = \frac{\Delta\omega^2(\boldsymbol{\rho})}{\omega_0^2(\boldsymbol{\rho})} = 2\,\frac{\min\limits_{\boldsymbol{k}} \omega_{j+1}^2(\mathbf{k},\boldsymbol{\rho}) - \max\limits_{\boldsymbol{k}} \omega_j^2(\mathbf{k},\boldsymbol{\rho})}{\min\limits_{\boldsymbol{k}} \omega_{j+1}^2(\mathbf{k},\boldsymbol{\rho}) + \max\limits_{\boldsymbol{k}} \omega_j^2(\mathbf{k},\boldsymbol{\rho})} \right\}. \quad (24)$$

This is a "dirty" objective function in the sense that it is a max-min problem with varying critical points (the $\boldsymbol{k}$-vector(s) for the critical frequencies may change during the optimization) and it may have several multiple eigenvalues. Interestingly, however, there is no need for a volume fraction constraint in the problem since neither a pure phase one structure produces a band gap (Fig. 7.10a) and nor does a pure phase two structure (see e.g. Fig. 7.10b). Somewhere in between there must be a volume fraction that results in the biggest band gap. Another interesting observation is that due to the missing volume constraint, the usual power-law interpolation (c.f. (10)) becomes useless in ensuring black-and-white designs. However, this is not a big problem since by experience, the optimized designs tend to be mostly black and white anyway. Finally, the mesh-independent filtering technique (19) works badly due to the missing volume constraint. Therefore, the regularized penalty function method [54] is used to ensure black-and-white and mesh-independent designs for this design problem.

The optimization problem may now be written as

$$\begin{aligned}
&\max_{\boldsymbol{\rho}} \ \frac{\Delta\omega^2(\boldsymbol{\rho})}{\omega_0^2(\boldsymbol{\rho})} \\
&\text{s.t.}: \ \left[\boldsymbol{K}(\boldsymbol{k}) - \omega^2\boldsymbol{M}\right]\boldsymbol{u} = \boldsymbol{0}, \quad \boldsymbol{k} \in [\Gamma - X - M - \Gamma]\,, \\
&\qquad 0 \le \rho_e \le 1, \quad e = 1, \ldots, N\,,
\end{aligned} \tag{25}$$

and may be rewritten in the more convenient bound formulation

$$\begin{aligned}
&\max_{\boldsymbol{\rho}} \ = \beta \\
&\text{s.t.}: \ \left[\omega_{j+1}(\boldsymbol{k})\right]_m \ge \beta, \quad m = 1, \ldots, M\,, \\
&\qquad \left[\omega_j(\boldsymbol{k})\right]_m \le \beta \qquad m = 1, \ldots, M\,, \\
&\qquad \left[\boldsymbol{K}(\boldsymbol{k}) - \omega^2\boldsymbol{M}\right]\boldsymbol{u} = \boldsymbol{0}, \quad \boldsymbol{k} \in [\Gamma - X - M - \Gamma]\,, \\
&\qquad 0 \le \rho_e \le 1, \quad e = 1, \ldots, N,
\end{aligned} \tag{26}$$

where the two first constraints take the M most critical values into account. This problem may efficiently be solved using MMA [36].

Results from optimizing the epoxy/duraluminum structures from Fig. 7.10 are shown in Fig. 7.11. The first example maximizes the relative band gap size between the first and the second band. The result is an almost square inclusion of a duraluminum in the epoxy matrix. The relative band gap size has increased from 0.32 for the circular inclusion in Fig. 7.10c to 0.65 for the square inclusion structure in Fig. 7.11a. The second example maximizes the relative band gap size between the second and the third band. In this case, the resulting structure consists of diamond and circular inclusion of duraluminum inclusions in the epoxy matrix. The relative size of the second band gap has increased from 0.11 for the circular inclusion in Fig. 7.10c to 0.61 for the structure in Fig. 7.11b.

Although this section has proved that increased band gap sizes can be obtained by optimizing *periodic* structures, it is not given that periodic structures

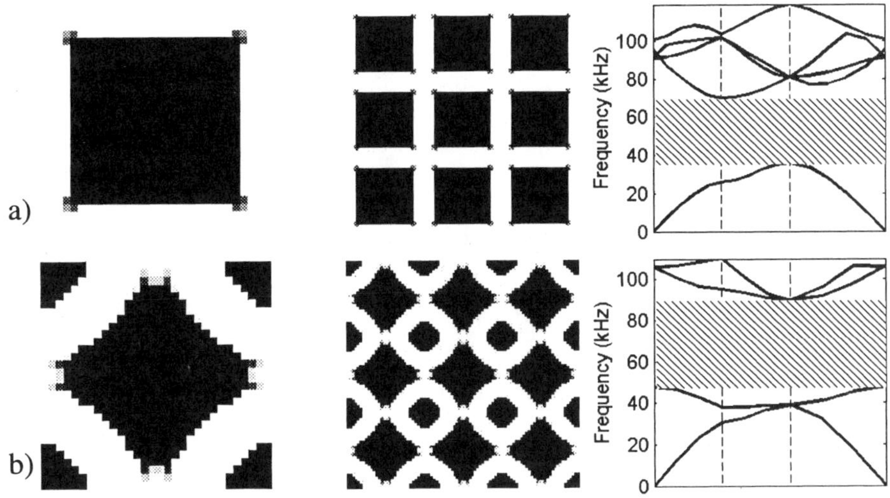

Figure 7.11. Maximization of relative band gap size between a) first and second band and b) second and third band.

with defects (Fig. 7.8) are the ideal wave guides. In fact, recent studies in [28, 29], indicate that non-periodic or semi-periodic structures are better for wave-guiding purposes.

There is a wealth of examples of photonic band gap materials and wave guides manufactured by microprocessing techniques [55, 56], however, there are very few papers apart from the above mentioned devoted to the systematic synthesis of the same.

6. Concluding remarks

In this chapter, we have demonstrated the principle of synthesis of PMMs by topology optimization. The examples were concentrated on linear elastic and thermoelastic structures. A multitude of extensions and application of the method can be thought of.

One possible direction is the inclusion of material and geometrical non-linearity. An example of the latter is negative Young's modulus PMMs [32]. Here, built-in prestressed members may cause buckling to happen for certain inputs, in turn reversing the output stress yielding locally negative Young's modulus behaviour. The proposed structures [32] are designed by intuitive means and may obviously be improved by systematic methods[4]. PMMs may

[4]Topology optimization studies for geometrically non-linear single mechanisms and snapping structures have recently been published [57–59].

also be designed to maximize energy absorption properties for crashworthiness improvement and the automotive industry. Another direction is multiphysics problems like magnetomechanical, optomechanical or other coupled problems.[5] An example could be strain sensitive fiber-optic sensors.

Acknowledgments

Parts of this work was sponsored by the projects "Systematic design of MEMS" and "Design band gap materials and structures with optimized dynamic properties" from the Danish Technical Research Council (STVF).

References

[1] G. F. Hawkins. Augmenting the mechanical properties of materials by embedding simple machines. *Journal of Advanced Materials - SAMPE*, 34(3):16–20, 2002.

[2] M. P. Bendsøe and N. Kikuchi. Generating optimal topologies in structural design using a homogenization method. *Computer Methods in Applied Mechanics and Engineering*, 71(2):197–224, 1988.

[3] M. P. Bendsøe and O. Sigmund. *Topology Optimization - Theory, Methods and Applications*. Springer Verlag, Berlin Heidelberg, Xiv+370 pp., 2003.

[4] O. Sigmund. *Design of material structures using topology optimization*. PhD thesis, Department of Solid Mechanics, Technical University of Denmark, December 1994.

[5] O. Sigmund. Materials with prescribed constitutive parameters: an inverse homogenization problem. *International Journal of Solids and Structures*, 31(17):2313–2329, 1994.

[6] O. Sigmund. Tailoring materials with prescribed elastic properties. *Mechanics of Materials*, 20:351–368, 1995.

[7] O. Sigmund. Some inverse problems in topology design of materials and mechanisms. In D. Bestle and W. Schielen, editors, *Symposium on optimization of mechanical systems*, pages 277–284, Netherlands, 1996. IUTAM, Kluwer.

[8] K. Terada and N. Kikuchi. Microstructural design of composites using the homogenization method and digital images. *Mat. Sci. Res. Int.*, 2:65–72, 1996.

[5]Topology optimization studies for coupled electrothermomechanical MEMS actuators have recently been published [60, 61].

[9] C. C. Swan and I. Kosaka. Voigt-Reuss topology optimization for structures with linear elastic material behaviours. *International Journal for Numerical Methods in Engineering*, 40(16):3033–3058, 1997.

[10] P. S. Theocaris and G. E. Stavroulakis. Multilevel optimal design of composite structures including materials with negative Poisson's ratio. *Structural Optimization*, 15(1):8–15, 1998.

[11] O. Sigmund. A new class of extremal composites. *Journal of the Mechanics and Physics of Solids*, 48(2):397–428, 2000.

[12] L. V. Gibiansky and O. Sigmund. Multiphase elastic composites with extremal bulk modulus. *Journal of the Mechanics and Physics of Solids*, 48(3):461–498, 2000.

[13] S. Hyun and S. Torquato. Designng composite microstructures with tageted properties. *Materials Research Society*, 16(1):280–285, 2000.

[14] M. M. Neves, H. C. Rodrigues, and J. M. Guedes. Optimal design of periodic linear elastic microstructures. *Computers & Structures*, 76:421–429, 2000.

[15] J. M. Guedes, H. C. Rodrigues, and M. P. Bendsøe. On the computational approximation of energy bounds for multiload cases with one level of microstructure. In G. D. Cheng, Y. Gu, S. Liu, and Y. Wang, editors, *WCSMO-4 - Proc. Second World Congress of Structural and Multidisciplinary Optimization*, Dalian, 2001. Liaoning Electronic Press. CD-rom, ISBN 7-900312-69-2.

[16] R. H. W. Hoppe and S. I. Petrova. VI large-scale computations for mechanical engineering problems - structural optimization of biomorphic microcellular ceramics by homogenization approach. *Lecture Notes in Computer Science*, 2179:353–360, 2001.

[17] J. Haslinger and J. Dvořák. Optimum composite material design. *RAIRO-Mathematical Modelling and Numerical Analysis-Modelisation*, 29(6):657–686, 1995.

[18] O. Sigmund and S. Torquato. Composites with extremal thermal expansion coefficients. *Applied Physics Letters*, 69(21):3203–3205, 1996.

[19] O. Sigmund and S. Torquato. Design of materials with extreme thermal expansion using a three-phase topology optimization method. *Journal of the Mechanics and Physics of Solids*, 45(6):1037–1067, 1997.

[20] D. Fujii, B.-C. Chen, and N. Kikuchi. Composite material design of two-dimensional structures using the homogenization design method. *International Journal for Numerical Methods in Engineering*, 50:2032–2051, 2001.

[21] E. C. N. Silva, J. S. O. Fonseca, and N. Kikuchi. Optimal design of piezoelectric microstructures. *Computational Mechanics*, 19(5):397–410, 1997.

[22] E. C. N. Silva, J. S. O. Fonseca, and N. Kikuchi. Optimal design of periodic piezocomposites. *Computer Methods in Applied Mechanics and Engineering*, 159(2):49–77, 1998.

[23] O. Sigmund and S. Torquato. Design of smart composite materials using topology optimization. *Smart Materials and Structures*, 8(3):365–379, 1999.

[24] C. C. Swan and J. S. Arora. Topology design of material layout in structured composites of high stiffness and strength. *Structural Optimization*, 13(1):45–59, 1997.

[25] Y. Yi, S. Park, and S.-K. Youn. Design of microstructures of viscoelastic composites for optimal damping characteristics. *International Journal of Solids and Structures*, 37:4791–4810, 2000.

[26] S. J. Cox and D. C. Dobson. Maximizing band gaps in two-dimensional photonic crystals. *SIAM Journal for Applied Mathematics*, 59(6):2108–2120, 1999.

[27] S. J. Cox and D. C. Dobson. Band structure optimization of two-dimensional photonic crystals in *h*-polarization. *Journal of Computational Physics*, 158(2):214–224, 2000.

[28] O. Sigmund and J. S. Jensen. Systematic design of phononic band gap materials and structures by topology optimization. *Philosophical Transactions of the Royal Society: Mathematical, Physical and Engineering Sciences*, page to appear, 2003.

[29] O. Sigmund and J. S. Jensen. Topology optimization of elastic band gap structures and waveguides. In H. A. Mang, F. G. Rammerstorfer, and J. Eberhardsteiner, editors, *Proceedings of the Fifth World Congress on Computational Mechanics*, page http://wccm.tuwien.ac.at. Vienna University of Technology, Austria, 2002.

[30] O. Sigmund. On the design of compliant mechanisms using topology optimization. *Mechanics of Structures and Machines*, 25(4):495–526, November 1997.

[31] S. J. Hollister, R. D. Maddox, and J. M. Taboas. Optimal design and fabrication of scaffolds to mimic tissue properties and satisfy biological constraints. *Biomaterials*, 23(20):4095–4103, 2002.

[32] R. S. Lakes and W. J. Drugan. Dramatically stiffer elastic composite materials due to a negative stiffness phase? *Journal of the Mechanics and Physics of Solids*, 50(5):979–1009, 2002.

[33] U. D. Larsen, O. Sigmund, and S. Bouwstra. Design and fabrication of compliant micromechanisms and structures with negative poisson's ratio. *IEEE Journal of Microelectromechanical Systems*, 6(2):99–106, 1997.

[34] A. Bensoussan, J.-L. Lions, and G. Papanicolaou. *Asymptotic analysis of periodic structures*. North-Holland Publ., 1978.

[35] R. M. Christensen. *Mechanics of Composite Materials*. Krieger, Amsterdam, 1991.

[36] K. Svanberg. The method of moving asymptotes - A new method for structural optimization. *International Journal for Numerical Methods in Engineering*, 24:359–373, 1987.

[37] O. Sigmund and J. Petersson. Numerical instabilities in topology optimization: A survey on procedures dealing with checkerboards, mesh-dependencies and local minima. *Structural Optimization*, 16(1):68–75, 1998.

[38] T. Borrvall. Topology optimization of elastic continua using restriction. *Archives of Computational Methods in Engineering*, 8(4):351–385, 2001.

[39] O. Sigmund. A 99 line topology optimization code written in MATLAB. *Structural and Multidisciplinary Optimization*, 21:120–127, 2001.

[40] A. G. Kolpakov. Determination of the average characteristics of elastic frameworks. *PMM Journal of applied Mathematics and Mechanics, U.S.S.R.*, 49:739–745, 1985.

[41] R. F. Almgren. An isotropic three-dimensional structure with Poisson's ratio = -1. *Journal of Elasticity*, 12:839–878, 1985.

[42] R. Lakes. Foam structures with negative Poisson's ratio. *Science*, 235:1038, 1987.

[43] G. W. Milton. Composite materials with Poisson's ratios close to -1. *Journal of the Mechanics and Physics of Solids*, 40(5):1105–1137, 1992.

[44] L. Rothemburg, A. A. Berlin, and J. Bathurst. Microstructures of isotropic materials with negative poisson's ratio. *Nature*, 354:470–472, 1991.

[45] R. A. Schapery. Thermal expansion coefficients of composite materials based on energy principles. *Journal of Composite Materials*, 2(3):380–404, 1968.

[46] B. W. Rosen and Z. Hashin. Effective thermal expansion and specific heat of composite materials. *International Journal of Engineering Science*, 8:157–173, 1970.

[47] L. V. Gibiansky and S. Torquato. Thermal expansion of isotropic multiphase composites and polycrystals. *Journal of the Mechanics and Physics of Solids*, 45(7):1223–1252, 1997.

[48] M. Avellaneda and P. J. Swart. Calculating the performance of 1-3 piezoelectric composites for hydrophone applications: An effective medium approach. *Journal of the Acoustical Society of America*, 103(3):1449–67, 1998.

[49] O. Sigmund, S. Torquato, and I. A. Aksay. On the design of 1-3 piezocomposites using topology optimization. *Journal of Materials Research*, 13(4):1038–1048, 1998.

[50] J. Mathews and R. Walker. *Mathematical Methods of Physics*. Addison-Wesley, Redwood City, Calif., 1964.

[51] C. Kittel. *Solid State Physics*. John Wiley & Sons, New York, 1986.

[52] L. Brillouin. *Wave propagation in Periodic Structures*. McGraw-Hill, New York, 1946.

[53] J. O. Vasseur, P. A. Deymier, G. Frantziskonis, G. Hong, B. Djafari-Rouhani, and L. Dobrzynski. Experimental evidence for the existence of absolute acoustic band gaps in two-dimensional periodic composite media. *Journal of Physics: Condensed Matter*, 10(27):6051–64, 1998.

[54] T. Borrvall and J. Petersson. Topology optimization using regularized intermediate density control. *Computer Methods in Applied Mechanics and Engineering*, 190:4911–4928, 2001.

[55] E. Yablonovitch, T. J. Gmitter, and K. M. Leung. Photonic band structures: The face-centered cubic case employing non-spherical atoms. *Physical Review Letters*, 67:2295, 1991.

[56] J. D. Joannopolis, R. D. Meade, and J. N. Winn. *Photonic Crystals: Molding the Flow of Light*. Princeton University Press, 1995.

[57] T. E. Bruns and O. Sigmund. Topology design of bistable mechanisms. In G. D. Cheng, Y. Gu, S. Liu, and Y. Wang, editors, *WCSMO-4 - Proc. Second World Congress of Structural and Multidisciplinary Optimization*, Dalian, 2001. Liaoning Electronic Press. CD-rom, ISBN 7-900312-69-2.

[58] T. E. Bruns, O. Sigmund, and D. A. Tortorelli. Numerical methods for the topology optimization of structures that exhibit snap-through. *International Journal for Numerical Methods in Engineering*, 55:1215–1237, 2002.

[59] T. E. Bruns. Topology design of thermomechanically actuated snap and bistable mechanisms. In H. A. Mang, F. G. Rammerstorfer, and J. Eberhardsteiner, editors, *Proceedings of the Fifth World Congress on Computational Mechanics*, page http://wccm.tuwien.ac.at. Vienna University of Technology, Austria, 2002.

[60] O. Sigmund. Design of multiphysics actuators using topology optimization - Part I: One-material structures. *Computer Methods in Applied Mechanics and Engineering*, 190(49-50):6577–6604, 2001.

[61] O. Sigmund. Design of multiphysics actuators using topology optimization - Part II: Two-material structures. *Computer Methods in Applied Mechanics and Engineering*, 190(49-50):6605–6627, 2001.

Chapter 8

PROCESS SYNTHESIS

Carlos Mastrangelo
University of Michigan and
Corning IntelliSense
mastrangch@corning.com

Edwin Carlen
Corning IntelliSense
carlenet@corning.com

Synopsis This chapter describes a systematic method for the automatic generation of fabrication processes of thin film devices. The method uses a partially ordered set (poset) representation of the device topology describing the order between its various components in the form of a directed acyclic graph. The component fabrication sequence is determined from the poset exensions. The component sequence is then expanded into a corresponding process flow. The graph-theoretic synthesis method is powerful enough to establish existence and multiplicity of flows thus creating a design space $\mathcal{D}$ suitable for optimization. The cardinality $||\mathcal{D}||$ for a device with N components is large with a worst case $||\mathcal{D}|| \leq (N-1)!$ yielding in general a combinatorial explosion of solutions. The number of solutions is controlled through *a priori* estimates of $||\mathcal{D}||$ and condensation of the device graph. Algorithms to construct process flows from schematics of thin film devices have been developed and implemented in the computer program MISTIC (Michigan Synthesis Tools for Integrated Circuits) which calculates specific process parameters using an internal database of process modules and materials. Currently MISTIC includes process modules for deposition, lithography, etching, ion implantation, coupled simultaneous diffusions, and reactive growth.

1. INTRODUCTION

In the brief history of IC manufacturing, design automation tools have played a major role in the rapid manufacturing of improved VLSI circuits and devices. Most of these tools have focused on ECAD (Electronic Computer Aided Design) for physical design of VLSI circuits, layout generation, simulation, testing, and verification. With the proliferation of VLSI ECAD tools in recent years, computer-aided design (CAD) of integrated circuits has become one of the most important design tools in the IC industry.

Like integrated circuit design, semiconductor device process development is a large and complex task that can benefit from design automation tools. Technology CAD (TCAD) is essential for the simulation of device performance, fabrication, and yield estimation. Although several specialized process simulation tools such as SUPREM [1], DEPICT [2], SAMPLE [3], SIMPL [4], and MEMCAD [5] are widely used, TCAD has not seen the equivalent parallel growth of ECAD. Most of this work has been highly focused on the accurate simulation of specific processes with very little work done in automated process flow synthesis. One of the reasons for this lag is the fact that until just a few years ago there were only a few commercial device processes available, namely CMOS and bipolar, which were developed by a combination of trial and error and extensive simulations.

Today there is an increasing need for application specific process design as opposed to process design problems which are focused on scaling of known structures. These application specific processes are essentially customized device processes that vary in structure for each different application [6]. Fabrication processes for EERAM, EPROM, DRAM, and MEMS devices are typical examples of these. This wide variety of devices leads to longer processes with increased complexity. It is now common to use process flows with 20 to 25 mask levels compared to the conventional 6-mask CMOS process. In longer device processes, trial and error methods are too time consuming and inefficient; hence a faster structured methodology is desirable.

The use of a structured representation has been suggested [6]. In this scheme, formal representations of wafer state and individual process steps are used to describe the effects of individual steps on the wafer thus providing a unifying framework for the specification of process flows. A system of this kind (SPEC) was implemented [7, 8] as part of the Microelectronics Manufacturing Science and Technology program [9] at Texas Instruments Incorporated. SPEC composes process flows from machine independent equipment, step, sequence, and wafer specification hierarchies. Research has also been performed in partitioning the task of process flow design into a number of abstraction levels [10]. Frameworks provide the basic glue connecting fundamentally different steps but the proper selection of steps and the overall assembly of the flow requires an expert operator. For more rapid process development, the operator must be eliminated and these two operations must be automated.

Recently [11, 12, 13], a unified method for the automatic generation of application specific process flows has been developed. This procedure, referred as process compilation, utilizes the basic device topology as its input and generates complete process flows as output. This methodology has been implemented in the MISTIC process compiler. In MISTIC, the step selection and flow assembly are determined from the nature of the device structure, the projection constraints of the lithographic process, and the processing capabilities of a particular laboratory.

2. STATE-SPACE REPRESENTATION

Our discussion begins with some useful definitions and a mathematical description for the planar microfabrication process. In planar processing, the starting material is a blank substrate. This substrate can be modified by a finite set of m operators, ϕ, such as lithography, etching, deposition, implantation, diffusion, etc. The operators are applied to a sample in a specific sequence, and the last operator yields the desired device structure. In this sense, the fabrication process is represented by the discrete system:

$$x_{i+1} = \phi_i(x_i) \tag{8.1}$$

where x_i are discrete states representing the current state of the sample, x_o the initial substrate, and x_n the complete device. This state abstraction is commonly used in process modeling frameworks [14]. We shall soon demonstrate that n is related to device complexity and the number of layers. For example, the fabrication process of a two-layer polysilicon surface micromachined MEMS device, shown in Fig. 8.1, consists of 12 operations performed in sequence with 11 intermediate states. Each intermediate cross-section can be considered as a point in a state space as shown in Fig. 8.2, and each operator ϕ_i represents a vector that connects subsequent states. In this state-space context, the synthesis problem consists of finding a finite path that connects the initial x_o and final x_n states. This path defines a sequence of operators with composition

$$x_n = \prod_{i=0}^{n} \phi_i(x_i) \equiv \mathcal{F}(x_o) \tag{8.2}$$

where $\mathcal{F}$ is the process flow characteristic to device x_n. Flow $\mathcal{F}$ consist of the sequence

$$\mathcal{F} = [\phi_o, \phi_1, ..., \phi_n] \tag{8.3}$$

operating on x_o. In this notation, brackets denote sequences and braces denote sets.

Searches for the path $\mathcal{F}$ may be accomplished in different ways. The number of possible paths however is large. If the operator set consists of m distinct operators that can be applied to each of the states, there are m^n possible paths of radius n originating from x_o that can be explored! For a typical device with $n = 10$ intermediate steps, and $m = 5$ operators there are $m^n = 5^{10} \approx 10^7$ possible paths. Thus random type searches are not practical.

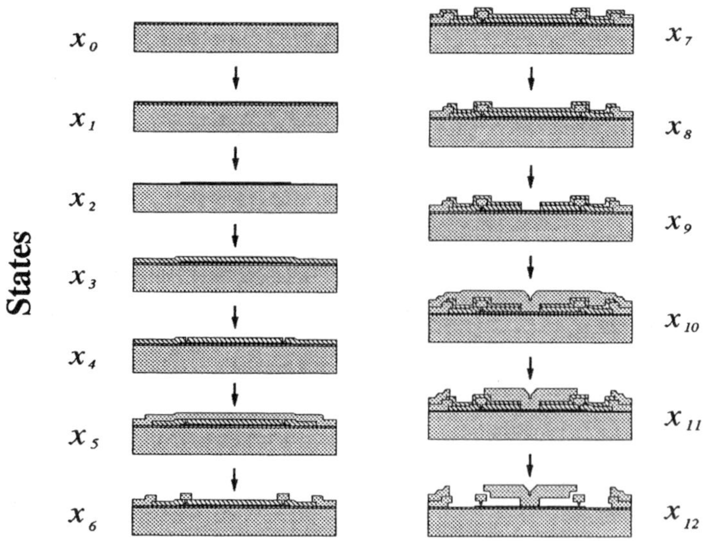

Figure 8.1. Cross-sections of a surface micromachined MEMS device at different steps in the fabrication process.

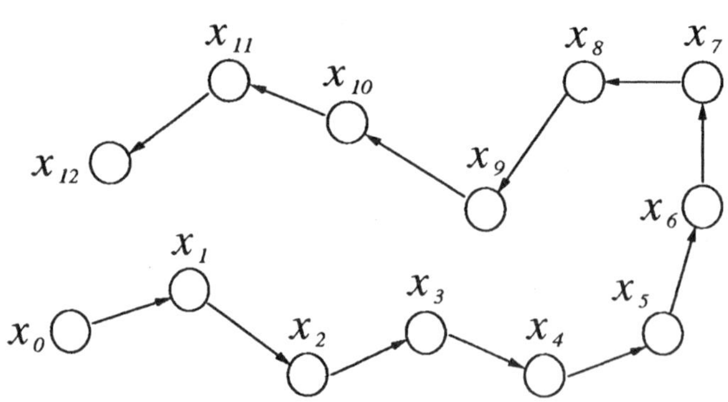

Figure 8.2. State-space abstraction of fabrication process.

The exponentially growing number of paths results in a combinatorial explosion which is a mixed blessing. The large path number leads in general to multiple solutions. The collection of feasible flows

$$\mathcal{D} = \{\mathcal{F}_1, \mathcal{F}_2, ..., \mathcal{F}_k\} \tag{8.4}$$

is defined as the design space $\mathcal{D}(x_n)$ of cardinality $||\mathcal{D}|| = k$. Since in general $||\mathcal{D}|| > 1$, the design space constitutes a basis for process flow optimization. However, a complete calculation of $\mathcal{D}$ may not be reached in finite time. The path is determined by extracting flow information from device topology, and imposing restrictions on $\mathcal{F}$ from approximate estimates of $\|\mathcal{D}\|$. In the next section a systematic procedure for efficiently finding a path $\mathcal{F}$ is described.

3. PLANAR DEVICE REPRESENTATION

The process flow $\mathcal{F}$ is intimately connected with the specific device topology; hence information leading to $\mathcal{F}$ is found by analyzing the device structure. We begin our discussion with some definitions relevant to the device itself. A device d consists of a set of components $\mathcal{C}(d)$ organized in a specific order $\mathcal{O}(c)$.

$$\begin{aligned} d &= \{\mathcal{C}, \mathcal{O}\} \\ \mathcal{C} &= \{c_o, c_1, ..., c_l\} \end{aligned} \tag{8.5}$$

Each c_i component represents a polygonal object in d of fixed attributes s_i such as material type, dimensions, and doping type. For instance, the drain, source, and gate are components of a MOSFET. Fig. 8.3 shows the components of a floating well diode.

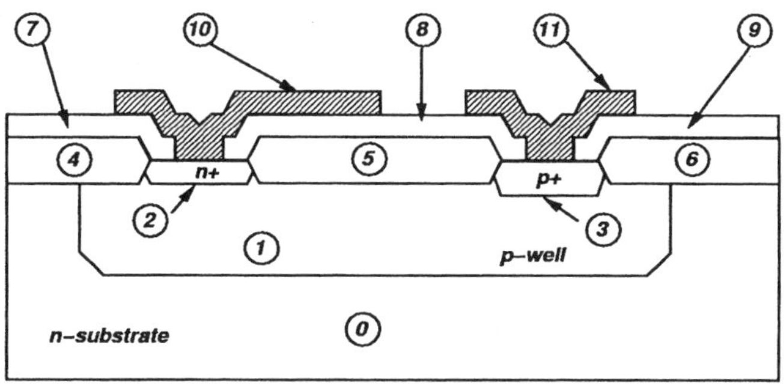

Figure 8.3. Example of floating well diode components.

Some of these components may belong to the same physical layer. For the moment we shall treat each of these as individual components. The order $\mathcal{O}(c)$ specifies the organization of these components in the device through binary relations as follows. Component c_i *rests above* c_j whenever c_i is in direct contact and covering or partially covering c_j. This relationship implicitly dictates that the fabrication of c_j precedes that of c_i, or in short notation $c_j \prec c_i$ or $c_i \succ c_j$. Similarly, if component c_i is diffused onto c_j then $c_j \prec c_i$. Additional precedence relations are established based upon operator compatibility. For example if c_i requires an operator which damages c_j then $c_j \prec c_i$. These precedence relations are irreflexive, ($c_i \not\prec c_i$) antisymmetric, and transitive (if $c_i \prec c_j$ and $c_j \prec c_k$ then $c_i \prec c_k$).

The device topology described by Eq. (8.5) can be represented as a directed graph where vertices describe its component and arcs their mutual order as shown in Fig. 8.4. An arc between c_i and c_j indicates $c_i \prec c_j$. This *digraph*

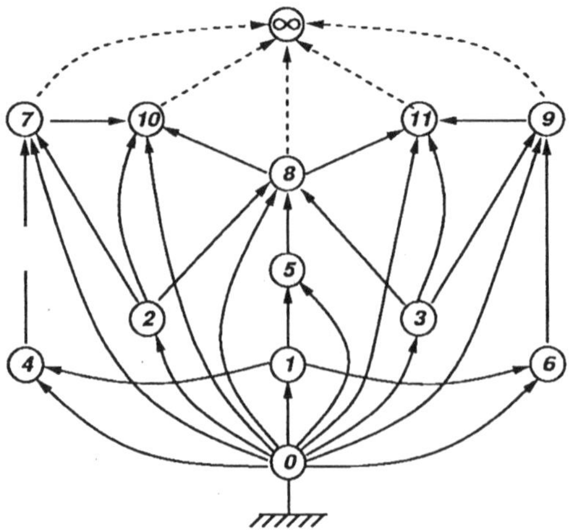

Figure 8.4. Digraph representation of diode component order.

representation is convenient since it eliminates all dimensional attributes of c_i.

The device digraph holds a number of interesting properties. In the graph, a vertex is a source or *root* if every other vertex in the graph can be reached from it. A *receiver* vertex is one that has outdegree zero and indegree > 0, and a *sink* is a vertex which can be reached by all of the rest. The device graph is rooted [15] since every element is in term connected to the source vertex representing the common substrate. The graph receiver vertices correspond to the device topmost components. Receiver vertices are in terms connected to a unique vacuum level sink c_∞ which is usually omitted.

In general, the device graph has no loops therefore is acyclic. Acyclic digraphs with a single root and sink are known as *st-graphs* [16] and play a central role in VLSI routing problems [17, 18]. While components c_i are organized in a plane, in general, the device graph is not planar if diffused layers are present because, in general, no order may be established between diffusions which share the same physical space without prior knowledge of diffusion conditions.

At this point we shall impose the additional restriction that the device digraph has no cycles. This is not a severe restriction since any n-vertex digraph with cycles can be expanded into an (n+1)-vertex acyclic digraph. Acyclic directed graphs are representations of partially ordered sets or *posets* [19, 20, 21, 22, 23, 24]. A poset is a combinatorial object consisting of a set of elements and a partial order between them. In the acyclic digraph, the arcs represent a partial order between the device components.

The poset is compactly stored in the digraph adjacency matrix $A(d)$. Matrix A is a square, Boolean (0,1)-matrix with $a_{i,j} = 1$ iff $c_i \succ c_j$ and zero elsewhere.

For example, the adjacency matrix for the structure shown in Fig. 8.3 is

$$A = \begin{bmatrix} 0 & 1 & 1 & 1 & 1 & 1 & 1 & 1 & 1 & 1 & 1 & 1 \\ 0 & 0 & 0 & 0 & 1 & 1 & 1 & 1 & 1 & 1 & 1 & 1 \\ 0 & 0 & 0 & 0 & 0 & 0 & 0 & 1 & 1 & 0 & 1 & 0 \\ 0 & 0 & 0 & 0 & 0 & 0 & 0 & 0 & 1 & 1 & 0 & 1 \\ 0 & 0 & 0 & 0 & 0 & 0 & 0 & 1 & 0 & 0 & 0 & 0 \\ 0 & 0 & 0 & 0 & 0 & 0 & 0 & 0 & 1 & 0 & 0 & 0 \\ 0 & 0 & 0 & 0 & 0 & 0 & 0 & 0 & 0 & 1 & 0 & 0 \\ 0 & 0 & 0 & 0 & 0 & 0 & 0 & 0 & 0 & 0 & 1 & 0 \\ 0 & 0 & 0 & 0 & 0 & 0 & 0 & 0 & 0 & 0 & 1 & 1 \\ 0 & 0 & 0 & 0 & 0 & 0 & 0 & 0 & 0 & 0 & 0 & 1 \\ 0 & 0 & 0 & 0 & 0 & 0 & 0 & 0 & 0 & 0 & 0 & 0 \\ 0 & 0 & 0 & 0 & 0 & 0 & 0 & 0 & 0 & 0 & 0 & 0 \end{bmatrix} \tag{8.6}$$

The component with zero column is the root, and all components with zero rows are receivers. Boolean matrices [25] are manipulated in much the same way as real matrices using Boolean AND, OR, and NOT operations, and they hold a number of useful properties [26, 27]. The *Hadamard element product*

$$C = A \times B \tag{8.7}$$

is defined such that $c_{i,j} = a_{i,j}\, b_{i,j}$. Boolean matrices are idempotent under the Hadamard product since $A \times A = A$. Hadamard products are useful to compare two device graphs [28]and in determining their process compatibility.

The specific form of A is dependent on the component and vertex labeling. Thus, the device matrix A is not unique since the component labeling is arbitrary. Two distinct matrices A and A' which represent the same device structure are *isomorphic*. Isomorphic matrices are related by the similarity transformation

$$A = P^T A' P \tag{8.8}$$

where P is a permutation matrix with a single one in each row and column and zeros elsewhere. In fact, for a device with N components, there are $N!$ permutation matrices and isomorphic adjacency matrix representations.

The binary order specified in A is transitive. The transitive closure matrix T specifies all transitive order relations.

$$T = \sum_{i=1}^{N-1} A^i \tag{8.9}$$

where the sum and power are Boolean AND and Boolean OR operations. Matrices A and T contain all order relations extracted from the device topology.

The graph-theoretic device representation provides a rigorous basis for the construction of a synthesis method as well as a means of comparison between different kinds of device topologies. We shall return to this topic when considering device integration problems.

4. BASIC SYNTHESIS METHOD

The flow path $\mathcal{F}$ joining x_o and x_n is constructed as follows. The operators in the flow $\mathcal{F}$ must form all k device components. The flow operators ϕ are thus intimately related to the device topology and the attributes of its components. Since each c_i is distinct we shall assume that there exists a specific operator group

$$\Phi(c_i) \;=\; [\phi_\alpha, ..., \phi_\beta](c_i) \tag{8.10}$$

required to form component c_i. The operator group $\Phi(c_i)$ is *selective* if it does not disturb the components existent in a state. Selective operator groups are the most useful since they decouple the operators required to form the device components. Using these definitions, we can establish a flow construction procedure where each component is constructed one at a time yielding the more restrictive state equation

$$x_{i+1} \;=\; x_i + c_{i+1} \tag{8.11}$$

Eq. (8.11) uniquely defines all intermediate states x_i and serves as a roadmap for finding the path. Since selective $\Phi(c)$ are known then from Eq. (8.1)

$$x_{i+1} \;=\; \Phi(c_{i+1}, x_i) \tag{8.12}$$

and the flow $\mathcal{F}$ is

$$\mathcal{F} \;=\; [\Phi(c_1), \Phi(c_2), ..., \Phi(c_n)] \tag{8.13}$$

From this discussion it follows that $||\mathcal{F}|| \geq k$. The flow $\mathcal{F}$ is hence an expansion of the component sequence $\mathcal{S}$

$$\mathcal{S} \;=\; [c_1, ..., c_k] \tag{8.14}$$

dictated by the order relationships $\mathcal{O}(c)$.

The component sequence $\mathcal{S}$ (and flow $\mathcal{F}$) is found by "stretching" the two-dimensional device graph into a linear chain or *linear extension*. Figure 5 shows a few linear extensions for the graph of Fig. 8.4. This procedure is known as *topological sorting* [29] and has many applications in computer science. The existence of $\mathcal{F}$ is thus assured by the existence of a linear extension in the device graph.

Theorem 1 (Fundamental) *Let d be a device of c_i components with order $\mathcal{O}$ represented by an acyclic directed graph. If there exists a selective operator group $\Phi(c)$ for each device component, there is at least one process flow for device d.*

Proof: Under the conditions stated above, the existence of a process flow is essentially assured by that of the linear extension of the digraph poset. The

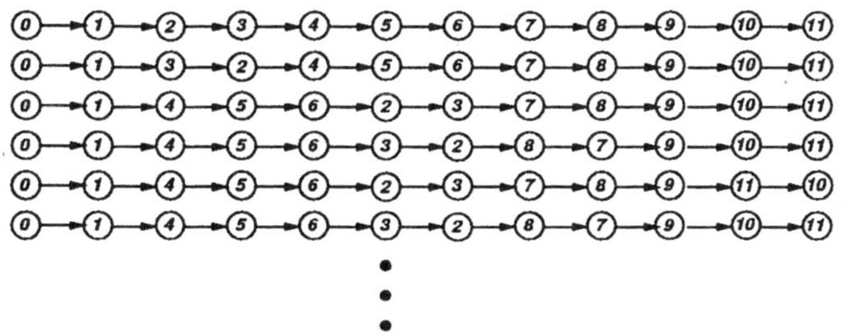

Figure 8.5. A few linear extensions.

existence of linear extensions was established by a well known theorem of Szpilrajn [30].

The fundamental theorem assures the existence of $\mathcal{F}$ in the topological sense. It does not assure the existence of selective Φ. We shall show later that good approximations for Φ can be reached. Furthermore the theorem can be slightly modified to accommodate operators which are not selective but distributed (such as thermal diffusion cycles).

Theorem 2 (Cardinality) *A device d of N components with r receivers and unique selective operator group Φ has a design space $\mathcal{D}$ of cardinality $r! \leq \|\mathcal{D}\| \leq (N-1)!$*

Proof: Since the receiver components have no mutual order, there are $r!$ possible linear chain arrangements. Szpilrajn's theorem assures the existence of a linear chain for the rest $N-r$ components. Thus there are are at least $r!$ linear chains of length N. The upper bound corresponds to a device with $N-1$ receivers and one root.

Theorem 2 states that devices with sparse transitive closure matrices are likely to have a very large number of flows $\mathcal{F}$. The multiplicity in $\mathcal{F}$ is beneficial since it allows the selection of an optimal flow from a large pool. However, in very complex devices, the factorial growth of $\|\mathcal{D}\|$ becomes intractable.

This combinatorial explosion is however avoided by carefully controlling the degrees of freedom in d through quick estimates of $\|\mathcal{D}\|$ and graph reduction techniques. These topics as well as specific methods for extracting the linear extensions are discussed in the sections below.

5. CARDINALITY OF DESIGN SPACE

A computationally inexpensive *a priori* measure of the design space cardinality $\|\mathcal{D}\|$ is an essential tool for the detection of a combinatorial explosion. This quantity is used to introduce reduction techniques and additional constraints if $\|\mathcal{D}\|$ is too large before the sequencing of the components is attempted.

A suitable measure of $||\mathcal{D}||$ is the number of linear extensions. The enumeration and generation of linear extensions has been studied extensively [31, 32, 33, 34, 35, 36]. For the general case, the generation of all linear extensions is an NP-complete problem [37]. It is well known [31] that the the number of linear extensions for a graph with N vertices is

$$n = ||\mathcal{D}|| = N! \operatorname{vol}(B(T)) \tag{8.15}$$

where $B(T)$ is an N-dimensional convex polytope imbedded within the unit hypercube, and vol() is the polytope volume. The polytope faces are hyperplanes defined by M constraints of T (ie. the number of ones in T), and $2N$ constraints from the hypercube faces. It has been shown that computing this volume is an NP-complete [36, 37] process. Therefore, counting the linear extensions exactly [38] is computationally equivalent to finding $\mathcal{D}$ through topological sorting. The volume however can be determined approximately at much smaller expense by several methods. The most common is the Monte Carlo technique [39, 40]. In graphs with a large number of nodes the volume of the polytope can be quite small; hence Monte Carlo techniques are computationally ineffective due to the high accuracy requirement at the lower range in vol().

A computationally efficient lower bound for vol() is found approximating the polytope volume by that of the largest hypersphere that will fit inside it [41]. Each of the $M + 2N$ bounding hyperplanes has the form

$$v_k^T q \leq b_k \tag{8.16}$$

where v_k is a unit vector pointing outward to hyperplane k, and b_k is a measure of the unit distance from the origin. The distance from an interior point q^* of the polytope to hyperplane k is

$$\gamma_k = v_k^T q^* - b_k \tag{8.17}$$

Therefore, the radius of the maximum hypersphere is found from the hyperplane constraints

$$v_k^T q^* + \rho \leq b_k \tag{8.18}$$

This is a standard linear programming problem solved in the worst case in $O(M^2)$ operations. The approximate $||\mathcal{D}||$ is then [42]

$$||\mathcal{D}|| \approx \frac{N!\, \pi^{N/2} \rho^N}{\Gamma(N/2 + 1)} \tag{8.19}$$

This bound is suitable for small M . The maximum M occurs when T is a total order isomorphic to an upper triangular matrix of ones [15] with $M = N^2/2$.

For large M, a less expensive upper estimate of $||\mathcal{D}||$ is found by relaxing some of the order constraints as follows. From the transitive closure matrix T, a *position restrictor matrix* R can be defined. The rows in R represent the device components, and its columns the possible positions (or bins) in the sequence S. Since there are as many positions as components, R is a square matrix. The elements $r_{i,j} = 1$ when c_i can occupy position j and zero elsewhere. For example in the structure shown in Fig. 8.6, component c_1 has three elements above it and one below; thus only $r_{1,2}$, and $r_{1,3}$ are ones. Component c_2 has two components above and one below thus $r_{2,1}$, $r_{2,2}$, and $r_{2,3}$ are ones, and so on. The number of components below c_i is the column count of T and the

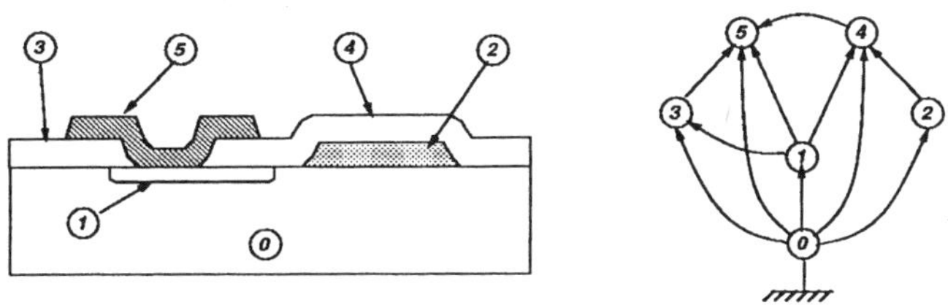

Figure 8.6. Position restrictor example.

number of components above c_i is its row count. The position restrictor matrix thus indicates the bins where a particular component may be placed as shown below.

$$T = \begin{bmatrix} 011111 \\ 000111 \\ 000011 \\ 000001 \\ 000001 \\ 000000 \end{bmatrix} \Longrightarrow R = \begin{bmatrix} 100000 \\ 011000 \\ 011100 \\ 001110 \\ 000110 \\ 000001 \end{bmatrix} \tag{8.20}$$

An estimate for $\|\mathcal{D}\|$ is found by counting the number of configurations of the N components under some conditions. No two components occupy the same position in the sequence (column), and each component is placed at a single position. Therefore in a valid configuration each row and column has one component resembling the placement of N non-attacking rooks in a restricted $N \times N$ chessboard. A few of these arrangements are shown below. Dashes represent forbidden bins, II components, and circles empty slots.

$$\begin{bmatrix} \text{II} & - & - & - & - & - \\ - & \text{II} & \bigcirc & - & - & - \\ - & \bigcirc & \bigcirc & \text{II} & - & - \\ - & - & \text{II} & \bigcirc & \bigcirc & - \\ - & - & - & \bigcirc & \text{II} & - \\ - & - & - & - & - & \text{II} \end{bmatrix}, \begin{bmatrix} \text{II} & - & - & - & - & - \\ - & \bigcirc & \text{II} & - & - & - \\ - & \text{II} & \bigcirc & \bigcirc & - & - \\ - & - & \bigcirc & \text{II} & \bigcirc & - \\ - & - & - & \bigcirc & \text{II} & - \\ - & - & - & - & - & \text{II} \end{bmatrix}, \begin{bmatrix} \text{II} & - & - & - & - & - \\ - & \bigcirc & \text{II} & - & - & - \\ - & \text{II} & \bigcirc & \bigcirc & - & - \\ - & - & \bigcirc & \bigcirc & \text{II} & - \\ - & - & - & \text{II} & \bigcirc & - \\ - & - & - & - & - & \text{II} \end{bmatrix}, \dots \tag{8.21}$$

The enumeration of these configurations is a classical problem in combinatorics quantified by the Nth coefficient of the rook polynomial $r_N(R)$ [43]. Thus

$$\|\mathcal{D}\| \leq r_N(R) \tag{8.22}$$

Since components are more constrained in T than in R, then r_N is an upper bound for $\|\mathcal{D}\|$.

The position restrictor matrix has an additional combinatorial interpretation. Components c_i belong to the device set d. Each bin position in sequence S is a group. Each configuration in R is a set of distinct $\{c_\alpha, ..., c_\beta\}$ where each c_i belongs to a particular group (column). Therefore each N-rook configuration is a *singular distinct representative* (S.D.R) of matrix R [44]. The number of S.D.R's and $r_N(R)$ in a Boolean matrix

$$||S.D.R|| = r_N(R) = \text{per}(R) \tag{8.23}$$

is equal to the *permanent*, per(R), of matrix R [45, 27, 46]. The permanent is a scalar quantity that often appears in many enumeration and combinatorial problems. For square matrices, the permanent is calculated as a 'plus-only' determinant [47]. The evaluation of per(R) requires in general $N!$ steps. Fortunately there exists an upper bound for per(). The quantity

$$||\mathcal{D}|| \leq r_N(r) = \text{per}(R) \leq \prod_{i=1}^{N} (b_i!)^{1/b_i} \tag{8.24}$$

is known as Bergman's bound [48] for the permanent, and b_i is the number of ones in row i of R. This bound is computationally inexpensive and can be calculated in N steps. Simpler estimates of per() are found as follows. First, the number of available bins in the first row, b_1, of T is counted. Then the first column and row are deleted from matrix T, and the procedure is repeated recursively on the reduced matrix. The permanent is approximately

$$\text{per}(R) \approx \prod_{i=1}^{N} b_i \tag{8.25}$$

and it is essentially a reduced factorial measure of the component combinations. This estimate reported in [49, 48] is known as the Lazyman's permanent. In practice the bound of Eq. (8.24) offers the most accuracy at the lowest computational cost.

6. GRANULARITY CONTROL THROUGH CONDENSATION

In practice, if $||\mathcal{D}|| > 10^4$, further simplifications are necessary to find $\mathcal{D}$ in a reasonable time. The cardinality $||\mathcal{D}||$ is a strongly increasing function of the number of vertices in the device graph. Therefore, substantial reductions are made by reducing the vertex count. This operation involves clustering of similar components under a single vertex now representing a layer. The resulting grouping operation in the graph is referred as a condensation. Components belonging to the same physical layer can be condensed into a group. The condensation is repeated recursively until either $||\mathcal{D}||$ is below an acceptable threshold or until no further condensation is possible. The scheme outlined above results in a drastic reduction in $||\mathcal{D}||$.

Theorem 3 (Condensation) *Let T and T^* be the transitive closure of a graph before and after the condensation of two components, then* $\|\mathcal{D}(T)\|/(N-1) \leq \|\mathcal{D}(T^*)\| \leq \|\mathcal{D}(T)\|/2$.

Proof: these bounds correspond to the two extreme cases of Fig. 8.7. In (a), the two components, c_1 and c_2, have identical lower and upper components. Therefore each linear extension has a subsequence of type $[..., c_1, c_2, ...]$ or $[..., c_2, c_1, ...]$. After the condensation, these two subsequences types reduce to one. In (b), one of the components has only order relations with the root. Therefore this component is placed in the linear extension at any of the $N-1$ positions above the root. After condensation each of these $N-1$ linear extensions reduces to just one.

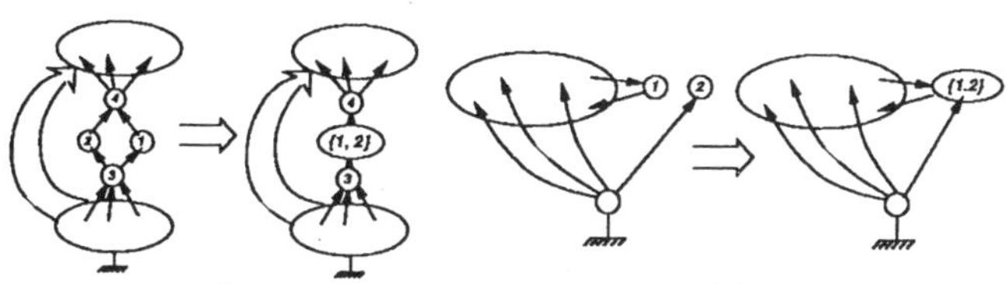

Figure 8.7. Condensation of components.

Each of the vertices in the condensed graph T^* represents a layer $l_j = \{c_1, ..., c_2\}$; hence the flow $\mathcal{F}$ is now specified in terms of a layer sequence

$$\mathcal{L} = [l_1, l_2, ..., l_m] \quad , \quad m \leq N \tag{8.26}$$

From the discussion it is evident that all components in a layer are made at the same time. Therefore only components which have identical attributes and hold no mutual order relationship can be condensed. This does not represent an obstacle since most devices have multiple components of the same material. For example in a MOSFET, both drain and source share the same attributes therefore are part of the same diffused layer. While straightforward, the condensation procedure can yield cyclic graphs if improperly used. This is illustrated in Fig. 8.8. In (b) and (c) c_1 and c_3 or c_2 and c_4 are condensed yielding graphs with 4 vertices. If the remaining components are condensed the resulting layer digraph in (d) contains a cycle. The cycle formation is eliminated however if the condensation is performed first between any two components that rest on a common component, and at each step, the transitive closure is recalculated. This sequential condensation procedure is somewhat restrictive as the condensed graph does not represent all possible mergings. This difficulty is avoided however if the entire collection of condensed graphs is used instead.

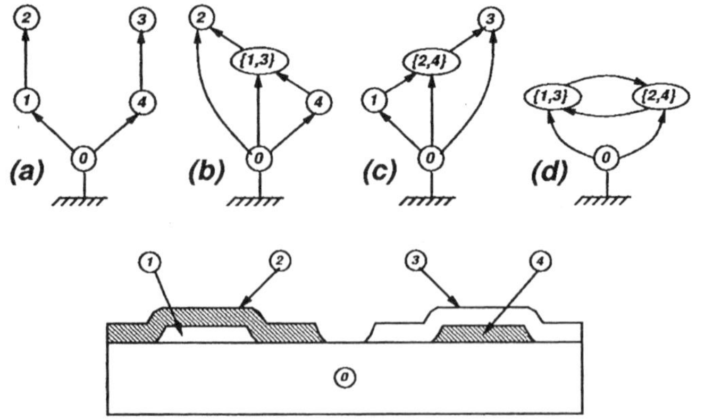

Figure 8.8. Condensation of components (a) uncondensed, (b), (c) acyclic condensed (d) cyclic condensation.

7. OTHER MISCELLANEOUS GRAPH-THEORETICAL RESULTS

The intrinsic connection between device structures and graph theory is very rich. Graph theory provides a wealth of theorems and corollaries directly applicable to planar processing and process design methodologies. In this section, a few of these results are discussed.

The graph theory framework allows the comparison of device structures in a relatively simple way. Two devices are compatible if they can be fabricated under a common flow $\mathcal{F}$.

Theorem 4 (Compatibility) *Given two devices d_o and $\|d_1\| < \|d_o\|$ with components $c(d_1) \in c(d_o)$ having identical labels and transitive closure matrices T_o and T_1, let T_o^* be the submatrix of T_o relating $c(d_1)$. Device d_1 is compatible with d_o iff the Hadamard product $T_o^* \times T_1^T = 0$.*

Proof: In order for d_1 to be compatible with d_o, the components of d_1 must be a subset of those in d_o. The order of these components specified in both T_o^* and T_1 must not be contradictory. This implies that if $(a_{ij})_o = 1$ then $(a_{ji})_1 \neq 1$, or alternatively $T_o^* \times T_1^T = 0$. Theorem 4 is also valid for devices which share a subset of their respective components.

In many occasions, the process flow for a particular device is fixed, and one is interested to know if a different device can be fabricated under the same process. This situation is most frequently encountered in integration of new devices with conventional processes such as CMOS, etc. Two devices are strictly compatible if they can be fabricated using a fixed flow $\mathcal{F}$.

Theorem 5 (Strict Compatibility) *Given two devices d_o and $\|d_1\| < \|d_o\|$ with components $c(d_1) \in c(d_o)$ having identical labels, let process flow $\mathcal{F}$*

specify a linear order for d_o *represented by transitive closure* $T_{\mathcal{F}}$. *Let* $T^*_{\mathcal{F}}$ *be the submatrix of* $T_{\mathcal{F}}$ *relating* $c(d_1)$ *according to the linear order. Device* d_1 *is strictly compatible with* d_o *and* $\mathcal{F}$ *iff* $T^*_{\mathcal{F}} + T_1 = T^*_{\mathcal{F}}$ *(or alternatively iff* $T^*_{\mathcal{F}} \times T_1^T = 0$*).*

Proof: Since flow $\mathcal{F}$ specifies a total linear order, the component labels can be rearranged such that $T_{\mathcal{F}}$ and $T^*_{\mathcal{F}}$ are full upper triangular. In order for d_1 to be strictly compatible with this total order, if $(a_{ij})_1 = 1$ then $(a_{ij})_{\mathcal{F}} = 1$. If this condition is violated, there exist additional order relations which are not included in $T^*_{\mathcal{F}}$ and d_1 cannot be fabricated using $\mathcal{F}$.

It is important to state the difference between the two theorems. Theorem 4 assures the existence of a common process which, in view of Theorem 1, is essentially a test for the presence of cycles in the merged $d_o + d_1$ graph. Theorem 5 imposes the same condition on the more dense (full upper triangular) $T_{\mathcal{F}}$ uniquely determined by $\mathcal{F}$; hence it includes Theorem 4 since $T_o \in T_{\mathcal{F}}$. The equivalence between the product and sum terms in Theorem 5 is easily demonstrated by taking the Hadamard product

$$\begin{aligned} T^*_{\mathcal{F}} \times (T^*_{\mathcal{F}} + T_1)^T &= T^*_{\mathcal{F}} \times (T^*_{\mathcal{F}})^T \\ T^*_{\mathcal{F}} \times T_1^T &= 0 \end{aligned} \tag{8.27}$$

which is exactly the condition of Theorem 4 with the dense matrix $T^*_{\mathcal{F}}$. Theorem 5 is useful to determine if a particular device can be imbedded onto another device process without changes.

A number of other interesting results stem from graph theory. Since devices are represented as graphs, then the number of different types of devices is related to the number of nonisomorphic, idempotent Boolean matrices. In general, the number of devices (or topologies) that can be fabricated with n layers of m different materials is approximately

$$\left(i\sqrt{2/\pi n}\right)\ m^n\, 2^{(n+1)^2/4+n} \tag{8.28}$$

Equation (8.28) is a modification of the formula obtained in [26, 50, 51, 52, 53, 54, 55] for unlabeled graphs. This number is very large but finite. Another interesting property of posets that may prove useful in microfabrication technology is their ability to be stored as a polynomial. It is well known that the order in labeled posets can be represented as polynomials [56, 57]. This fact may prove useful as a compact means for describing complex device topology.

8. PROCESS FLOW CONSTRUCTION

The first step in the construction of flow $\mathcal{F}$ consists of finding linear extensions of the condensed device graph. For finite graphs, the topological sort of its vertices can be performed in several ways. These sorting algorithms are based on (a) sequential generation of permutations [58, 29, 59, 60, 61, 62], (b) finding directed paths in acyclic graphs [63, 64, 65], and (c) triangulation of its adjacency matrix [66]. Efficient parallel computation algorithms [67, 17, 68, 69] have also been developed.

Since most device graphs contain a large number of vertices, it is necessary to use an algorithm which does not generate any redundant information. We have adopted a modification of a sequential algorithm due to Steiner [62]. This algorithm constructs the extensions using the ideals (feasible subsets) consistent with the poset order as shown in Fig. 8.9 for the graph of Figure i8.6. An initial

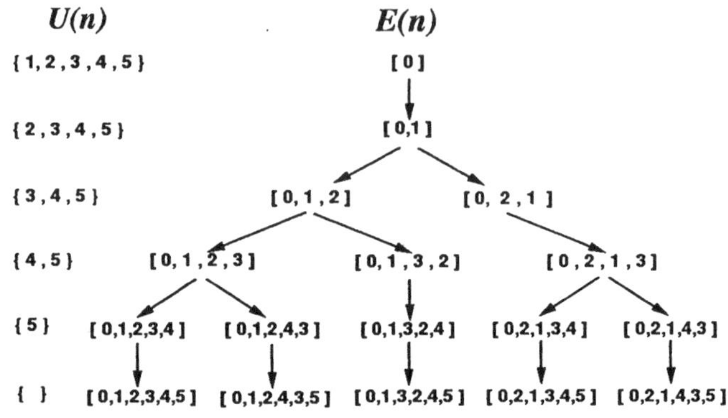

Figure 8.9. Steiner's topological sorting algorithm.

list of vertices $U(n)$ and an empty extension $E(0)$ are first formed. Any vertex from $U(n) \rightarrow U(n-1)$ is first moved to $E(0) \rightarrow E(1)$. In the next step, a second vertex from $U(n-1) \rightarrow U(n-2)$ is removed and added at all positions in the ideal consistent with the poset. This forms at most two distinct ideals $E_1(2)$ and $E_2(2)$. The procedure is repeated until $U = U(0)$ is empty yielding all linear extensions. Since all the ideals are distinct no repeated extensions are generated.

The layer sequences $[l_1, l_2, ..., l_N]$ are next expanded into flows using a selective operator set $\Phi(l_i)$ as in Eq. (8.12). In the discussion below we shall assume that these operators exist and defer the discussion on how to find these to a later section. The operator set Φ consist of a group of operators ϕ in a specific sequence. The structure of $\Phi(l_i)$ depends exclusively on the layer type. In the process construction, we identify three distinct classifications of layers with correspondingly different Φ. In our approach, deposited, diffused, and reactively grown layers are treated separately. The structure of Φ for two layers of the same category is the same; however the parameters of Φ are specific to the layer materials. The specific expansion procedure for each is outlined below.

a) Deposited Layers (Φ_D)*:* These layers are formed by the solidification of added components on top of the wafer surface. Examples of deposited layers are CVD, spin-casted, evaporated, and sputtered thin films. Components constructed from these layers can be formed as follows. (a) In the subtractive approach, a blanket layer is deposited (ϕ_D) onto the substrate. This layer is next patterned by a subsequent lithography (ϕ_L) and etch (ϕ_E). (b) In the additive approach, the lithography (ϕ_L) is performed first, the layer is next deposited (ϕ_D), and the resist is next removed along with the layer above it. The latter is known as a liftoff process (ϕ_F), and it is restricted to low-temperature deposition processes that do not damage the resist (such as metallizations).

b) Diffused Layers (Φ_F)*:* In most modern devices, diffused layers are formed by ion implantation. Under this process, a layer of resist is first patterned (ϕ_L), the dopants are next ion implanted (ϕ_I), and diffused by a drive-in anneal (ϕ_A). Doped layers can also be formed by first depositing and patterning a high-temperature mask (ϕ_D, ϕ_L, ϕ_E) followed by in-situ doping (ϕ_N) from the gas phase, drive-in (ϕ_A), and mask removal (ϕ_E). Diffused layers are subject to all temperature steps present in the process. Their final junction depths and surface concentrations are functions of the sample cumulative thermal history. The inherently interacting nature of thermal treatments however does not pose a difficulty if solved simultaneously to achieve the desired final diffusion characteristics. This topic will be discussed further in the parameter calculation section.

c) Reactively-Grown Layers (Φ_G)*:* These layers are formed by the chemical reaction between a reacting substrate material and added components. Silicon dioxide is the most common example of a reactively-grown film by the reaction of O_2 with solid silicon. These films grow on regions where the reacting substrate is exposed. Reactively-grown layers are formed by the following mechanisms. In the one approach, the material is first grown (ϕ_G) everywhere and selectively etched (ϕ_L, ϕ_E) in much the same way as deposited layers. Selective growth can be accomplished by blocking the surface with a high-temperature mask followed by its removal. The selective grow process is very important as it provides a basis for self alignment of patterns. Reactively grown layers are affected by thermal cycles in the presence of reacting gases. In general, the effect of growth operators is cumulative; hence growth operators are subject to the same considerations as diffused layers.

From the discussion above, the three type of layers are formed by the operator groups

$$
\begin{aligned}
\Phi_D(l_i) &= \begin{cases} [\phi_D, \phi_L, \phi_E] & \text{etching} \\ [\phi_L, \phi_D, \phi_F] & \text{lift-off} \end{cases} \\
\Phi_F(l_i) &= \begin{cases} [\phi_L, \phi_I, \phi_A] & \text{implant} \\ [\phi_D, \phi_L, \phi_E, \phi_N, \phi_A, \phi_E] & \text{in-situ doping} \end{cases} \\
\Phi_G(l_i) &= \begin{cases} [\phi_G, \phi_L, \phi_E] & \text{reactive growth} \\ [\phi_D, \phi_L, \phi_E, \phi_G, \phi_E] & \text{localized growth} \end{cases}
\end{aligned}
\qquad (8.29)
$$

therefore a maximum of 2^n processes are formed from each layer linear extension. This however rarely occurs as the second operator group from each category is used under very restricted conditions.

d) Implant-Through Layers: Dopants can be ion implanted through thin dielectric layers; therefore under proper conditions implants are not subject to visibility constraints. This special property of implants can be accounted by considering the transparency of device components. Dielectric layer l_i is transparent to an implant when its thickness is less than a threshold value $t(l_i) \leq d_t$. The set $C(l_j)$ of layers that immediately cover diffusion l_j is first found from its adjacency matrix. Let $W(C)$ represent the subset of roots in the subgraph containing vertices $l_i \in C$. If there is only one root and this vertex is transparent to the implant then the arc connecting l_j with it is converted to a soft-link indicated as a dashed arc in the device graph as shown in Fig. 8.10(b). The dashed line defines the subgraph of elements above the diffusion. In order to allow the deposition of these transparent layers to occur prior to the

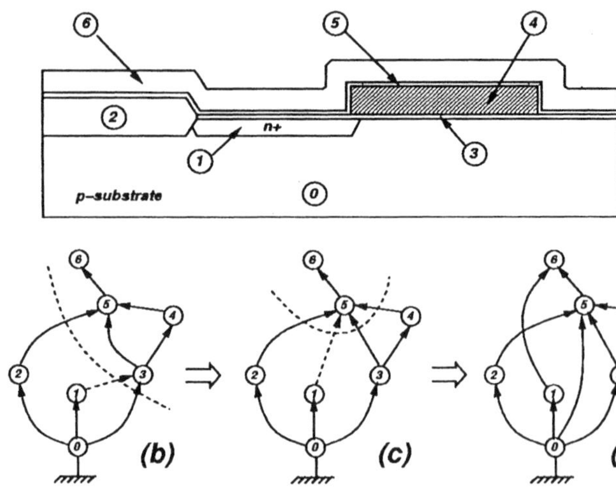

Figure 8.10. Transparency of thin layers to ion implants.

diffusion, the device graph is modified as follows. First, all incoming arcs in the diffusion vertex are extended to all vertices immediately above it. The soft link is removed, and arcs are added from the diffusion vertex to those above the transparent layer blocking the diffusion as in Fig. 8.10(c). This procedure breaks the mutual order between thin layers and diffusions while maintaining the rest of the component order intact. The procedure is extended to stacked transparent layers by keeping track of a reduced threshold $d_t = d_t - t(l_i)$ and repeating the procedure iteratively as in Fig. 8.10(d). If however, there are several roots in W, the graph modifications are repeated for each root vertex in C. In the flow $\mathcal{F}$, only processes with stacked transparent layers that fully cover the diffusion opening are generated to prevent the formation of a stepped implant.

e) Process Assembly: A process flow may now be assembled as a direct expansion of Eq. (8.12) using Eq. (8.29). This procedure is however overly restrictive since it does not include the possible self-alignment of device components or the use of multiple masks for the same layer. These two important methods substantially increase the process yield and reduce the overall number of steps. We shall show that these two processing methods are accounted through the permutation, elimination, and expansion of ϕ in $\mathcal{F}$.

The process flow $\mathcal{F}$ is assembled in three phases. First, the operators that define layers $(\phi_D, \phi_I, \phi_G, \phi_A)$ are inserted in $\mathcal{F}$. Thus

$$[l_1, l_2, ..., l_n] \Longrightarrow [\phi_D(l_1), \phi_D(l_2), ..., \phi_I(l_j), \phi_A(l_j), ..., \phi_G(l_n)] \tag{8.30}$$

Next, etching steps are inserted. The mutual order between the deposition operators and $\phi_E(l_i)$ is determined by its characteristic mask set $\mathcal{M}_o(l_i)$. The mask set consist of a collection of openings or *gaps* $g_{i,j}$ representing the areas where the layer is removed (for positive resists).

$$\mathcal{M}_o(l_i) = \{g_{i,1}, g_{i,2}, ..., g_{i,k}\}. \tag{8.31}$$

Self-aligned processes are obtained if gaps from different masks are permutated in $i\mathcal{F}$ followed by reconstruction of $\mathcal{M}$. The gaps and layers hold a specific order. Layers patterned by regular etching must be present before definition therefore $g_{i,k} \succ l_i$. Furthermore the segment of layer l_i corresponding to $g_{i,k}$ must be visible. The layer visibility is determined by examining the blocking graph of the layer sequence [70, 71]. If this gap is covered by components from layer l_j then $g_{i,k} \prec l_j$. Precedence constraints between $g_{i,k}$ from different mask sets are established when $l_j \succ l_i$ and $g_{i,k}(l_i) \notin g_{j,m}(l_j)$ implying that $g_{j,m} \succ g_{i,k}$. These order relations are compactly stored in the L_G and G_G matrices.

$$L_G = \begin{array}{c} \\ \\ l_1 \\ \vdots \\ l_n \end{array} \begin{array}{c} \overbrace{g_{1,1}\ g_{1,2} \cdots g_{1,k}}^{\mathcal{M}_o(l_1)} \quad \overbrace{g_{2,1}\ g_{2,2} \cdots g_{2,m}}^{\mathcal{M}_o(l_2)} \quad \overbrace{\cdots\cdots g_{h,j}}^{\mathcal{M}_o(l_n)} \\ \left[\begin{array}{ccc} \ddots \cdots\cdots & \cdots\cdots & \ddots \\ \vdots \cdots\cdots & \vdots \cdots\cdots & \ddots \\ \vdots \cdots\cdots & \vdots \cdots\cdots & \ddots \end{array}\right] \end{array} \tag{8.32}$$

$$G_G = \begin{array}{c} \\ g_{1,1} \\ \vdots \\ g_{m,j} \end{array} \begin{array}{c} g_{1,1}\ g_{1,2} \cdots g_{1,k} \quad g_{2,1}\ g_{2,2} \cdots g_{2,m} \quad \cdots\cdots g_{h,j} \\ \left[\begin{array}{ccc} \ddots \cdots\cdots & \cdots\cdots & \ddots \\ \vdots \cdots\cdots & \vdots \cdots\cdots & \ddots \\ \vdots \cdots\cdots & \vdots \cdots\cdots & \ddots \end{array}\right] \end{array} \tag{8.33}$$

Gaps $g_{i,k}$ are next inserted into $\mathcal{F}$ consistent with this order. Therefore different gaps for the same layer can be etched at different times. This is in fact required for a large number of device processes. Since the deposition operator order is fixed by the layer order, this procedure is equivalent to merging of two ordered sets [38].

The insertion of $g_{i,k}$ in $\mathcal{F}$ takes place according to the matrix order. The procedure uses the same topological sorting but now with the additional constraint that each $g_{i,k}$ must operate on the corresponding layer. This is determined by the visibility constrains specified in the G_G matrix. This condition is dynamically captured efficiently by a modification of the sequencing algorithm. The resulting $\mathcal{F}$ is a stream of ϕ_D, ϕ_I, ϕ_A, and g_i.

$$\begin{aligned}[]
[\phi_D(l_1), g_{1,1}, g_{1,2}, ..., \phi_D(l_2), g_{2,1}, ...g_{2,k}, g_{1,k}, g_{1,l}...,\\
+ \; g_{j,1}, g_{j,2}, ..., \phi_I(l_j), \phi_A(l_j), ...]
\end{aligned} \tag{8.34}$$

In Eq. (8.34), adjacent $g_{i,j}$ operating on the same layer are grouped into corresponding masks $\mathcal{M}_k(l_i) \in \mathcal{M}_o(l_i)$.

$$\begin{aligned}[]
[\phi_D(l_1), \mathcal{M}_1(l_1), ..., \phi_D(l_2), \mathcal{M}_1(l_2),\\
+ \; \mathcal{M}_2(l_1), ..., \mathcal{M}_1(l_j), \phi_I(l_j), \phi_A(l_j), ...]
\end{aligned} \tag{8.35}$$

If layer l_i is patterned by etching using $\mathcal{M}(l_i)$, the mask is replaced by a corresponding etching operator $\phi_E(\mathcal{M}(l_i))$.

$$\begin{aligned}[]
[\phi_D(l_1), \phi_E(\mathcal{M}_1(l_1)), ..., \phi_D(l_2),\\
+ \; \phi_E(\mathcal{M}_1(l_2)), \phi_E(\mathcal{M}_2(l_1)), ...,\\
+ \; \mathcal{M}_1(l_j), \phi_I(l_j), \phi_A(l_j), ...]
\end{aligned} \tag{8.36}$$

This procedure generates multiple etching steps and corresponding lithographic steps for a single layer. If adjacent etching operators operate on different layers of the same material; these in term are condensed into a single etch.

In the last phase of the assembly, the lithography operators ϕ_L are added to $\mathcal{F}$. At every state in $\mathcal{F}$, the device outline is determined. If all gaps $g_{i,j} \in \mathcal{M}_k(l_i)$ conform to the exposed components it may be possible to eliminate the lithography step completely yielding a self-alignment. The resulting $\mathcal{F}$ consists of a stream of operators specifying the various processes that the wafer must undergo to fabricate the device. Each of these processes contains a set of parameters which must be calculated.

9. DETERMINATION OF SELECTIVE OPERATORS

The idealization of a universally selective $\Phi(l_i)$ is rarely achievable in practice. Most processes used in VLSI technology, for IC's or MEMS, will interact with other layers to certain extent. In general, there exist two types of interactions. Thermal interactions occur when ϕ exposes the sample to high temperatures. For deposited layers, these interactions are destructive or non-destructive.

Destructive interactions are *a priori* detected and eliminated through appropriate changes in the device transitive closure matrix. Non destructive thermal interactions however are easily accounted by considering the cumulative thermal effect of all operators in $\mathcal{F}$. Thermal interactions are inherently distributed phenomena of particular relevance to diffused layers. A detailed discussion of distributed thermal interactions is found in the flow parameter section.

Chemical interactions occur in general for all ϕ. For deposition and reactive growth operators, these interactions are either negligible, well known, or unknown, Unknown interactions invalidate a flow. In feasible flows, layer materials are either immune or reacting through a known set of rate equations. For example, the chemical interaction of reactive grow operators is well known while lithographic operators have nearly none. The most significant chemical interactions in a process occur during etching steps. Improper choice of etchants can lead to process failure; therefore proper etchant selection is essential. Etching operators can be selective to some materials but highly interacting with others. For example, many plasma etchants used for silicon nitride etch do not attack oxide layers but attack polycrystalline silicon aggressively. On the other hand hot phosphoric acid does not attack oxide or silicon but strips resist. To compound these difficulties the response of many etchants on materials is often unknown. Therefore the task of the compiler is to construct selective ϕ from an imperfect and incomplete data set.

The precarious knowledge of ϕ poses severe restrictions. The first step toward finding selective Φ's is restricting the possible interactions. The operator set $\Phi(l_i)$ forms all components $c_i \in l_i \in x_i$ without disturbing $l_j \prec l_i, l_j \in x_{i-1}$. This causality on l_i relaxes the selectivity restrictions since $\Phi_i = \Phi(l_i, x_{i-1})$. Chemical interactions are further restricted to components which are exposed to the *wafer surface*. This component subset of x_i form the outline set $\Gamma_i \subset x_i$ hence $\Phi_i = \Phi(l_i, \Gamma_{i-1})$. The number of components in Γ_i is in general small because most device components are separated by uniform passivation dielectric layers. While a good choice of layer order minimizes interactions, these may not be negligible. In etching operators, the etch selectivity quantifier

$$S(\phi_E(l_i), x) = \min_{l_j \in \Gamma_{i-1}} \left(\frac{R(\phi_E, l_j)}{R(\phi_E, l_i)} \right) \tag{8.37}$$

where $R(\phi_E, l_j)$ is the etch rate of ϕ_E on layer l_j, is an indication of the operator attack on the outline components. This parameter is related to device yield. In the process expansion, etching operators are selected such that the film is etched at a reasonable rate yet having a negligible attack on other exposed materials. A lower limit on $S \geq S_{min}$ is also imposed. If no etchant meets these conditions, the flow is discarded. Experimentally, a value of $S_{min} = 5$ results in good yields. A second consideration in ϕ_E selection is its etching time. Etchants with etching times fitting inside a window

$$t_{min} \leq t(\phi_E) \leq t_{max} \tag{8.38}$$

are accepted to assure a reasonable process time. In Eq. (8.38), t_{min} is in the order of a few seconds and t_{max} of a few hours. The operator ϕ_E that best fits this criteria is selected. Similar time limits exist for both deposition and reactive growth operators.

The above procedure chooses a reasonable approximation to a selective operator set as a function of the state x_i and its outline Γ_i. This fact implies that flow operators may change for different intermediate states x_i resulting from distinct layer orders. It is the large variety of these orders that in general allows us to find a reasonable Φ which matches the information known about operator interactions with x_i. Many of the linear extensions are hence discarded resulting in general in a small design space $\mathcal{D}$.

10. PROCESS FLOW PARAMETERS

Each of the flow operators $\phi(l_i) \in \mathcal{F}$ contains a set of parameters specific to the attributes of l_i. These parameters depend on the type of operator. For example, for a polysilicon deposition process, the SiH_4 flow, tube temperature, pressure, and deposition time are some of its parameters.

In view of the complexity of most operators, we have adopted a *recipe-based* approach. This simplification is necessary because many processes are laboratory and machine specific. Furthermore, the operator behavior under widely varying parameter ranges is in general unknown. Recipe parameters however are finely tuned for satisfactory performance through many experiments yielding predictable results. Most recipes are specific with few parameters that can be adjusted. Operator recipes are stored in a database containing general [72] as well as lab specific data.

a) Lithography: The lithography recipe consist of a dehydration cycle, resist application, soft bake, exposure and hard bake (for etching). The photoresist thickness is determined from the roughness of the surface topography. Good resist coverage is in generally found when

$$t_{resist} \geq \frac{1}{3} t_{step}(\Gamma) \tag{8.39}$$

where t_{step} is the maximum step on the surface profile at a particular time. Increasing the resist thickness improves its coverage but degrades the sharpness of the patterns. Therefore the minimum acceptable thickness is used unless the resist is severely attacked by an etchant. Appropriate spin-speeds, exposure and development times are calculated internally as a function of this thickness.

b) Deposition and Etch: Deposition and etch recipes characterize all operators in terms of linear deposition and etch rates. The actual times are proportional to the layer thickness and inversely proportional to their rates. The etchant selection is made according to the principles established above maximizing selectivity and fabrication yield.

c) Reactive Growth: The thickness of reactively-grown layers is affected by subsequent temperature steps. For example, in thermal oxide growth, the field oxide thickness is changed when the gate oxide is regrown and invariably at any time that the oxide surfaces are exposed to a high temperature oxidizing environment. For thermal oxidation, the thickness grow of the oxide layer from u_n to its final state u_{n+1} by a temperature cycle at temperature T_n and Δt_n duration is

$$h(u_{n+1}, u_n, T_n) = \frac{(u_{n+1}^2 - u_n^2)}{B(T_n)} + \frac{(u_{n+1} - u_n)}{(B/A)(T_n)} = \Delta t_n \tag{8.40}$$

where B and B/A are the quadratic and linear rate constants [73]. For other type of reactions, Eq. (8.40) has the general form

$$u_{n+1} = f(u_n, T_n, \Delta t_n) \tag{8.41}$$

Therefore interacting reactive growth steps are solved simultaneously to achieve all the desired thicknesses at the process end. Suppose the process contains k interacting reactive growth processes with k final thicknesses $u_{i,f}$. At any given time, u_i are easily calculated if Eq. (8.41) is rewritten in vector form as

$$\widehat{U}_n = \widehat{g}(\widehat{U}_{n+1}, T_n, \Delta t_n) \tag{8.42}$$

where

$$\widehat{U} = (u_1, u_2, ..., u_k)^T \tag{8.43}$$

Eq. (8.42) is solved by noting that the ith layer is only affected by subsequent $j \geq i$ steps. Therefore, by causality, the last reactive growth step is

$$(u_k)_k = f((u_k)_{k-1}, T_{k-1}, \Delta t_{k-1}) \tag{8.44}$$

since $(u_k)_k = u_{k,f}$ and $(u_k)_{k-1} = 0$ are both known, then Δt_{k-1} is solved from Eq. (8.40).

$$\Delta t_{k-1} = h(u_{k,f}, 0, T_{k-1}) \tag{8.45}$$

and the growth of layer k no longer needs to be considered. Since Δt_{k-1} is now known the thicknesses for all layers $i < k$ at the $k-1$ step are determined from Eq. (8.42)

$$\widehat{U}_{k-1} = \widehat{g}(\widehat{U}_k, T_{k-1}, \Delta t_{k-1}) \tag{8.46}$$

Equations (8.45) and (8.46) form the basis for a backward recursion where each Δt_i is found from Δt_{k-1} to Δt_o. This method of solution is very general and can be applied when only numerical forms of $h()$ exist.

The above solution procedure is first carried out on a fixed set of temperatures. If the Δt_i are too short or long, the temperature for these steps are increased or decreased correspondingly to fit Δt_i within an acceptable time window. If any Δt is negative, the process is invalidated.

d) Diffusion and Implantation: Doped layers are affected by all temperature cycles in the process; therefore, all diffusion and implantation parameters are calculated simultaneously. In the present implementation, diffused layers are accepted in terms of two parameters: junction depth x_j and surface (top) concentration N_o. The junction depth specifies the diffused region thickness. The parameters that must be calculated are the implant dose Q and energy E_I, diffusion temperature T and diffusion time t.

As an initial estimate, diffusions are approximated as Gaussians profiles, and implants are considered to be shallow. Therefore junction depths are computed from intersection between two Gaussians or a Gaussian and a constant background. The junction depth constraints thus form a set of equations

$$N_1\, e^{-x_{j1}^2/4(\beta_T)_1} = N_2\, e^{-x_{j1}^2/4(\beta_T)_2} \;, \quad N_3\, e^{-x_{j3}^2/4(\beta_T)_3} = N_4 \;, \;\ldots \tag{8.47}$$

These relations yield the final straggle, $(\beta_T)_i$, for each profile. The Gaussian straggle is the cumulative D product of all successive steps

$$(\beta_T)_i \;=\; \sum_{j\geq i} D_i(T_j)\, t_j \tag{8.48}$$

and $D_i(T_j)$ is the diffusion coefficient for the ith diffused layer [73]

$$D_i(T_j) \;\approx\; (D_i)_0 \cdot e^{-\frac{E_A}{kT_j}} \tag{8.49}$$

where we have assumed a simple Arrhenius form for D. The buildup of $(\beta_T)_i$ occurs in small increments distributed throughout the process as in Eq. (8.48).

In general, each diffusion is first formed by an implantation step followed by an intentional drive-in, and any other subsequent high-temperature steps. The diffusion schedule solver of the compiler manipulates the temperature, time of the drive-in cycles, and implant dose to achieve desired junction depth and surface concentration specifications. Eq. (8.48) is rewritten in terms of partial increments $(\beta_p)_i$ due to each drive-in.

$$(\beta_T)_i = C_i + \sum_{j=i}^{N} \alpha_{ij}(T_j)\cdot(\beta_p)_j \;, \quad (\beta_p)_i \;=\; D_i(T_i)t_i \tag{8.50}$$

Where N is the number of drive-in annealing steps, C_i accounts for all fixed thermal cycles following diffusion i, and α is a ratio of diffusion coefficients

$$\alpha_{ij} = \frac{D_i(T_j)}{D_j(T_j)} \tag{8.51}$$

The formulation of Eq. (8.50) is convenient since it removes the drive-in temperatures from the unknown. The above equation can be rewritten in the following matrix form:

$$\tilde{\alpha}\,[\widehat{T}]\,\widehat{\beta}_p + \widehat{C} = \widehat{\beta}_T \tag{8.52}$$

Eq. (8.52) requires an appropriate selection of $\widehat{T}$. Our implementation uses an initial choice of 1000°C for all steps. Since the ith diffusion is only affected by subsequent steps $j \geq i$, then matrix $\tilde{\alpha}$ is upper triangular, and is solved easily by back substitution. If any β_p is negative, the process is discarded. If all $\beta_p \geq 0$, corresponding drive-in times and implant doses are calculated from

$$\begin{aligned} t_i &= \frac{(\beta_p)_i}{D_i(T_i)} \\ Q_i &= \sqrt{\pi(\beta_p)_i}\, N_i \end{aligned} \tag{8.53}$$

While this procedure yields in general a feasible diffusion schedule, the diffusion times may not be adequate. In order to assure reasonable drive-in times, these must fit within an acceptance time window. If the corresponding times are too long or too short, T_i is increased or decreased and Eq. (8.52) is recalculated.

The Gaussian approximation gives good values for Q and t in low concentration diffusions when the material remains essentially intrinsic at the diffusion temperature. However, in high concentration diffused layers, the diffusion coefficient is a function of concentration and dopant migration is affected by local electric fields. In this regime, the concentration of species C_i obeys [74, 75]

$$\frac{\partial C_i}{\partial t} = \nabla \cdot D_i(\nabla C_i \,+\, Z_i f_i \nabla C_{net}) \tag{8.54}$$

where f_i is the electric field enhancement factor

$$f_i = \frac{0.5N_i}{[(0.5C_{net})^2 \,+\, n_i^2]^{0.5}} \tag{8.55}$$

and

$$C_{net} = -\sum_{i=1}^{n} Z_i\, N_i \tag{8.56}$$

In this regime, the Gaussian approximation is a very crude guess to the actual profile. This difficulty however is eliminated if diffusion profiles are solved numerically. Numerical solution of Eqs. (8.54)-(8.56) yields an expression

$$(\widehat{N}_i)_{n+1} = \widehat{g}((\widehat{N}_i)_n, Q_n, T_n, t_n) \tag{8.57}$$

subject to similar causality constraints as those present in the treatment of reactive growth. Diffusion k is primarily affected by subsequent drive-in cycles

$j \geq k$, and somewhat affected by electric fields induced by profiles of diffusions $j < k$. If a good guess of the profiles is available, most of the noncausal contribution to the motion of diffusion j will be accounted for. Therefore errors contributed by this approximation are in general minor perturbations which can be treated as second order corrections. The diffusion solver thus starts with a Gaussian based guess for the profiles and iteratively refines this guess until convergence is achieved.

In this scheme, Q and t in Eq. (8.57) are solved using a recursive backward loop with non-causal corrections in a forward loop as shown in Fig. 8.11. The entire procedure is repeated until appropriate convergence is achieved. In the

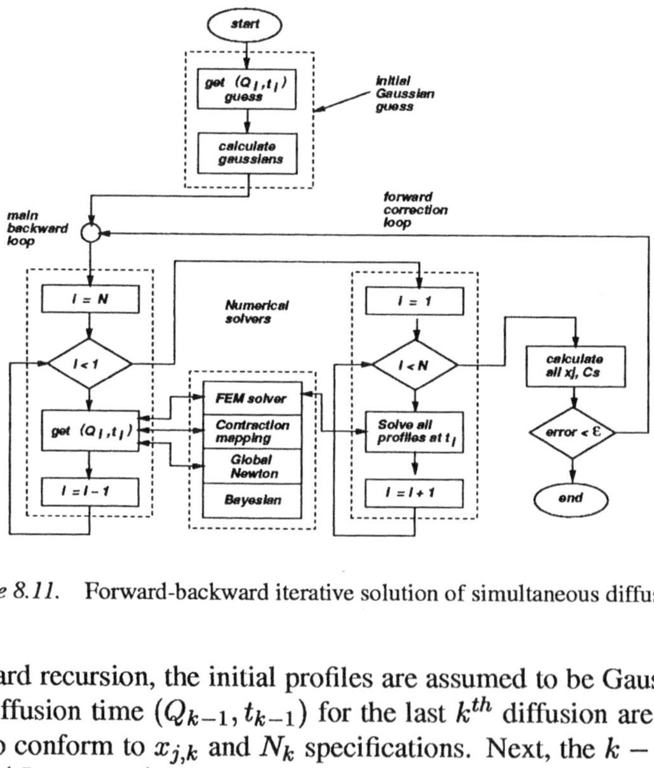

Figure 8.11. Forward-backward iterative solution of simultaneous diffusions.

first backward recursion, the initial profiles are assumed to be Gaussians. The dose and diffusion time (Q_{k-1}, t_{k-1}) for the last k^{th} diffusion are solved numerically to conform to $x_{j,k}$ and N_k specifications. Next, the $k-1$ diffusion parameters (Q_{k-2}, t_{k-2}) are solved with the updated (t_{k-1}, Q_{k-1}) and corresponding numerical profile. The backward recursion is continued until (Q_o, t_o) is found. An inherent error exists in these parameters due to the approximate nature of the initial guesses. In the forward loop, new numerical guesses are calculated using the updated (Q_j, t_j). These new guesses are used again in the backward loop to obtain second order corrections. The backward-forward recursion loop is repeated until errors in x_j and N_j in the forward loop are negligible. The main virtue of this scheme is that at any given time, parameters for a single diffusion layer are determined.

The calculation of (Q_j, t_j) for each diffusion drive-in step is the most computationally intensive of the procedure. Numerical solutions of Eq. (8.54) provide (N_j, x_j) in terms of (Q_j, t_j), but not vice versa. Therefore correct (Q_j, t_j) are arrived through iterative error minimization in $\epsilon_{x_j} = (x_j - x_{j,f})$, and $\epsilon_{N_j} = (N_j - N_f)$, where $(N_f, x_{j,f})$ are its desired values. Several iteration schemes for (Q_j, t_j) have been implemented. These include a globally convergent Newton method, a contraction mapping, and a Bayesian global optimization [76]. The most robust scheme is a globally convergent Newton method described in [77]. The main disadvantage of Newton-type codes is in the numerical evaluation of gradients which may confuse the solver ending the iteration in a local minima rather than a root. The fastest convergence is accomplished by the Gaussian-based iteration

$$\begin{aligned} t_{n+1} &= t_n \left(\frac{x_{j,f}}{x_j}\right)^2 \\ Q_{n+1} &= Q_n \left(\frac{N_f\, x_{j,f}}{N_j\, x_j}\right) \end{aligned} \tag{8.58}$$

Eq. (8.58) is in general a contraction which converges very quickly. In general, approximately 5-10 iterations are required for convergence in each diffusion. Once the first diffusion of the process is reached, the forward loop is initiated again with these new values. Typically 5-25 backward-forward loops are necessary for full convergence within 3% of specifications. Since the existence of solutions is not warranted, if both algorithms fail, the Bayesian global optimizer finds the best possible fit. The Bayesian scheme constructs a probability density of the cost function from each evaluation. This scheme is particularly efficient for the computationally expensive cost functions considered here.

e) Yield and Figure of Merit: The process yield is affected by both deterministic and random factors. Systematic errors reduce the process yield through known deterministic nonuniformities in the process operators causing certain areas of the chip to fail. Yield is also reduced by the presence of random distribution of point defects on the wafer as well as random variations in the process operators. Random effects on yield have been studied by numerous authors [78, 79, 80] for fine tuning of well known semiconductor processes. Since the compiler generates a large number of tentative processes with limited selectivities, we have adopted to estimate the deterministic yield instead.

The most important deterministic factor on yield is the loading effect. The loading effect results in nonuniformities in the radial distribution of deposition and etch rates caused by a balance between diffusion of fresh reactants and depletion of deposited or used species. The loading effect is primarily destructive during etching where it is manifested as a propagating front delineating areas of the wafer where the etch is not complete. The actual front propagation depends on the density of patterns on the wafer, but in most instances, the front propagates from the outer edge toward its center forming a noticeable 'bullseye' pattern.

If the reacting species diffuses from the wafer's outer edges, and if the depletion is uniform, in radial coordinates, the rate R is approximately determined by

$$r^2 \frac{d^2R}{dr^2} + r\frac{dR}{dr} - \left(\frac{r^2}{L_d^2}\right) R = 0 \tag{8.59}$$

with $R(r_w) \approx R_\infty = R_o$ where r_w is the wafer radius. In the above equation L_d is the characteristic depletion length

$$L_d = \sqrt{\frac{D\delta}{R_o}} \tag{8.60}$$

D is the species diffusivity, R_o the reaction velocity, and δ the thickness of a boundary layer above it. The solution of Eq. (8.59) is

$$R(r) = \frac{R_o I_o(r/L_d)}{I_o(r_w/L_d)} \tag{8.61}$$

where $I_o()$ is the zero order modified Bessel function of the first kind.

If a uniform film of thickness t_f is etched, the rate $R(r_w)$ is larger than that at the center $R(0)$ therefore the film is clearing from the outside of the wafer toward its center. On the cleared areas, the films beneath are being attacked by the etchant with a selectivity S. We can now define a cutoff radius r_c over which the fraction of overetched film is below a critical value α_c. Areas where the overetch fraction exceeds α_c are lost as shown in Fig. 8.12. Therefore, the

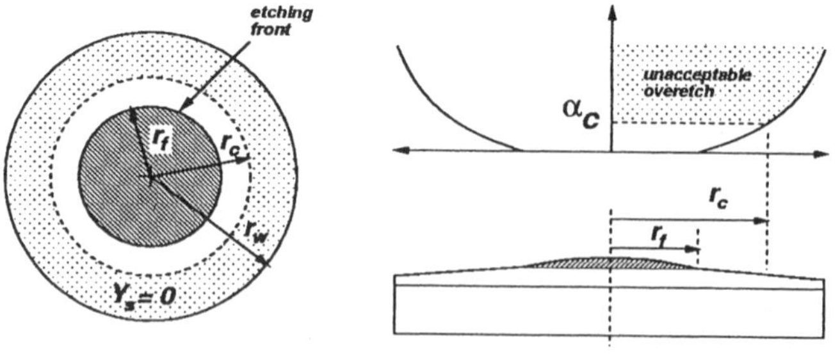

Figure 8.12. Loading effect decreases the yield.

nonuniformity in $R(r)$ is responsible for a systematic yield loss. In order to clear the film on the entire wafer, the areas near the outer edge will be overetched. The overetch time is a function of radius

$$T_{ov}(r) = \frac{t_f}{R(r)}\left(\frac{R(r)}{R(0)} - 1\right) \tag{8.62}$$

therefore the fraction of film beneath etched away is

$$\alpha = \frac{R_o}{S\,t_b} T_{ov}(r) \tag{8.63}$$

where t_b is the thickness of the film below. The cutoff radius corresponding to α_c is

$$r_c = \begin{cases} L_d I_o^{-1}\left[\frac{I_o(r_w/L_d)}{I_o(r_w/L_d) - aS}\right] & r_c \leq r_w \\ r_w & \text{otherwise} \end{cases} \tag{8.64}$$

where $a = (\alpha_c t_b / t_f)$. The systematic yield for the etching step is then

$$Y_s = \min_{l_j \in \Gamma} \left(\frac{r_c^2}{r_w^2} \right) \tag{8.65}$$

for all layers exposed and beneath the etched layer. Equation (8.64) can be used for all etching steps in the process in a sequence providing an overall systematic yield for the entire process. Under normal conditions, a small overetch time is always used to assure that the top film is completely gone. Therefore

$$T_{etch} = (1 + \epsilon) \left(\frac{t_f}{R(0)} \right) \tag{8.66}$$

where ϵ is a deliberate overetch fraction of the total etching time, typically set between 10 - 25 %. Since $I_o(x) \approx 1 + x^{2/4}$, then for each etching step

$$r_{n+1} = \begin{cases} \min \sqrt{\frac{4L_{d,n}^2 (4a_n S_n L_{d,n}^2 - \epsilon r_w^2)}{4(1-a_n S_n) L_{d,n}^2 - (1+2\epsilon) r_w^2}} \\ r_n \end{cases} \tag{8.67}$$

whichever of these is the lowest. While Y_s is not a true indication of the overall yield Y_t, it is an upper bound hence a suitable figure of merit (FOM). For a process with k etching steps

$$\text{FOM} = \frac{r_k^2}{r_w^2} \geq Y_t \tag{8.68}$$

The figure of merit indicates the precision over which patterns can be defined due to finite selectivities. The figure of merit has a maximum of one and can only decrease with low selectivity etching processes.

A second factor affecting the figure of merit is the number of alignments performed in the sample. Processes with self-aligned features are immune to registration errors; therefore their figure of merit is higher than that of processes where there is none. In this implementation, this is accounted as a gradual degradation of the FOM for every instance an alignment is required.

11. SOFTWARE IMPLEMENTATION

The structure of MISTIC is shown in Fig. 8.13. The compiler consists of four major software modules: (a) an input device builder, (b) a compilation core, (c) an output processor, and (d) a database of materials and laboratory

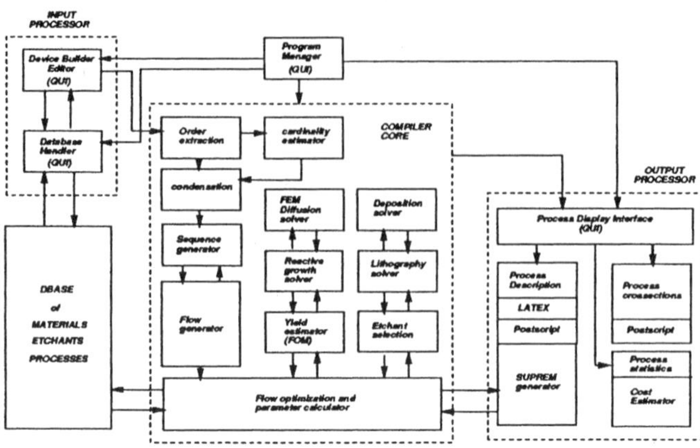

Figure 8.13. Internal organization of MISTIC.

processes. The four parts are supervised by a program manager which controls the flow of data among the various parts.

The input to the compiler core is generated by the device builder. This tool consists of a graphical user interface for drawing and editing device cross-sections using the materials stored in the database. In addition to regular graphical editing commands, the device builder has special drawing routines for quick calculation of conformal outlines and merging of common device macros (such as MOSFET's). After completion of editing, device structures are saved under a common format. The current implementation of the device builder accepts Manhattan geometry. Future work will include the incorporation of polygonal geometries.

The compilation core consists of several submodules. First, device cross section files are read and the component order is determined thus generating adjacency and transitive closure matrices. These matrices are examined by a cardinality estimator. If $||\mathcal{D}||$ is too high, the device components are recursively condensed until an acceptable limit is reached.

The condensed device is next sequenced using a low-memory budget modification of Steiner's algorithm [62]. For each of the generated sequences, the flow generator generates the L_G and G_G matrices of Eqs. (8.32)-(8.33) necessary for the flow expansion. The flow generator then inserts the gaps in the layer sequence consistent with these matrices. Typically more than one flow is generated for a given sequence. Gaps are condensed into masks, and lithographies are inserted when self alignment is not possible. The output of the flow generator is a flow with all the instructions necessary to fabricate the sample. The parameters of the flow are next determined.

The parameter calculator interacts with six other submodules. First specific etchants are selected for all etching steps. The device outline and the list of materials which are exposed during and after each etch is calculated. The etchant selection module scans the database for etchants that attack the film under consideration within a specified time window yet having the highest selectivities possible respect to the rest of the exposed materials on the outline and beneath it. Deposition parameters, lithography parameters, and figure of merit are next calculated from lab-dependent recipes stored in the database. The parameter calculator also interacts with the reactive growth solver that determines the reactive growth schedule. The diffusion doses, times, and temperature are determined by a one-dimensional finite element diffusion solver implementing the forward-backward iteration described above.

The finally assembled process is next graded by a process optimizer in terms of the figure of merit and process cost. A design space of a small, user-specified number of best processes is stored. Since the time required by the numerical diffusion solver can be substantial, it is customary to perform the optimization with the Gaussian guesses, and the numerical solver is used for refinement of the best processes. The compilation core generates a set of files containing design space statistics and each of its final processes. Process files contains a machine description of the flow interpreted by the output processor, and a text description of the final process with a complete set of instructions and associated parameters.

The database consists of an array of records of different types. These records include material entries describing their properties, recipes for deposition and growth conditions, and equipment specifications; wet and plasma etchants entries with corresponding recipes and selectivity data; diffusion coefficient entries; lithography and cleaning cycle information. The database is organized in two parts, a general library of processes, and a lab specific library. The general library consists mainly of wet etchant data [72] since depositions and plasma etches are very much machine and lab specific. The database contents are accessed by all the parts of MISTIC. These contents are included in the output files of the various modules. The database contents can be manipulated through a graphical interface for inspection, printing, editing, and porting of specific laboratory data.

The output interface extracts information from the compilation files and generates output in various formats. Its graphical interface displays device cross section at each step during the process as well as process statistics. In addition, the output interface generates SUPREM files for simulation of the process flow.

12. COMPILER TESTING

In order to determine the effectiveness of the compilation approach, the method was initially tested with the two devices shown in Fig. 8.14. First a

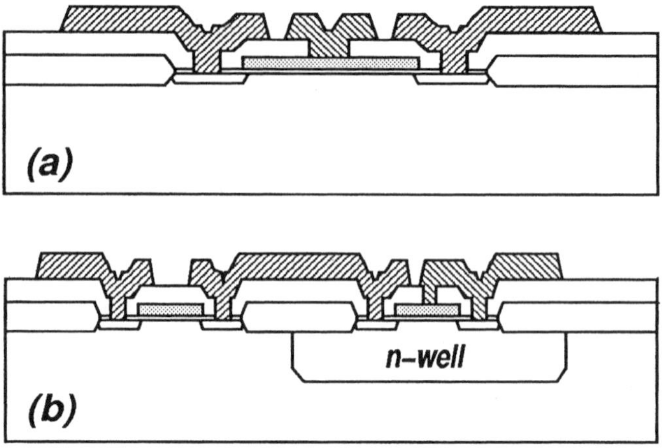

Figure 8.14. Test devices (a) NMOS, (b) CMOS.

16-component NMOS transistor was input to the compiler. The design space cardinality for this device was approximately 1.8×10^9 out of $15! \approx 10^{12}$ combinations. After condensing the gate oxide, field oxide, PSG passivation, and metal polygons onto corresponding layers, the vertex number was reduced to 7. The cardinality for the condensed graph was reduced to $\|\mathcal{D}\| \approx 6$. This dramatic decrease in $\|\mathcal{D}\|$ is typical of devices containing many components formed by the same layers. After adding an additional order for the reactive growths the cardinality was $\|\mathcal{D}\| \approx 2$. The arcs between the diffusions and the gate oxide were next eliminated according to the implant-through transparency rules. The device cardinality increased back to $\|\mathcal{D}\| \approx 6$. The compiler generated 12 process flows for this device. The best process requires 23 steps, 83 substeps, 4 lithographies, and 111 hours of process time. The worst process requires 34 steps, 7 lithographies, and 119 hours of process time. The best process included the localized oxidation of the field oxide, self alignment for the source and drain diffusions, and the same mask for the contact holes etching through the PSG passivation layer and gate oxide indicative of the mask reconstruction and elimination scheme implemented in the compiler.

The 29-component CMOS structure of Fig. 8.14(b) was next input to the compiler. The initial cardinality estimate was $\|\mathcal{D}\| \approx 2 \times 10^{26}$. After 22 condensations, the cardinality was reduced to $\|\mathcal{D}\| \approx 46$. The 9-vertex condensed graph is shown in Fig. 8.16. Additional orders were introduced between the reactive growth and the well and source diffusions yielding $\|\mathcal{D}\| \approx 3$. After

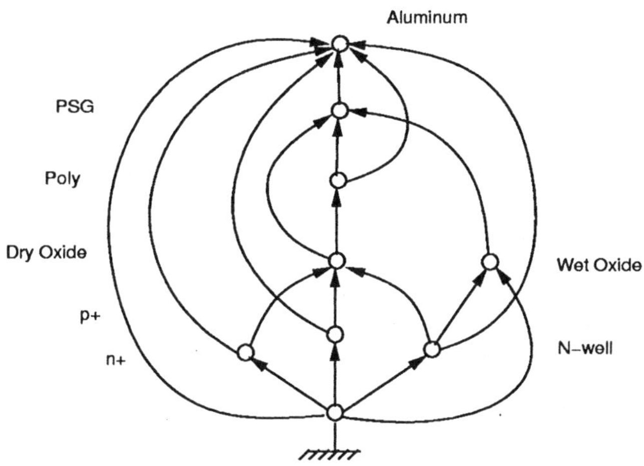

Figure 8.15. 8-vertex condensed CMOS digraph.

modification of the graph for implant-through layers, the cardinality increased to $\|\mathcal{D}\| \approx 29$. The compiler generated 18 complete CMOS processes.

The best process requires 36 steps, 85 substeps, 7 lithographies, and 278 hours of process time. The worst process requires 44 steps, 105 substeps, 9 lithographies, and 284 hours of process time. Due to the computational demands of the diffusion schedule solver, the initial processes for both devices are initially generated with the Gaussian guesses. Accurate dopant doses and diffusion times are solved only on the best process. Figure 8.17 shows comparison of doping profiles for the CMOS structure in both the NMOS and PMOS devices at the source calculated by the compiler FEM diffusion solver, and those from SUPREM-III using the MISTIC compiler specified doses, temperatures, and time. The diffusion solver was able to find the implant doses, drive-in temperatures and times to meet the specified junctions and surface concentrations for three simultaneous diffusions within 2% of specifications. There is a small difference in the diffusion profiles provided by MISTIC and SUPREM due to their slightly different diffusion coefficient models. These discrepancies are easily eliminated through matching of the internal models.

Figure 8.16 shows simultaneous diffusion profiles for the source/drain regions of both MOSFETs in good agreement with SUPREM-III simulations. The diffusion drive-in cycles ranged from 1175°C for the N-well for 4.5 hours to 1000°C for 2 hrs for the boron p+ source drain and 1000°C for the phosphorus n+ diffusion.

One of the most attractive virtues of the systematic compilation procedure is that its methods are independent of the number of components in the device; hence the compiler is capable of handling very complex structures. The

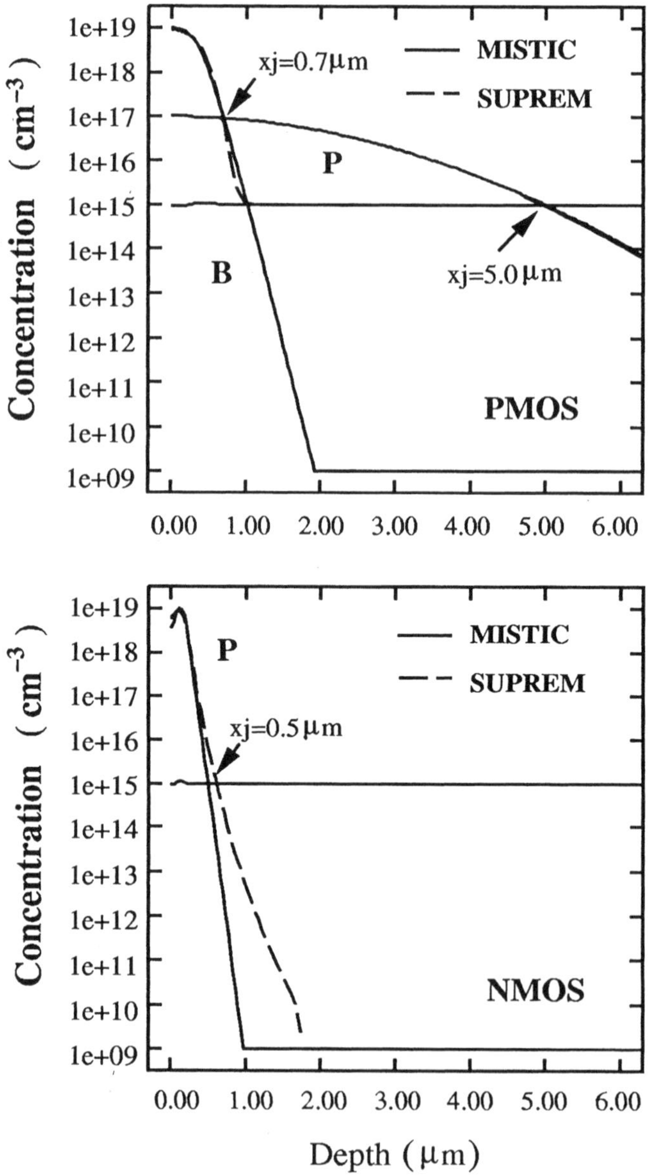

Figure 8.16. (a) PMOS well and source junctions, (b) NMOS source junction.

compiler was also tested with the double metal twin-well BiCMOS structure with its associated 64-vertex graph shown in Fig. 8.17. Despite its initial 135 order constraints, the cardinality for this structure is an immense number $\|\mathcal{D}\| \approx 5 \times 10^{73}$. After 44 condensations, the graph is reduced to the 20-vertex graph shown in Fig. 8.18. Despite its 47 order relations, the cardinality for this graph is approximately $\|\mathcal{D}\| \approx 2.6 \times 10^7$. Due to the very large number of processes, additional constraints are added between its three reactive growths reducing $\|\mathcal{D}\| \approx 7 \times 10^5$ and its eight diffusions $\|\mathcal{D}\| \approx 81$. The cardinality increases to $\|\mathcal{D}\| \approx 860$ after the graph modification for implant-through layers. It is found that for very large structures, the order between the diffusions must be included to complete the process assembly in a finite time. For example if there is no a-priori order set for its eight diffusions, the compiler must calculate each of its $8! = 40{,}320$ possible permutations. The a-priory order is easily established by assuring that for any two given diffusions, the β_p are positive. The vector C in Eq. (8.52) can only make β_p decrease thus further restricting the diffusion orders. After proper selection of etch steps, the compiler gener-

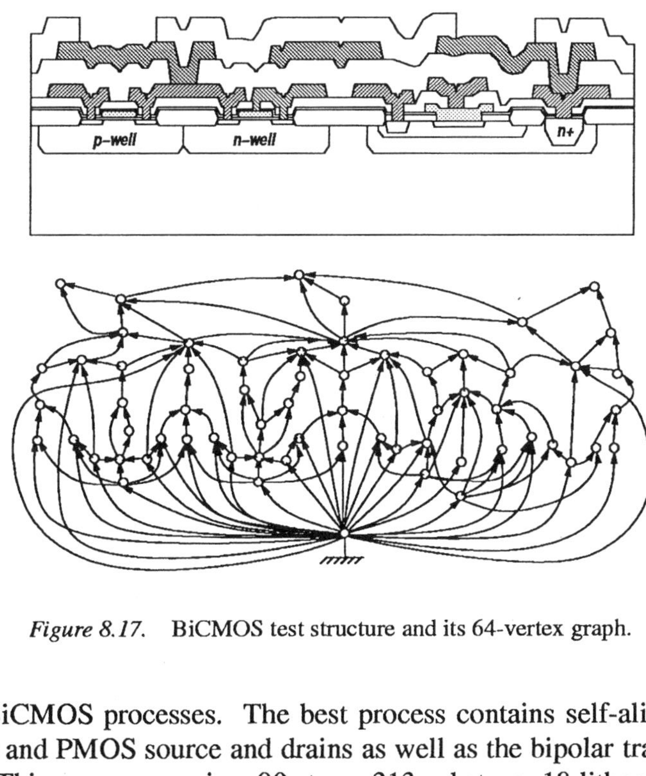

Figure 8.17. BiCMOS test structure and its 64-vertex graph.

ated 168 BiCMOS processes. The best process contains self-alignments for the NMOS and PMOS source and drains as well as the bipolar transistor base diffusion. This process requires 90 steps, 213 substeps, 18 lithographies, and 723 hours of process time. Most of this time is spend in the transit required by its 8 out of lab implants. The worst process requires 106 steps, 253 substeps, 22 lithographies, and 747 hours of process time. The process compiler predicts the correct order of first well, base, collector contact, base contact, emitter, and source diffusions as well as the correct order of reactive growths with first the passivation wet oxide, next the bipolar intermediate oxide, and last the gate oxide growth.

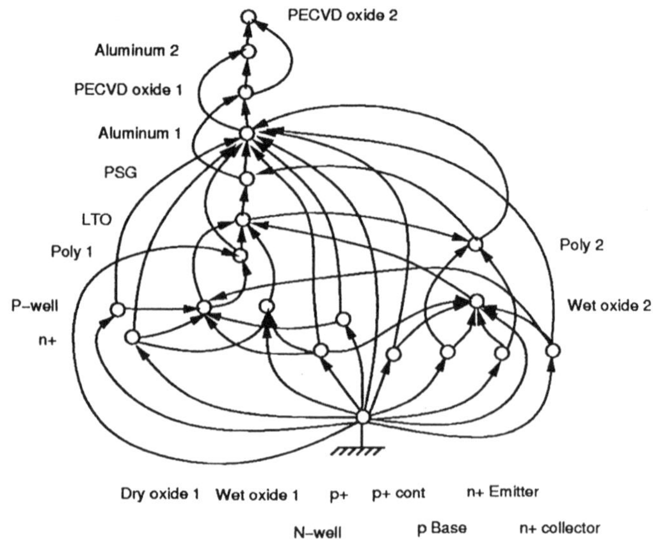

Figure 8.18. 20-vertex condensed BiCMOS graph.

Device	Components	Processes	S/A	Masks	Total steps
NMOS	16	12	1	4	83
CMOS	26	18	2	7	85
BiCMOS	64	168	3	18	213

The effectiveness of the diffusion schedule solver was tested for the BiCMOS device shown above. The backward-forward diffusion solver converged to the correct specifications within 3% after 5 loops. The resulting profiles of the source/drain regions of the NMOS and PMOS transistors with their corresponding wells are shown in Fig. 8.19. The net concentration of the dopants in the bipolar transistor is shown in Fig. 8.20. Note the good agreement of MISTIC with SUPREM-III simulations. The least important diffusions for the base and collector contacts also agree within target specifications but are not shown here. Drive-in temperatures ranged from 1175 °C for the 3-μm boron well to 950 °C for the boron source of the PMOS transistor. The process statistics for the NMOS, CMOS, and BiCMOS structures are summarized in Table 1. Since the Gaussian guesses are used to determine an order between the various diffusions, in some instances, the diffusion solver times are not as short as they could be. A more precise and fast approximation for the diffusions is

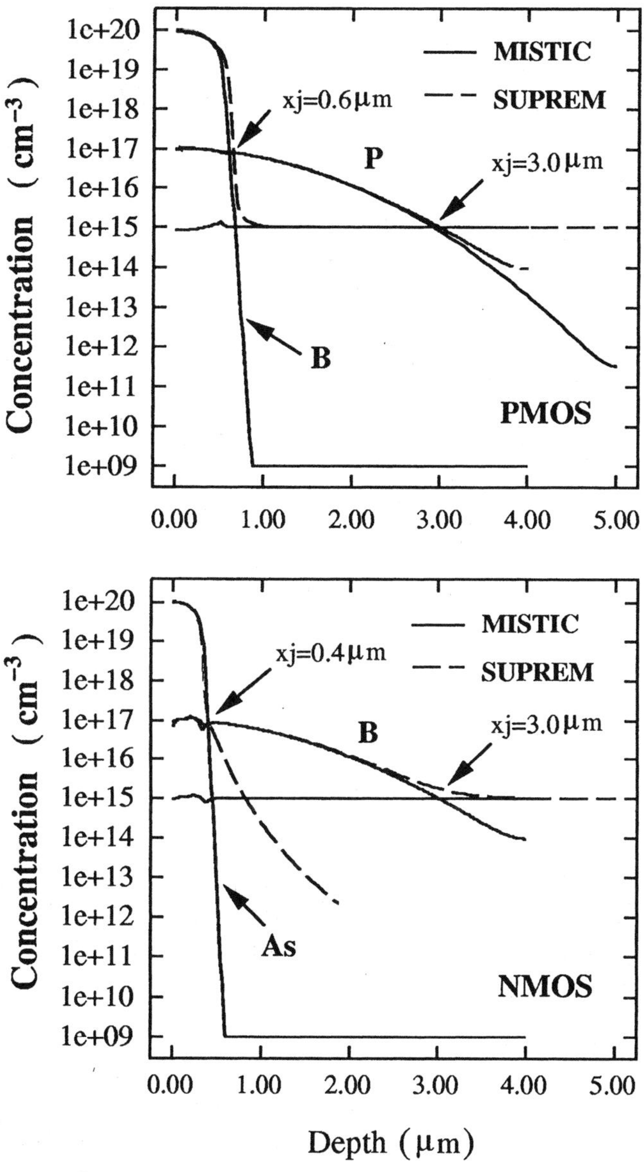

Figure 8.19. BiCMOS PMOS and NMOS source and well areas.

required. For optimal performance, the order can be *a priori* established using the nonlinear diffusion solver at a moderate computational expense.

The rigorous theoretical basis of the compiler provides a sound framework for the addition of numerous improvements. The current compiler version lacks some of the features available in common simulation tools. In particular, the

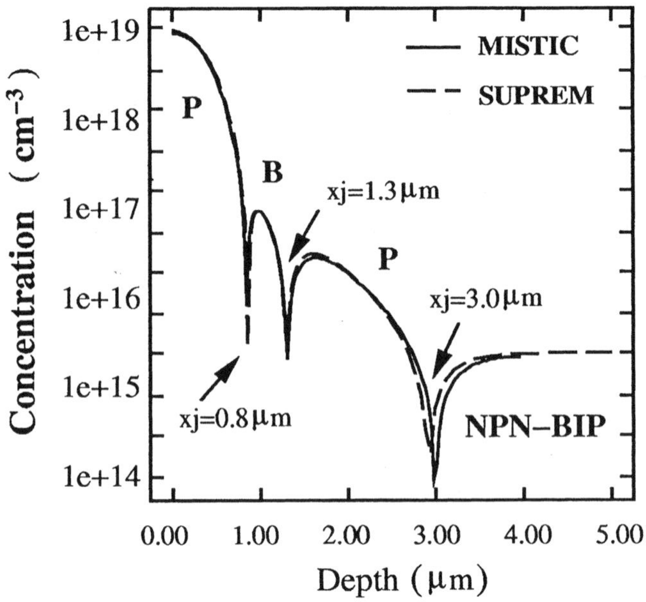

Figure 8.20. BiCMOS npn bipolar transistor concentration profile.

compiler has no provision for selecting an adequate implant energy. In this implementation, these energies are fixed, and the initial profiles are assumed as truncated Gaussians. These and many other refinements will be included in future versions.

13. SUMMARY

A method for the automatic generation of fabrication process sequences of thin film devices has been developed. The method uses a two-dimensional device description as input and using topological sorting techniques find feasible flows for the device. It is shown that the set of feasible flows is large; hence, an optimal flow is selected which maximizes the device yield. The method was implemented in a compiler program which uses a graphical interface to input the device cross-section and outputs the process flow data. The fabrication process flow is described by fundamental processing steps as deposition, lithography, etching, reactive growth, and diffusion. The method was successfully applied to the compilation of NMOS, CMOS, and double metal twin-well BiCMOS structures.

ACKNOWLEDGMENTS

We thank L. Mockus for providing the source code of the Bayesian optimizer. We also thank Dr. S. B. Crary for numerous helpful discussions. This research was sponsored by the National Science Foundation under grant ECS-9309229.

REFERENCES

[1] M. E. Law and R. W. Dutton. Verification of analytic point defect models using SUPREM-IV. *IEEE Trans. Computer-Aided Design*, CAD-7:181–190, 1988.

[2] K. Toh. *Two-dimensional images with effects of lens aberrations in optical lithography*. PhD thesis, University of California, Berkeley, 1975.

[3] W. G. Oldham, S. N. Nandgaonkar, A. R. Neureuther, and M. M. O'Toole. A general simulator for VLSI lithography and etching processes: Part I-application to projection lithography. *IEEE Trans. Electron Devices*, ED-26:717–722, 1979.

[4] K. Lee. *SIMPL-2 simulated profiles from the layout - version 2*. PhD thesis, University of California, Berkeley, 1985.

[5] S. D. Senturia, R. M. Harris, and *et. al.* A computer-aided design system for microelectromechanical systems (MEMCAD). *IEEE J. Microelectromechanical Syst.*, 1:3–13, 1992.

[6] D. S. Boning. *Semiconductor process design: representations, tools, and methodologies*. PhD thesis, Massachusetts Institute of Technology, 1986.

[7] D. Durbeck, J.-H. Chern, and D. S. Boning. A system for semiconductor process specification. *IEEE Trans. Semi. Manuf.*, 4:297–305, 1993.

[8] P. R. Kirstoff and D. P. Nunn. The process specification system for MMST. *IEEE Trans. Semi. Manuf.*, 8:262–271, 1995.

[9] H. McGehee, J. Heley, and J. Mahaffey. The MMST computer-integrated manufacturing system framework. *IEEE Trans. Semi. Manuf.*, 7:107–116, 1994.

[10] S. Saxena, P. K. Mozumder, A. Unruh, and R. Burch. A methodology for the top-down synthesis of semiconductor process flows. In *Intl. Symp. on Semi. Manuf.*, pages 36–40, 1995.

[11] M. Hasanuzzaman and C. H. Mastrangelo. Process compilation of thin-film microdevices. *IEEE Trans. Comp. Aid. Des.*, 30:1438–1453, 1996.

[12] M. Hasanuzzaman, E. T. Carlen, and C. H. Mastrangelo. Automatic generation of thin film process flows - Part I: Basic algorithms. *IEEE Trans. Semi. Mfg.*, 12(1):1–12, Feb. 1999.

[13] M. Hasanuzzaman, E. T. Carlen, and C. H. Mastrangelo. Automatic generation of thin film process flows - Part II: Recipe generation, flow evaluation, and system framework. *IEEE Trans. Semi. Mfg.*, 12(1):13–22, Feb. 1999.

[14] D. S. Bonning, M. B. McIlrath, P. Penfield, and E. M. Sachs. A general semiconductor process modeling framework. *IEEE Trans. Semic. Manuf.*, 5:266–280, 1992.

[15] F. Harary, R. Z. Norman, and D. Cartwright. *Structural Models: An Introduction to the Theory of Directed Graphs.* Wiley, New York, 1965.

[16] G. Di Battista and R. Tamassia. Algorithms for plane representations of acyclic digraphs. *Theor. Comp. Sci.*, 61:175–198, 1988.

[17] R. Tamassia and J. Vitter. Parallel transitive closure and point location in planar structures. *Siam J. Comput.*, 20:708–725, June 1991.

[18] P. Rosenstiehl and R. E. Tarjan. Rectilinear planar layouts and bipolar orientations of planar graphs. *Discrete Comp. Geom.*, 1:343–353, 1986.

[19] B. Dushnik and E. W> Miller. Partially ordered sets. *Am. J. Math.*, 63:600–610, 1941.

[20] P. C. Fishburn. *Interval Orders and Interval Graphs.* Wiley, New York, 1985.

[21] L. Budach. *Algebraic and Topological Properties of Finite Partially Ordered Sets.* Teubner, Leipzig, 1988.

[22] I. Rival. *Ordered Sets.* Kluwer, Boston, 1981.

[23] I. Rival. *Algorithms and Order.* Kluwer, Dordrecht, 1989.

[24] W. T. Trotter. *Combinatorics and Partially Ordered Sets.* Johns Hopkins University Press, Baltimore, 1992.

[25] R. A. Brualdy and H. J. Ryser. *Combinatorial Matrix Theory.* Cambridge Univ. Press, New York, 1991.

[26] K. H. Kim. *Boolean Matrix Theory and Applications.* M. Dekker, New York, 1982.

[27] R. A. Brualdy and H. J. Ryser. *A Survey of Finite Mathematics.* Cambridge Univ. Press, New York, 1991.

[28] A. Kaufmann. *Graphs, Dynamic Programming, and Finite Games.* Academic Press, New York, 1967.

[29] D. E. Knuth and J. L. Szwarcfiter. A structured program to generate all topological sorting arrangements. *Information Processing Letters*, 2:153–157, 1974.

[30] E. Szpilrajn. Sur l'extension de l'ordre partiel. *Fund. Math*, 60:175–186, 1930.

[31] N. Linial. Hard enumeration problems in geometry and combinatorics. *SIAM. J. Alg. Disc. Meth.*, 7:331–335, April 1986.

[32] S. G. Mohanty. *Lattice Path Countings and Applications*. Academic Press, New York, 1979.

[33] I. Rival. Linear extensions of finite ordered sets. *Annals of Discrete Mathematics*, 23:355–370, 1984.

[34] G. R. Brightwell. Linear extensions of infinite posets. *Discrete Mathematics*, 70:113–136, 1988.

[35] M. Pouzet and I. Rival. Which ordered sets have a complete linear extension? *Can. J. Math.*, XXXIII:1245–1254, 1981.

[36] V. Bouchitte and M. Habib. NP-completeness properties about linear extensions. *Order*, 4:143–154, 1987.

[37] G. R. Brightwell and P. Winkler. Counting linear extensions is NP-complete. In *Proceedings of the Twenty Third Annual ACM Symposium, Theory of Computing*, pages 175–181, 1991.

[38] M. D. Atkinson and H. W. Chang. Computing the number of mergings with constraints. *Information Processing Letters*, 24:289–292, 1987.

[39] M. Dyer, A. Frieze, and R. Kannan. A random polynomial time algorithm for approximating the volume of convex bodies. In *Proceedings of the Twenty First Annual ACM Symposium, Theory of Computing*, pages 375–381, 1989.

[40] D. Applegate and R. Kannan. Sampling and integration of near log-concave functions. In *Proceedings of the Twenty Third Annual ACM Symposium, Theory of Computing*, pages 156–163, 1991.

[41] S. W. Director and G. D.Hatchel. The simplicial approximation approach to design centering. *IEEE Trans. Cir. Syst.*, CAS-24:363–372, 1977.

[42] H. S. M. Coxeter. *Regular Polytopes*. Dover, New York, 1973.

[43] V. Bryant. *Aspects of Combinatorics*. Cambridge Univ. Press, New York, 1993.

[44] M. Marcus. *A Survey of Finite Mathematics*. Dover, New York, 1993.

[45] M. Marcus and H. Minc. *A Survey of Matrix Theory and Matrix Inequalities*. Dover, New York, 1992.

[46] W. Wan-di. Permutations with restricted positions and k-permanent. *Acta Math. Appl. Sinica*, 6:177–182, 1983.

[47] H. Minc. *Permanents*. Addison-Wesley, Reading, 1978.

[48] N. Alon, J. H. Spencer, and P. Erdos. *The Probabilistic Method*. Wiley, New York, 1992.

[49] B. Gogoi, R. Yuen, and C. H. Mastrangelo. The automatic synthesis of planar fabrication process flows for surface micromachined devices. In

MEMS 94, International Conference on Microelectromechanical Systems, pages 153–157, 1994.

[50] D. J. Kleitman and B. L. Rothschild. Asymptotic enumeration of partial orders on a finite set. *Trans. Amer. Math. Soc.*, 205:205–221, 1975.

[51] D. J. Kleitman and B. L. Rothschild. The number of finite topologies. *Proc. Amer. Math. Soc.*, 25:276–282, 1970.

[52] R. L. Davis. The number of structures of finite relations. *Proc. Amer. Math. Soc.*, 4:486–495, 1953.

[53] K. K. Butler. The number of partially ordered sets. *J. Comb. Theor.*, 13:276–289, 1972.

[54] F. Harary. The number of linear, directed, rooted and connected graphs. *Trans. Amer. Math. Soc.*, 78:445–463, 1955.

[55] M. H. El-Zahar. Enumeration of ordered sets. In I. Rival, editor, *Algorithms and Order*, pages 327–352. Kluwer, Dordrecht, 1989.

[56] J. Neggers. Representations of finite partially ordered sets. *J. Combin.*, 3:113–133, 1978.

[57] R. P. Stanley. Ordered structures and partitions. *Memoirs Amer. Math. Soc.*, 119:1–102, 1972.

[58] A. D. Kalvin and Y. L. Varol. On the generation of all topological sortings. *J. Algorithms*, 4:150–162, 1983.

[59] A. B. Kahn. Topological sortings of large networks. *Comm. ACM*, 5:558–562, 1962.

[60] D. E. Knuth. *The Art of Computer Programming*, volume 1. Addison-Wesley, Reading, Mass., 1968.

[61] Y. L. Varol and D. Rotem. An algorithm to generate all topological sorting arrangements. *The Computer Journal*, 24:83–84, 1981.

[62] G. Steiner. An algorithm to generate the ideals of a partial order. *Operations Research Letters*, 5:317–320, 1986.

[63] A. V. Aho, M. R. Garey, and J. D. Ullman. The transitive reduction of a directed graph. *SIAM J. Comput.*, 1:131–137, 1972.

[64] E. M. Reingold, J. Nievergelt, and N. Deo. *Combinatorial Algorithms: Theory and Practice*. Prentice-Hall, Englewood Cliffs, N. J., 1977.

[65] R. Tarjan. Finding dominators in directed graphs. *SIAM J. Comput.*, 3:62–89, 1974.

[66] A. Nijenhuis and H. S. Wilf. *Combinatorial Algorithms*. Academic Press, New York, 1975.

[67] M. C. Er. A parallel computation approach to topological sorting. *The Computer Journal*, 26:293–295, 1983.

[68] P. Chaudhuri. Fast parallel algorithms for a class of graph problems. *Jr. Comb. Inf. and Syst. Sci.*, 12(1):51–65, 1987.

[69] K. Culik and I. Fris. Topological transformations as a tool in the design of systolic networks. *Theoretical Computer Science*, 37:183–216, 1985.

[70] E. Welzl. Constructing the visibility graph for n-line segments in $O(n^2)$ time. *Information Processing Letters*, 20:167–171, 1990.

[71] S. Foldes, I. Rival, and J. Urrutia. Light sources, obstructions and spherical orders. *Discrete Mathematics*, 102:13–23, 1992.

[72] J. L. Vossen. *Thin Film Processes.* Academic Press, New York, 1978.

[73] S. M. Sze. *VLSI Technology.* McGraw-Hill, New York, 1988.

[74] B. R. Penumalli. A comprehensive two dimensional VLSI process simulation program, BICEPS. *IEEE Trans. Electron Devices*, ED-30:986–992, 1983.

[75] C. P. Ho, J. D. Plummer, S. E. Hansen, and R. W. Dutton. VLSI process modeling - SUPREM III. *IEEE Trans. Electron Devices*, ED-30:1438–1453, 1983.

[76] J. Mockus. *Bayesian Approach to Global Optimization : Theory and Applications.* Kluwer, Boston, 1989.

[77] W. H. Press, B. P. Flannery, S. A. Teulosky, and W. T. Vetterling. *Numerical Recipes in C.* Cambridge Univ. Press, New York, 1993.

[78] A. J. Strowjas. *Statistical Design of Integrated Circuits.* IEEE, New York, 1987.

[79] S. W. Director and W. Maly. *Statistical Approach to VLSI.* North-Holland, New York, 1994.

[80] D. Moore and H. Walker. *Yield Simulation For Integrated Circuits.* Kluwer, New York, 1987.

Chapter 9

MASK SYNTHESIS

With a focus on surface micromachined MEMS

Radha Sarma

Department of Mechanical Engineering, The University of Michigan, Ann Arbor, MI 48105, U. S. A.

G. K. Ananthasuresh

Department of Mechanical Engineering and Applied Mechanics, University of Pennsylvania, Philadelphia, PA 19104, U. S. A.

Synopsis: Photolithography masks provide the link between the design of a MEMS device and its microfabrication. Numerous techniques and computer software tools are available for creating a geometric model from the masks by geometrically emulating the microfabrication process. This is the *analysis* or the *forward* problem. The opposite of this, i.e., the *synthesis* or the *inverse* problem, entails direct generation of masks from a 3-D or 2-D model of a device. There are only a few techniques available for this important aspect of MEMS design and fabrication. After noting the current practice adopted by MEMS designers of today, approaches to mask synthesis of devices fabricated with bulk micromachining and surface micromachining are presented in this chapter. The latter kind is described in detail. A mathematical framework for geometric modeling, which was developed to streamline the mask synthesis problem, is also presented. It is shown, illustrated with examples, how this framework reduces the geometric problem of mask synthesis of surface micromachined devices to an over constrained linear system of equations that can be solved using singular value decomposition..

1. INTRODUCTION

Microelectromechanical Systems (MEMS) are manufactured by adapting existing VLSI (Very Large Scale Integration) microelectronic fabrication technology that is used to fabricate integrated circuit (IC) chip. While the IC fabrication uses relatively simple Manhattan geometries, the functionality of a MEMS device is highly dependant on its geometric sophistication. Currently, some common MEMS fabrication processes, which can achieve complex geometries, include bulk micromachining, surface micromachining, and LIGA [1, 2]. Some others are also underway. In most microfabrication processes, the photolithography masks are the links between the design and fabrication. The masks determine the shape of the device once a process is selected. Thus, the end product of a MEMS design effort is a set of mask layouts, which are simply 2-D geometric patterns of black and white. Ironically, however, MEMS design effort today often begins with this "end product". That is, the designers begin by drawing masks in a layout editor. These are used to create a 2-D or 3-D model of the device by geometrically emulating the process. The models are then used by the designers to perform behavioral simulations (often with numerical techniques such as finite element analysis, boundary element analysis, and others) to see if the design is satisfactory. As is often the case, the 2-D or 3-D models are re-designed to improve the performance. After a design is finalized, it is essential to go backwards to generate the mask layouts that will result in the finalized 2-D or 3-D geometry in the fabricated device or a system. In other words, the *synthesis* of masks is a crucial step in the Computer Aided Design (CAD) of MEMS.

It is interesting to compare the design and manufacture at the macro scale with those of MEMS. The designers of macro devices have the advantage of starting with a geometric model and being able to directly visualize or manipulate their designs. The geometric model is then queried to generate the process specific data, for example, the tool paths for Computer Numerically Controlled (CNC) machining. These tool paths are analogous to the masks layouts at the micro scale. Imagine how unnatural and difficult it would be for a designer to begin the design an object with tool paths! Yet, a MEMS designer is able to do it by beginning the design procedure with masks by taking advantage of tools and techniques of microelectronics and the VLSI technology. One approach is to use parameterized mask layout generation models [3] wherein the design process is limited to certain geometries. And it works well for those cases. At other times, designers manually design the masks, by a hit or miss fashion, to get the geometry that they want. But as the MEMS technology develops further, as judged from its current fast pace, the complexity of the geometry increases driven mainly by the functionality. Two issues arise then. First, parameterized layout models may prove to be overly restrictive for a designer. Second, the manual process of generating masks from the final design becomes increasingly challenging and time-consuming. Systematic procedures for mask synthesis are thus useful.

1.1 Terminology

MEMS devices and systems are made of multiple materials. Hence, the definition of the *geometric model* here is broad so that it also includes the material information in addition to the geometry. A model may be described in 2-D or 3-D. The *process* is a sequence of steps such as etching, deposition, doping, chemo-mechanical polishing (CMP), etc. *Masks* are of course the 2-D geometric patterns that determine where the material is selectively removed or doped. The interplay among these three is shown in Fig. 1. The generation of the geometric model from the process and mask information is called the *forward* or *analysis* problem. Many algorithms and software tools are available to solve this problem. The inverse of analysis, i.e., the *synthesis* problem is of two types depending on whether a process is assumed or synthesized along with the masks. The procedure for synthesizing the masks and the process for a given geometric model is shown with dashed curves in the figure. This is yet to be developed. The synthesis of masks for a given process is the focus of this chapter. Also included in the figure are *parameterized layout models*. They consist of 2-D geometrical patterns of fixed topology and shape but varying dimensions (the parameters). They usually link not only a geometric model but also a functional entity. The folded-beam suspension spring of an electrostatic comb-drive is an example of this. As shown in the figure, they are process-specific and, therefore, constrain the geometric modeling effort. They have the ability to directly synthesize masks because the geometry is not changed topologically or in shape during the design procedure.

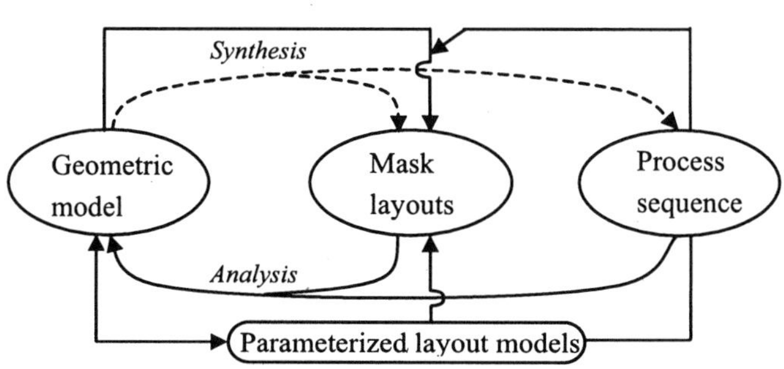

Figure 1. Interplay between the geometric model (design), masks, and process

In the remainder of the chapter, first, the current literature and practice are briefly reviewed (Section 2). This chapter focuses on mask synthesis for surface micromachined devices. Hence, Section 3 contains a mathematical framework for creating geometric models of surface micromachined devices. In Section 4, the synthesis problem and a solution technique are presented. Concluding remarks are in Section 5.

2. RELATED WORK

There is a large body of work related to the forward problem. The OYSTER project at IBM [4, 5] and the MEMCAD project at MIT [6] first recognized the need for geometric CAD in the micromechanical realm. Emphasis was placed on creating process simulations and 3-D CAD models that could be used to predict the performance of MEMS devices using numerical techniques. In these integrated CAD environments, three-dimensional models were created from mask and process data [7].

Given a set of masks, three-dimensional models of MEMS devices fabricated using a bulk micromachining process were generated using a crystal plane offset approach [8] and a cellular automata approach [9]. Image processing and morphological operations were also used on three-dimensional cellular representations to simulate etches and deposits based on etch and deposition rates [10]. MEMShapes, another three-dimensional simulator, uses solid modeling techniques to build models of MEMS devices [11]. Process algebra was developed that captured all the geometric and material transformations occurring to the MEMS device as it is being fabricated. Finally, there are numerous MEMS-CAD software companies [12] which have modules that make the forward problem easy for designers. Some of these rely on existing solid modeling software and some have developed their own solid modeling capabilities with varying degrees of flexibility and versatility. But none seem to have the true synthesis capability although some use the parameterized layout model approach.

Several projects related to process modeling are available. They include the SUPREM project at Stanford [13] and the MiSTIC project at Michigan [14-16]. The latter focuses on a systematic method to synthesize the process flow (i.e., sequence of deposition, etching, lithography, ion implantation, diffusion, reactive growth needed to build a device) for a two-dimensional cross- section of a device. This is described in detail in Chapter 8.

In contrast, the mask synthesis problem has received very little attention. The reason for this is perhaps the need for it did not arise in the initial stages of the development of MEMS. Or, perhaps many saw the use of parameterized layout models is adequate. Some may even feel that manually generating masks from refined designs is not difficult for the MEMS devices developed so far. Arguably, MEMS have become geometrically complex now. There are examples where laying out masks manually to get a desired shape that a designer has in mind is cumbersome. A notable example is the large array of mechanical transmissions from the Sandia research group [17] who work with a seven-layer surface micromachining process. The advances in microfluidics also warrant techniques for mask synthesis. In microelectronics arena, techniques have been developed and are in use to correct mask layouts for optical aberrations and inevitable fabrication limitations for submicron, ultra-deep lithography [18]. Corner compensation is an example of such a technique. In MEMS, there are two types of synthesis techniques: (i) for bulk micromachining, and (ii) for surface micromachining. They are described next.

2.1 Mask synthesis for bulk micromachining

The most widely used technique in bulk micromachining is the wet etching where the silicon wafer, covered with a suitable mask layer, is exposed to the chemical etchant in solution. In single crystal silicon, etch rates of an etchant are different in different directions. Some etches stop on certain crystal planes. Because of the variable etch rates, the resulting shapes in bulk micromachining are complex and do not bear an easily discernible relationship to the shapes in the mask openings. Several researchers have successfully developed simulation techniques for predicting the 2-D (top-view) and 3-D (shallow or deep pits left by etching) shapes. Some of these include [8, 9, 19-22]. More details and an overview can be found in [23].

The synthesis problem of determining the masks for a bulk-etched single layer using multiple etchants was done using genetic and evolutionary algorithms combined with computational techniques to predict the emerging shapes [24-26]. A comprehensive description of this method can be found in [23]. Thinking further about the synthesis problem, the following was also noted in that article:

> "Because for most applications the behavior or the function of the microsystem is of direct interest, rather than its three-dimensional (3-D) shape, the procedure described here can be expanded...to include a simulation of the fabrication of the device, followed by a simulation of its function (e.g., a finite element analysis of flow in channels). In this way, the *desired function* of a microdevice can be synthesized."

2.2 Mask synthesis for surface micromachining

In comparison with the bulk micromachining, surface micromachining has fewer complications with the resulting geometry of the parts. In surface micromachining, as is well known, the geometric complexity and versatility comes due to its layered construction by way of a sequence of deposits and etches. As the number of layers increases, the complexity quickly grows. Therefore, the Sandia MEMS research group developed a visualization tool for its seven-layers process [27] and a set of design rules for the designer to follow [28]. Others utilize a restricted set of geometries and use parameterized layout models for beams, anchors, and plates [3]. One of the first attempts at systematic mask synthesis for surface micromachining was described in [29-31]. The remainder of the chapter describes this approach in detail. A mathematical framework developed for this purpose is presented next.

3. MATHEMATICAL FRAMEWORK

A rigorous representation of the model and a set of unambiguously defined operators are deemed essential in systematic geometric modeling [32, 33]. Such a framework, which is presented below, is useful for both the forward and inverse (synthesis) problems involving surface micromachining.

At the highest level, a surface micromachining process is described in this chapter by the following identifiers: the number of layers, the number of process steps, a list of deposit descriptors, a list of etch descriptors, and a list of doping descriptors. This is shown algorithmically in the left column of Table 1 with more details in the right column.

Consider a surface micromachined device of J layers (that include the structural and sacrificial layers) that is fabricated in I process steps (sum of the total number of deposits J, etches K, and dopings L, i.e., $I = J + K + L$). The indices i, j, k, and l are henceforth used to count the corresponding steps. The model M of a MEMS device is a layered, multi-material point set that can be interpreted as a union of all J layers, i.e., $M = \cup L_j$, which are themselves multi-material point sets. Thus the modeling space comprises a point set $\{P\}$, a corresponding set of unique material tags $\{m\}$ at each point, and a corresponding unique layer tag LT at each point, i.e., $M = \{P, \{m\}, LT\}$. This modeling space allows for the representation of both doped and undoped regions, for example, in an undoped region a point will be associated with a single primary material tag,

Table 1 Algorithmic representation of surface micromachining process steps

``class ProcessBuild (nLayers, nSteps, StepsList) { int nLayers; int nSteps; struct_of_pointers *StepsList; // pointers to classes etch/deposit/doping ProcessBuild(int nl, int ns, struct_of_pointers *sl) { nLayers = nl; nSteps = ns; for (int i = 1; i < ns; i++) StepsList[i] = sl[i]; } }``	nLayers = the total number of layers including the structural and sacrificial layers nSteps = the total number of steps including deposits, etches, and dopings The model M_i is modified by i^{th} process step as follows: $M_{i+1} = \text{ProcessStep}(M_i, \textit{descriptors})$ The details of each process step (i.e., deposit, etch, dope) along with their descriptors appear below.

class Deposit (Type, Material, Thickness) {
char *Type;
char *Material;
double Thickness;
Deposit (char ty, char ma, double th) {
Type = ty;
Material = ma;
Thickness = th;
}
}
	Type = *type* = conformal, planar, via, or stack as described in [34].
	Material = m = material of deposit
	Thickness = t = thickness of the deposit
	$M_{i+1} = Deposit(M_i, type, m, t)$
	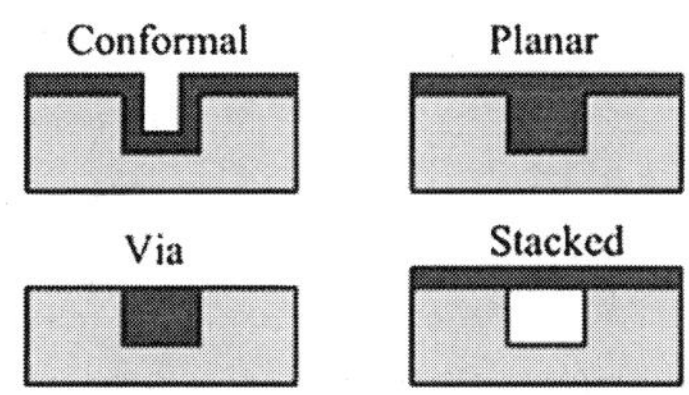

class Etch (Type, nAffectedLayers, AffectedLayerList,
FractionalDepth) {
char *Type;
int nAffectedLayers;
int *AffectedLayerList;
double FractionalDepth;
Etch (char ty, int nl, int *ll, double fd) {
Type = ty;
nAffectedLayers = nl;
AffectedLayerList = new int(nl);
for (int i = 1; i < nl; i++)
AffectedLayerList[i] = ll[i];
FractionalDepth = fd;
}
}
	Type = *type* = straight wall, undercut, or release
	Each etch may affect more than one layer. The number and the list of affected layers are specified with n_{AL} and L_A, respectively.
	An etch may end at a layer, at a material, or at a fractional depth of a material or layer. The parameter FractionalDepth, f_d, is multiplied by the layer thickness to get the depth of etch.
	The mask is denoted by $\overline{m}_i$.
	A release etch does not usually have a mask.
	$M_{i+1} = Etch(M_i, type, n_{AL}, L_A, f_d, \overline{m}_i)$

class Dope (Type, nAffectedLayers,
AffectedLayerList, FractionalDepth,
NewMaterial) {
char *Type;
int nAffectedLayers;
int *AffectedLayerList;
double FractionalDepth;
Dope (char ty, int nl, int *ll, double fd,
char ma) {
Type = ty;
nAffectedLayers = nl;
AffectedLayerList = new int(nl);
for (int i = 1; i < nl; i++)
AffectedLayerList[i] = ll[i];
FractionalDepth = fd;
NewMaterial = ma;
}
}
	Doping is similar to etching with the difference that the material in the affected portion is exchanged with a new material, m_{new}, rather than being removed.
	$M_{i+1} = Dope(M_i, type, n_{AL}, L_A, f_d, \overline{m}_i, m_{new})$

e.g., (p,m_p,lt), whereas in a doped region a point may be associated with additional secondary material tags corresponding to the doped material as in $(p,(m_p,m_{s1},m_{s2},...),lt)$.

As an example, consider a hypothetical one-layer surface micromachining process (involving a total of three deposited layers) shown in Fig. 2. The variables f_{e1}, f_{e2}, and f_{o1} indicate the fractional depths to which the last affected layer is etched/doped; t_1, t_2, and t_3 indicate the thicknesses of the deposited layers; and m_1, m_2, and m_3 indicate the material tags associated with a layer. This process sequence is used for the examples presented in this chapter.

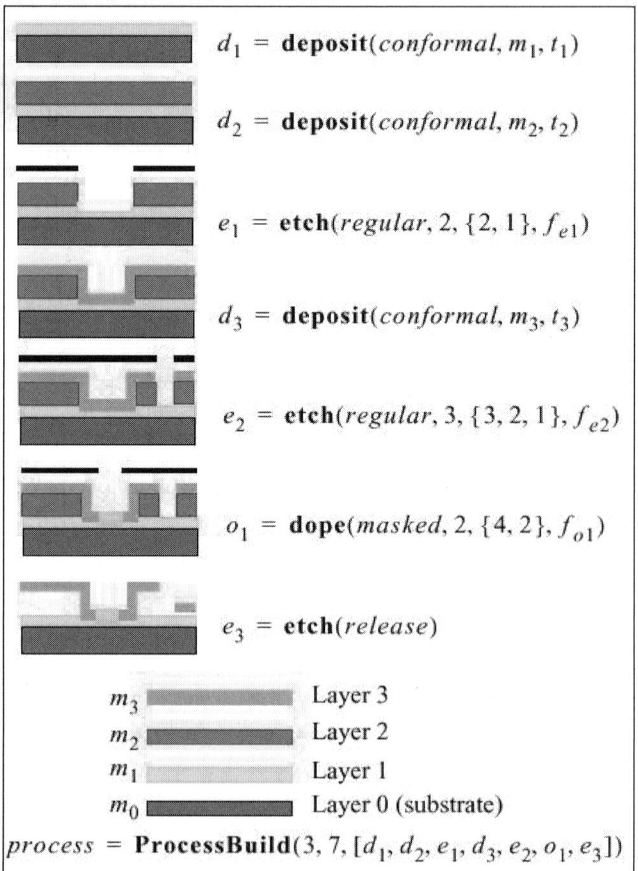

Figure 2. A hypothetical one-layer process

The exposed boundary of a model M is represented by the symbol $\partial_e M$. Each ith process step modifies the existing model M_i to create the updated model M_{i+1}. The modification of models in the modeling space is done by means of *state change operators* (Fig. 3) and *queries* (Fig. 4) following the usual approach in systematic geometric modeling [33]. Note that each operation is pictorially shown in addition to its mathematical representation. For instance, the multi-material union operator is defined such that the first material is overridden (consequently the operation is non-commutative). Note that the union, intersection, and subtraction operators are also non-commutative. The interested reader may find more detailed mathematical descriptions of the general multi-material modeling paradigm in [35]. The operators and queries are used to model each step in the surface micromachining process in terms of a set operations (e.g., unions, differences, Minkowski additions) on heterogeneous or multi-material point sets. Note that there are multiple ways to perform an operation. In this chapter, the mathematical model of heterogeneous point sets is represented by means of a cellular or voxel representation [36]. The processing steps are implemented as shown below. It should be noted that these are just one of the many ways of obtaining the desired effect for a process step.

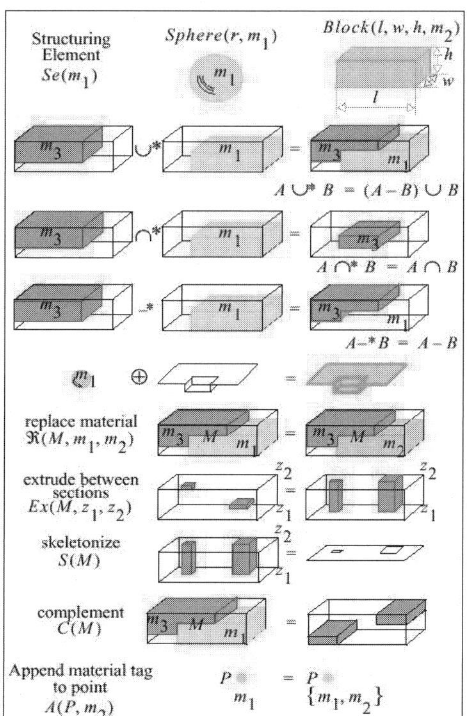

Figure 3. State change operators

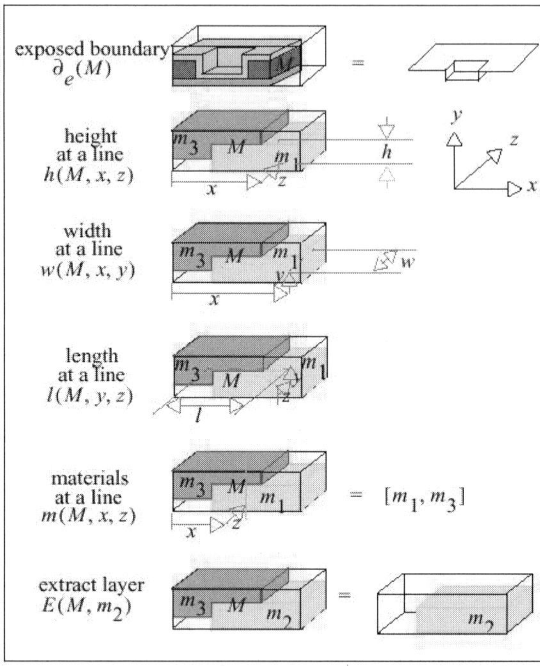

Figure 4. Queries on the mathematical and geometric models

Conformal deposit:
$$S = circle(r, m_j)$$
$$M_{i+1} = (S \oplus \partial_e M_i) \cup M_i \tag{1}$$

Planar deposit:
$$a = \max\{h(M_i, x)\} + t$$
$$b = \max\{w(M_{i-1}, y)\}$$
$$S = \text{rectangle}(a, b, m_j) \tag{2}$$
$$M_{i+1} = S \cup M_i$$

Via filling:

$$\begin{aligned} a &= \max\{h(M_i, x)\} \\ b &= \max\{w(M_{i-1}, y)\} \\ S &= \text{rectangle}(a, b, m_j) \\ M_{i+1} &= S \cup M_i \end{aligned} \tag{3}$$

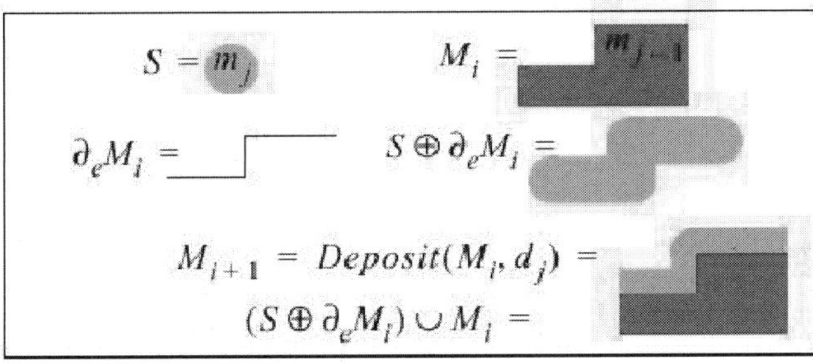

Figure 5. A conformal deposit in 2-D

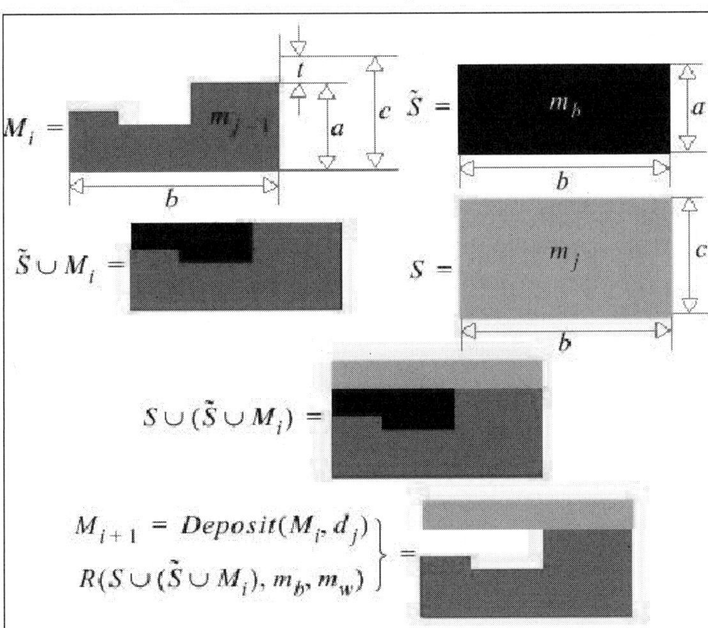

Figure 6. Modeling a stack deposit (e.g., wafer bonding)

Stack deposit:

$$\begin{aligned} a &= \max\{h(M_i, x)\} \\ b &= \max\{w(M_{i-1}, y)\} \\ \widetilde{S} &= \text{rectangle}(a, b, m_{black}) \\ c &= \max\{h(M_i, x)\} + t \\ S &= \text{rectangle}(c, b, m_j) \\ M_{i+1} &= R(S \cup (\widetilde{S} \cup M_i), m_{black}, m_{white}) \end{aligned} \tag{4}$$

The conformal and stack deposit operations are illustrated in Figs. 5 and 6 respectively.

Masks, which are used in etches and dopings, are of one dimension less than the dimension of the modeling used. That is, they are lines in 2-D modeling and planes in 3-D. All points on a mask are classified as open or closed as illustrated in Fig. 8. An open point exposes the layer directly underneath and allows the etchant/dopant to act on the exposed portions of the part. Conversely, a closed point protects the surface directly underneath. Any contiguous set of open points on the mask is referred to as a mask opening. A single mask may have multiple openings.

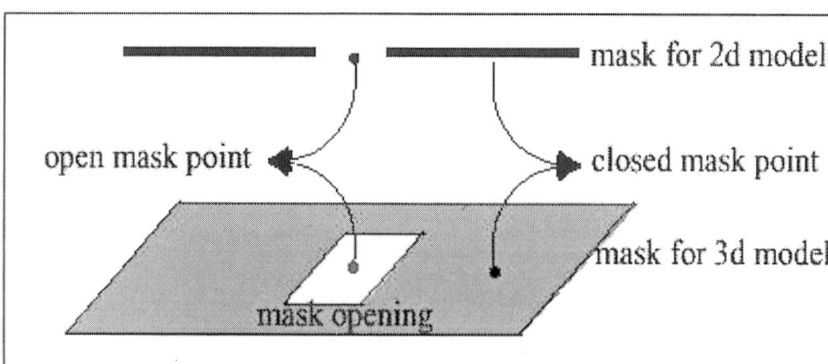

Figure 7. Masks in 2-D and 3-D

Figures 8 and 9 illustrate how a regular straight-wall etch and an undercut isotropic etch are modeled in this framework. At first sight, even though these operations may look involved, they are algorithmically convenient and are unambiguous in the way they ought to be implemented. Additionally, the definition of the operations is independent of the implementation, which is also an advantage, as each user may want to implement differently depending on the particular needs and desires.

The stage is now set to describe the synthesis problem.

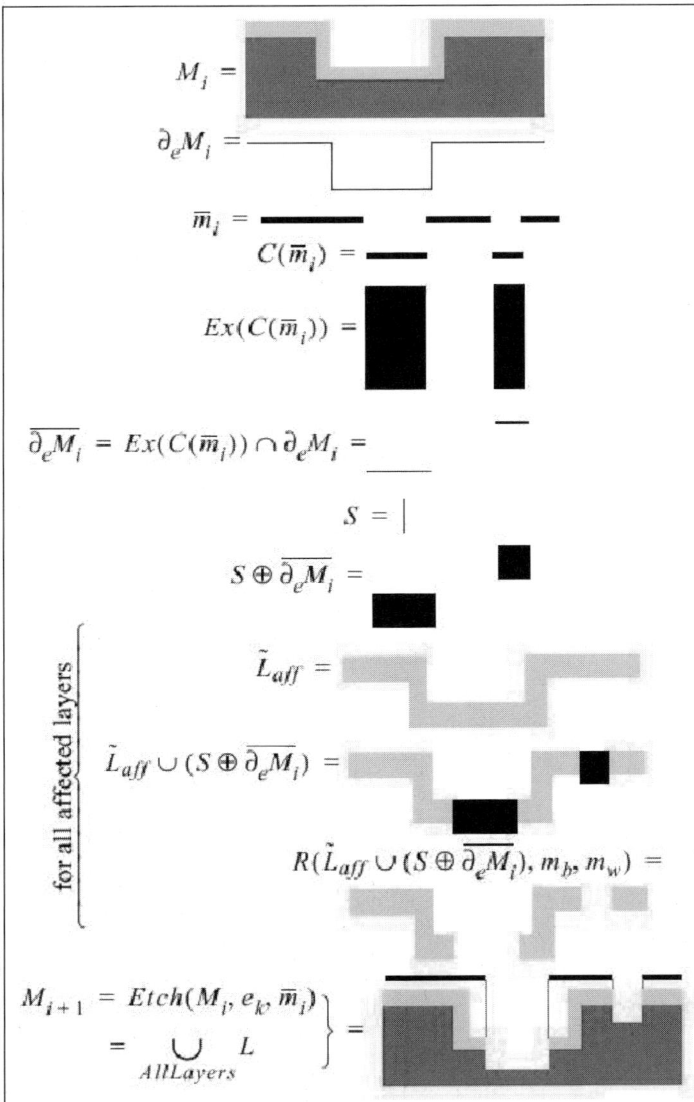

Figure 8. Modeling a straight-wall regular etch in 2-D

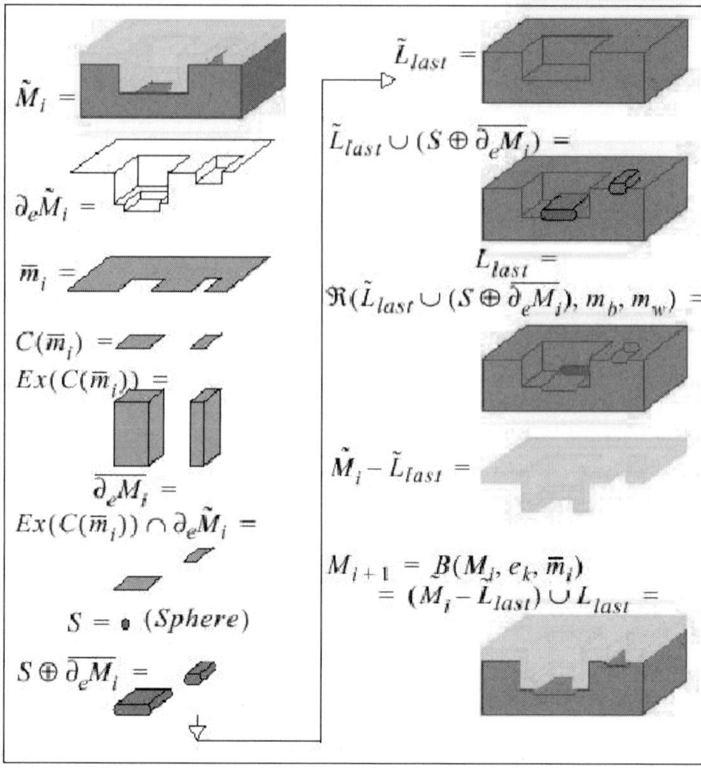

Figure 9. Modeling an undercut isotropic etch in 3-D

4. SYNTHESIS

The synthesis problem involves obtaining a set of masks for given a complete geometric model M of the MEMS device and a process sequence chosen by the designer. The technique presented here to solve the synthesis problem uses an *intermediate geometric model* that is reconstructed from the beginning and transformed gradually in stages using the known process sequence. The procedure starts with the initialization of an intermediate geometric model $\widetilde{M}_0$ that represents the substrate. The list of steps in the process sequence is then sequentially queried, and the corresponding intermediate geometric model $\widetilde{M}_i$ $(i = 1...I)$ is built. For each step i in the process sequence, a series of actions are undertaken depending on whether an etch, deposit, or doping is encountered. The detailed actions are described below and a pictorial sequence of the actions is shown in Fig. 10. Note that the description below is general enough to cover three-dimensional MEMS devices; only the illustration is two-dimensional.

4.1 Deposits

If a deposition step d_j is encountered, the intermediate geometric model $\widetilde{M}_i$ corresponding to the ith step in the process sequence is generated using the forward problem as shown in steps $d_j\ (j=1,2,3)$ in Fig. 10. Note that it is possible to build the intermediate geometric model since deposits do not use masks and all other information is known from the process sequence.

4.2 Etches

Three actions are undertaken for each etching step encountered. First, the intermediate geometric model $\widetilde{M}_i$ is set equal to $\widetilde{M}_{i-1}$. Second, the layer L of the complete geometric model (i.e., the part) corresponding to the first affected layer of etch e_k is extracted. Finally, the layer $\widetilde{L}$ of the intermediate geometric model corresponding to the first affected layer is extracted. The mathematical representation of these actions is shown below.

$$
\begin{aligned}
&\widetilde{M}_i = \widetilde{M}_{i-1} \\
&f = e_k \cdot AffectedLayerList[1] \\
&L = E(M, m_f) \\
&\widetilde{L} = E(\widetilde{M}, m_f)
\end{aligned}
\tag{5}
$$

The extracted layers L and $\widetilde{L}$ are then compared to generate a potential set of mask openings as explained in Section 4.4. These steps are summarized in the first column of step e_1 in Fig. 10. The mask openings obtained in this manner may not be correct as several etches could affect any given layer. Hence a procedure to detect and eliminate incorrect mask openings is undertaken as explained in Section 4.5, resulting in the correct set of masks $\overline{m}_i$. The intermediate geometric model is then updated appropriately. This procedure described here is shown in steps e_1 and e_2 in Fig. 11.

4.3 Doping

A doping step is handled in the same way as an etching step with the difference that in doping the material change is considered, as it remains unaltered geometrically. The step o_1 in Fig. 10 illustrates this.

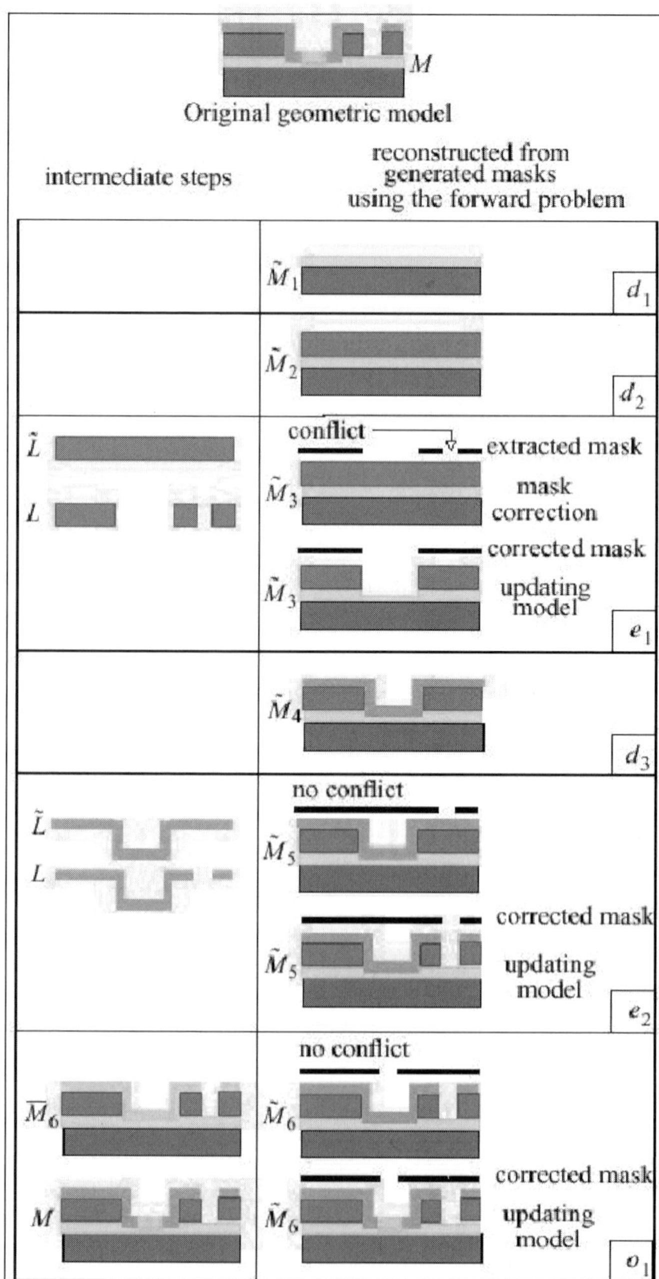

Figure 10. Illustrated procedure for the synthesis problem

All the steps in the inverse problem are best summarized algorithmically as follows:

InverseProblem (Model M, ProcessBuild P)
$\widetilde{M}_0 = substrate$;
i = 0 // A counter to indicate the current step in process sequence
for (i = 1; i <= P.nSteps; i++) {
   j = 0; // counter to indicate the current deposit step
   k = 0; // counter to indicate the current etch step
   l = 0; // counter to indicate the current doping step
   if P.StepsList[i] is a deposit {
      j = j + 1;
      d_j = P.StepsList[i]; // building the intermediate geometric model $\widetilde{M}_i = Deposit(\widetilde{M}_{i-1}, d_j)$;
   } else if P.StepsList[i] is an etch {
   k = k + 1;
   e_k = P.StepsList[i];
   $\widetilde{M}_i = \widetilde{M}_{i-1}$; // instantiate the current model
   $f = e_k \cdot AffectedLayerList[1]$; // find the first affected layer of the etch
   $L = E(M_i, m_f)$;// extract a layer from complete geometric model
   $\widetilde{L} = E(\widetilde{M}_i, m_f)$// extract a layer from intermediate geometric model
   $\overline{m}_i = GenerateMasks(\,)$; // generate a potential mask set
   $\overline{m}_i = SubdivideMasks(\overline{m}_i)$; // subdivide the potential mask set based on
      // 1. change in thickness of last affected layer
      // 2. layer in contact with the upper surface of the last affected layer.
   $\overline{m}_i = CorrectMaskOpenings(\overline{m}_i)$; // resolve conflicts; generate correct mask
$\widetilde{M}_i = Etch(\widetilde{M}_i, e_k, \overline{m}_i)$; // update the intermediate model with correct mask
}
   } else if P.StepsList[i] is a doping {
      $l = l + 1$;
      o_l = P.StepsList[i];
$\widetilde{M}_i = \widetilde{M}_{i-1}$; // instantiate the current model
$\widetilde{M}_i = Dope(\widetilde{M}_i, o_l, unmasked)$; // Perform unmasked doping
   $\overline{m}_i = GenerateMasks(\,)$; // generate a potential mask set
   ... //steps similar to etching in obtaining the correct masks
   $\widetilde{M}_i = Dope(\widetilde{M}_i, o_l, \overline{m}_i)$; // update the intermediate model with the correct mask
   }
}

4.4 Generating potential mask openings

Comparing the openings on corresponding layers of geometric models (complete and intermediate), gives an indication of the mask openings of the mask for a particular etch or doping step. However, not all the predicted mask openings need to belong to the mask associated with the step under consideration. This is because multiple etches/dopings can affect one layer. Hence they are termed as potential mask openings. This section outlines how these potential mask openings are generated. Sections 4.5 and 4.6 describe how to resolve whether a mask opening can be correctly associated with a particular etch step.

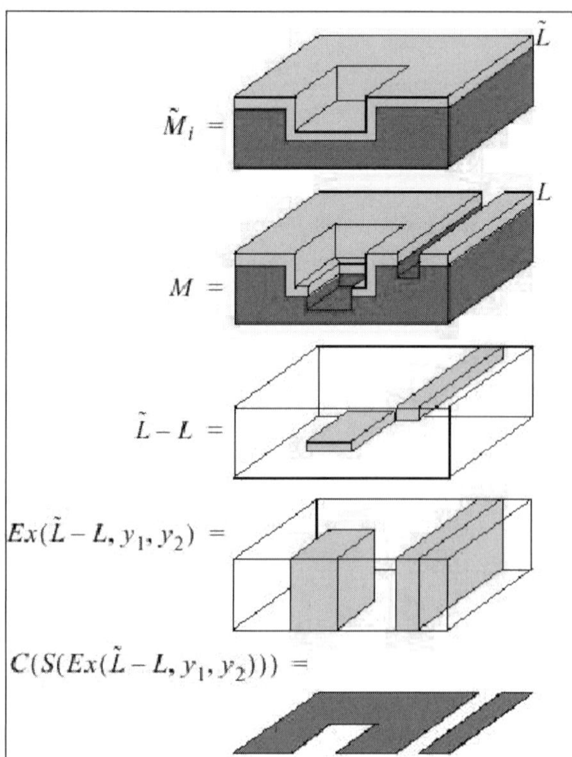

Figure 11. Generation of potential masks

Etch masks are generated by extracting the first affected layer of the kth etch. These layers are subtracted from each other to detect regions of difference in layers. These differences indicate the presence of mask openings. The mask openings are generated by extruding and skeletonizing the intersections in layers as illustrated in Fig. 11 and in the following equations.

$$\overline{m}_i = C(S(Ex(\widetilde{L} - L), y_1, y_2))) \quad (6)$$
$$\text{where } L = E(M, m_f) \text{ and } \widetilde{L} = E(\widetilde{M}_i, m_f)$$

The mask generation procedure for dopings is given below.

$$\widetilde{M}_i = Dope(\widetilde{M}_i, o_l, nullmask) \quad (7)$$
$$\overline{m}_i = C(S(Ex(E(M, m_{o_l}) \cap E(\widetilde{M}_i, m_{o_l}), y_1, y_2)))$$

Since the procedure for doping steps is almost similar to that of the etches, here forward, only etches are discussed.

4.5 Subdivision of mask openings

A single opening in a mask (as predicted by the procedure in Section 4.4) may be the result of more than one etches acting on the layer under consideration. Thus each mask opening needs to be unambiguously subdivided into distinct mask openings, which reflect a distinct series of etches acting on them. The cases where this occurs are enumerated below and illustrated in Fig. 12. Completely overlapping masks of two distinct etches acting on the same layer will yield a single potential mask opening. Partially overlapping masks of two distinct etches acting on the same layer will yield a single potential mask opening. Touching masks of two distinct etches acting on the same layer will yield a single potential mask opening. Redundant etches specified in the process sequence, e.g., specifying an etch of the same parameters two times in sequence, is disallowed. In all the cases mentioned above, the potential mask opening needs to be subdivided into more than one distinct mask openings. Two criteria are used for the subdivision.

i. The subdivision of a mask opening based on the thickness of the last affected layer of the etch: This criterion is illustrated by the "process flow 1" shown in Fig. 12, where the potential mask for etch e_1 is subdivided based on the thickness of the last affected layer. If the thickness of the last affected layer changes, it is indicative of than more than one etch has occurred. Hence there are as many subdivisions in the mask as the number of different thicknesses in the layer.

ii. The subdivision of a mask opening based on the layer which the last affected layer is immediately in contact with: This criterion is illustrated by "process flow 3" shown in Figure 12. In this case the openings of two different masks are adjacent to each other and have the same depth in the last affected layer making it impossible to subdivide based on thickness. Hence the layers immediately in contact with the top surface of the last affected layer are checked. The subdivision is based on regions of the last affected layer in contact with a distinct upper layer. The case shown in "process flow 2" is a combination of "process flow 1" and "process flow 3".

4.6 Validating Mask Openings

This step follows after the distinct potential mask openings have been generated using the procedures outlined in the previous sections. Each distinct mask opening has to be checked to detect whether the current etch has affected the fth layer. To check whether a given mask opening affects the fth layer, it is sufficient to check the complete geometric model along any given line (referred to as a check line) passing anywhere through the mask opening, e.g., line AA' or BB' in Fig. 13. Note that if the subdivision of masks was not undertaken every point in the mask opening has to be checked using the procedure described below whereas after the subdivision only one point in the mask opening need be checked. Coordinates are then given to the complete geometric model along the check line. The coordinates are simply the "measured" thicknesses of layers along the check line from top to bottom.

Figure 13 shows the coordinates along check lines AA' and BB'. The coordinates exist in J-dimensional space $\Re^J$ where J is the maximum number of deposited layers. Since the example illustrated in Fig. 2 has three deposited layers, the coordinates in Fig. 13 lie in $\Re^3$. Each etch and deposit is also given a coordinate depending on the thickness of the deposit and the depth of the etch. The coordinate of a deposit is a vector with a single positive entry corresponding to the deposit thickness. The coordinate of an etch is a vector with negative entries corresponding to the etch depth in each affected layer. Coordinates of the deposits of Fig. 2 are listed below.

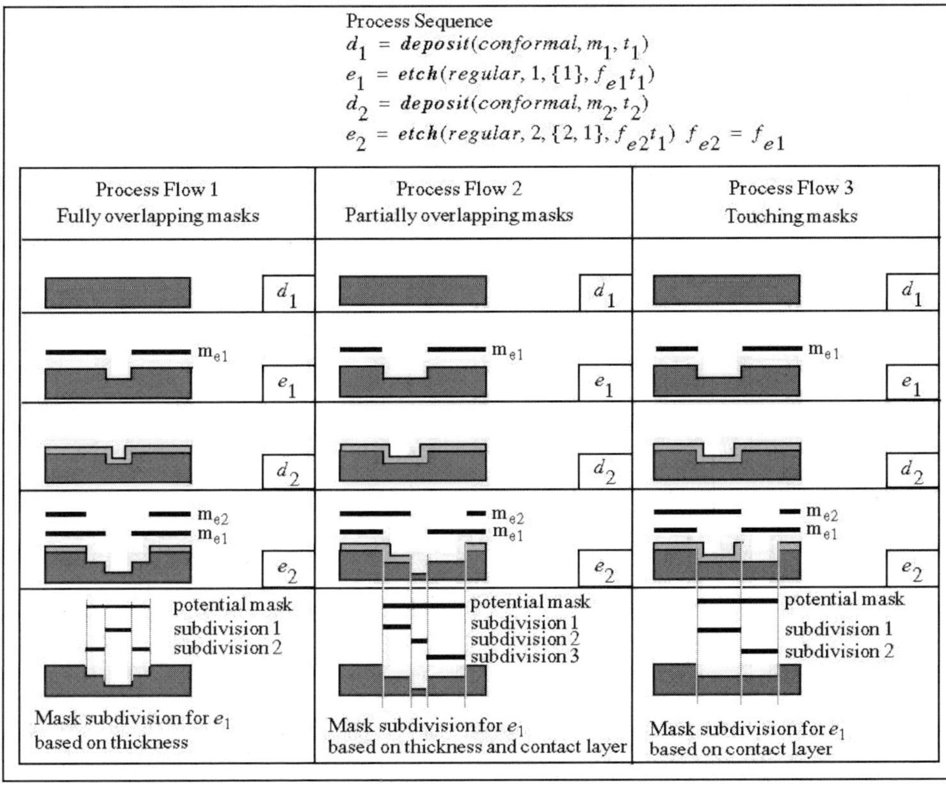

Figure 12. Subdivision of mask openings to associate an etch to a particular step

$$d_1 \to \begin{Bmatrix} 0 \\ 0 \\ t_1 \end{Bmatrix},\ d_2 \to \begin{Bmatrix} 0 \\ t_2 \\ 0 \end{Bmatrix},\ d_3 \to \begin{Bmatrix} t_3 \\ 0 \\ 0 \end{Bmatrix},\ e_1 \to \begin{Bmatrix} 0 \\ -t_2 \\ -f_{e1}t_1 \end{Bmatrix},\ e_2 \to \begin{Bmatrix} -t_3 \\ -t_2 \\ -f_{e2}t_1 \end{Bmatrix} \tag{8}$$

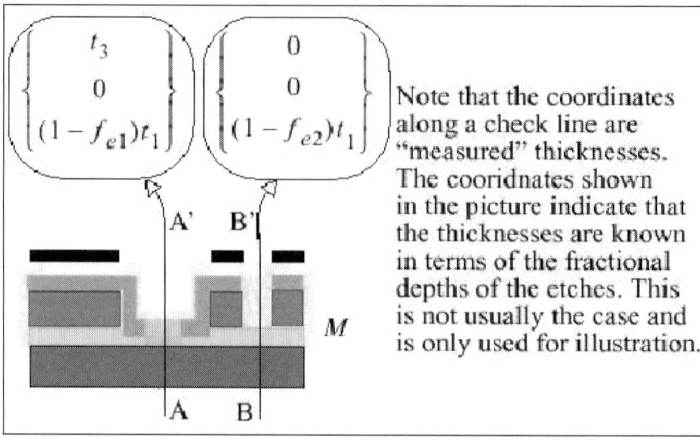

Figure 13. Check lines to validate mask openings

Each coordinate of the check line is a result of a linear combination of a sequence of deposits and etches as shown below for check lines *AA*' and *BB*' respectively:

$$\begin{Bmatrix} t_3 \\ 0 \\ (1-f_{e1})t_1 \end{Bmatrix} = c_1 \begin{Bmatrix} 0 \\ -t_2 \\ -f_{e1}t_1 \end{Bmatrix} + c_2 \begin{Bmatrix} -t_3 \\ -t_2 \\ -f_{e2}t_1 \end{Bmatrix} + c_3 \begin{Bmatrix} 0 \\ 0 \\ t_1 \end{Bmatrix} + c_4 \begin{Bmatrix} 0 \\ t_2 \\ 0 \end{Bmatrix} + c_5 \begin{Bmatrix} t_3 \\ 0 \\ 0 \end{Bmatrix} \tag{9a}$$

$$\begin{Bmatrix} 0 \\ 0 \\ (1-f_{e2})t_1 \end{Bmatrix} = c_1 \begin{Bmatrix} 0 \\ -t_2 \\ -f_{e1}t_1 \end{Bmatrix} + c_2 \begin{Bmatrix} -t_3 \\ -t_2 \\ -f_{e2}t_1 \end{Bmatrix} + c_3 \begin{Bmatrix} 0 \\ 0 \\ t_1 \end{Bmatrix} + c_4 \begin{Bmatrix} 0 \\ t_2 \\ 0 \end{Bmatrix} + c_5 \begin{Bmatrix} t_3 \\ 0 \\ 0 \end{Bmatrix} \tag{9b}$$

At a given check line, the unknown constants c_i $(i = 1..5)$ can only take values of 0 or 1 because an etch/deposit either is there or not. In particular, the constants related to deposits (i.e., c_i $(i = 3..5)$) are always 1 because deposits always exist. Thus to detect whether a particular etch occurred at a check line, one needs to find the values of the constants c_i $(i = 1, 2)$ related to the etches. For a general process with J deposits and K etches, this will yield J equations in K unknowns. The system of equations will be over-constrained, perfectly-constrained, or under-constrained depending on whether the number of deposits is greater than, equal to, or lesser than the number of etches. In all the cases, there can be multiple solutions, one solution, or no solutions depending on the rank of the above linear system of equations. Another notable point is that all the equations are independent of deposit thicknesses and only depend on the fractional depths of the etches. The equations above simplify to the form:

$$[A\}\{c\} = \{b\} \tag{10}$$

For check lines AA' and BB', respectively, Eq. (10) can be obtained as:

$$\begin{bmatrix} 0 & 1 \\ 1 & 1 \\ f_{e1} & f_{e2} \end{bmatrix} \begin{Bmatrix} c_1 \\ c_2 \end{Bmatrix} = \begin{Bmatrix} 0 \\ 1 \\ f_{e1} \end{Bmatrix} \text{ and } \begin{bmatrix} 0 & 1 \\ 1 & 1 \\ f_{e1} & f_{e2} \end{bmatrix} \begin{Bmatrix} c_1 \\ c_2 \end{Bmatrix} = \begin{Bmatrix} 0 \\ 1 \\ f_{e2} \end{Bmatrix} \tag{11}$$

The general solution of the system of equations involves using a singular value decomposition (SVD) to find the basic and free variables [37, 38]. If there are R free variables, there are 2^R potential solutions because the variables here can only take on the values 0 or 1. However not all the 2^R solutions need be feasible because the basic variables may not be only 1s and 0s. Thus the system of equations can potentially yield zero, one, or multiple solutions. The situation of no solutions implies that the given device cannot be fabricated with the specified process. Multiple solutions imply that the given mask opening can be fabricated by different sequences of etches. For the example shown above, the solutions are unique ($(c_1, c_2) = (1,0)$ for AA' and $(c_1, c_2) = (0,1)$ for BB') and correspond to the original mask openings in Fig. 2.

The above solution procedure thus gives an answer as to whether the given etch e_k is present or absent for a given mask opening. If the etch is present, then the mask opening can be kept, if not, the mask opening needs to be closed. If there are multiple feasible solutions, user input is required to select a given solution or multiple branches of solutions could be maintained. Additionally, if the status of a prior etch has been determined (i.e., the corrected mask openings of a previous etch have already been calculated and determined), there will be a reduction in the number of variables. In summary, the steps involved in detecting and eliminating incorrect mask openings are:

1. Select any check line that passes through the given mask opening.
2. Find the coordinates of the geometric model along the check line.
3. Find the coordinates of all the etches and deposits.

4. Express the coordinates along the check line as a linear combination of etches and deposits to get a system of linear equations.
5. Solve the system of linear equations using SVD, while taking care to ensure that some variables may have been assigned values previously.
6. If there are multiple solutions, prompt the designer to decide on whether or not the mask opening will remain.
7. Finally, retain or eliminate (close up) the potential mask openings depending on the solution.

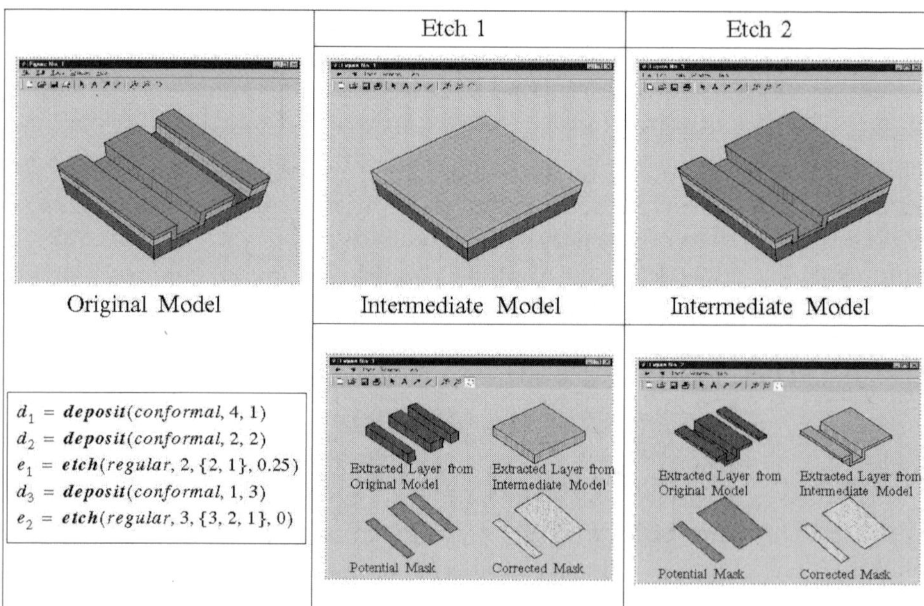

Figure 14. A hypothetical surface-micromachined device to illustrate the potential versus corrected masks.

5. EXAMPLES

The mathematical model, operations, and steps presented in the previous sections can be implemented in many different ways depending on how the geometric model is represented and manipulated in the computer. In this work, a volumetric model of MEMS devices was implemented using voxels

[36]. A voxel is a representation of a volume element of a solid body. A number of voxels are put together to form a volumetric model. Boolean and other operators have been implemented to create a computer model of the MEMS devices. The implementation of the voxel model was developed in C++ and the model was displayed by a program written in MATLAB®.

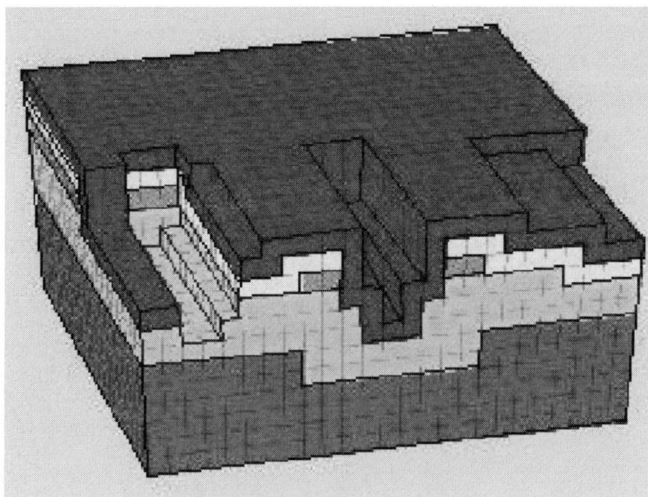

$$
\begin{aligned}
d_1 &= \textbf{\textit{deposit}}(conformal, 4, 1) \\
e_1 &= \textbf{\textit{etch}}(regular, 1, \{1\}, 0.5) \\
d_2 &= \textbf{\textit{deposit}}(conformal, 3, 2) \\
e_2 &= \textbf{\textit{etch}}(regular, 1, \{2\}, 0.5) \\
d_3 &= \textbf{\textit{deposit}}(conformal, 1, 3) \\
e_3 &= \textbf{\textit{etch}}(regular, 2, \{3, 2\}, 0) \\
d_4 &= \textbf{\textit{deposit}}(conformal, 1, 4) \\
e_4 &= \textbf{\textit{etch}}(regular, 2, \{4, 2\}, 0) \\
e_5 &= \textbf{\textit{etch}}(regular, 3, \{4, 3, 2\}, 0.25) \\
d_5 &= \textbf{\textit{deposit}}(conformal, 1, 5) \\
e_6 &= \textbf{\textit{etch}}(regular, 2, \{5, 2\}, 0.25)
\end{aligned}
$$

Figure 15. A five-layer part whose masks are to be synthesized for the process shown

6. CONCLUSIONS

In this chapter, the geometric mask synthesis problem for surface-micromachined devices was posed systematically and solved mathematically using singular value decomposition of an overconstrained linear system of equations. A systematic mask synthesis procedure enables a MEMS designer to automatically generate masks from a geometric model of the MEMS device. Therefore, the MEMS designer does not need to constantly check if a change in the design is compatible with a process when trying to improve the performance of a device. Future modeling work will focused on relaxing some of the assumptions so that the inverse problem can be undertaken for a larger domain of MEMS devices fabricated by processes other than surface micromachining. Finally, while the synthesis procedure is able to find all solutions for mask openings, it is not able to currently suggest a feasible process that can fabricate the device when no solutions are found. This is another practically useful extension of this work.

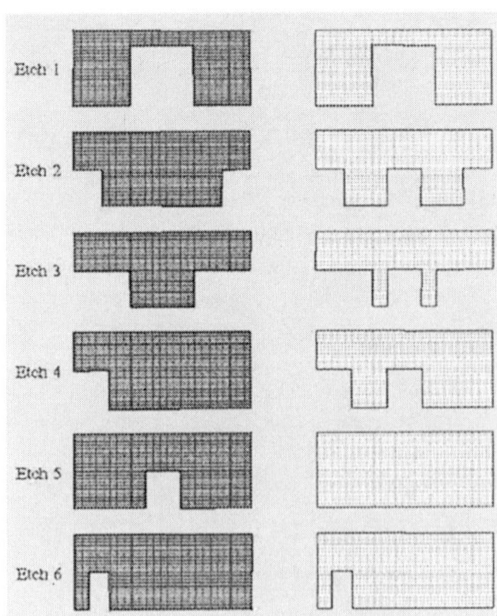

Figure 16. Two valid sets of masks synthesized for the part shown in Fig. 15.

ACKNOWLEDGMENTS

The grant support {DMI-9970021 (RS) and DMI-9970059 (GKA)} by National Science Foundation is gratefully acknowledged. The authors wish to acknowledge their respective students Venkat Ananthakrishnan (who did most of this work) and Andrew Perrin for their contributions to this work. Partial support for this work by the Iowa State University Carver Trust Grant is also gratefully acknowledged by RS.

REFERENCES

1. Madou, M. J., *Fundamentals of Microfabrication*, CRC Press, 1997.
2. Fatikow, S., and Rembold, U., *Microsystem Technology and Microrobotics*, Springer Verlag, 1997.
3. S. Iyer, T. Mukherjee and G.K. Fedder, "Multi-Mode Sensitive Layout Synthesis of Microresonators," *1998 International Conference on Modeling and Simulation of Microsystems, Semiconductors, Sensors and Actuators (MSM 98)*, Santa Clara CA, April 6-8, 1998.
4. Koppelman, G. M., and Wesley, M. A., "OYSTER: A Study of Integrated Circuits as Three-Dimensional Structures," *IBM Journal of Research and Development*, Vol. 27, No. 2, March 1983, pp. 149-163.
5. Koppelman, G. M., "OYSTER: A Three Dimensional Structural Simulator for Microelectromechanical Design," *Sensors and Actuators*, Vol. 20, 1989, pp. 179-185.
6. Senturia, S. D., Harris, R. M., Johnson, B. P., Kim, S., Nabors, K., Shulman, M. A., and White, J. K., "A Computer-Aided Design System for Microelectromechanical Systems," *IEEE/ ASME Journal of Microelectromechanical Systems*, Vol. 1, No. 1, March 1992, pp. 3-13.
7. Senturia, S. D., "CAD Challenges for Microsensors, Microactuators, and Microsystems," *Proceedings of the IEEE*, Vol. 86, No. 8, 1998, pp. 1611-1626.
8. Hubbard, T., and Antonsson, E. K., "Emergent Faces in Crystal Etching," *Journal of Microelectromechanical Systems*, Vol. 3, No. 1, 1994, pp. 19-28.
9. Hubbard, T., and Antonsson, E. K., "Cellular Automata in MEMS Design," *Sensors and Materials*, Vol. 9, No. 7, 1997, pp. 437-448.
10. Strasser, E., and Selberherr, S., "Algorithms for Cellular Based Topography Simulation," *IEEE Transactions on Computer- Aided Design of Integrated Circuits and Systems*, Vol. 14, No 9, 1995, pp. 1104-1114.
11. Dixit, H., Kannapan, S., and Taylor, D. L., "3D Geometric Simulation of MEMS Fabrication Processes: A Semantic Approach," *Proceedings of the 4th ACM Symposium on Solid Modeling and Applications*, 1997, pp. 376-387.
12. MEMS-CAD companies: (incomplete list) CFD-GEOM, CFD Research Corporation, www.cfdrc.com; CoventorWare, Coventor, www.memcad.com; IntelliSuite, Corning Intellisense Inc., www.intellisense.com; MEMSCAP, Memscap, www.memscap.e-sip.com; MEMS-Pro, Tanner Research, www.tanner.com.

13. Ho, C. P., Plummer, J. D., Hansen, S. E., and Dutton, R. W., "VLSI Process Modeling - SUPREM III," *IEEE Transactions on Electron Devices*, Vol. 30, No. 11, pp. 1438-1453.
14. Hasanuzzaman, M., and Mastrangelo, C. H., "Process compilation of thin film microdevices," *IEEE Transactions on Computer-Aided Design of Integrated Circuits and Systems*, Vol. 15, No. 7, 1996, pp. 745-764.
15. Zaman, M. H., Carlen, E. T., and Mastrangelo, C. H., "Automatic Generation of Thin Film Process Flows: Part I - Basic Algorithms," *IEEE Transactions on Semiconductor Manufacturing*, Vol. 12, No. 1, 1999a, pp. 116-128.
16. Zaman, M. H., Carlen, E. T., Mastrangelo, C. H., "Automatic Generation of Thin Film Process Flows: Part II - Recipe Generation, Flow Evaluation, and System Framework," *IEEE Transactions on Semiconductor Manufacturing*, Vol. 12, No. 1, 1999b, pp. 129-138.
17. Rodgers, M.S. and Sniegowski, J.J., "Designing Microelectromechanical Systems-on-a-Chip in a Five-Level Surface Micromachining Technology," *Proceedings of the International Conference on Engineering Design Automation*, 1998.
18. Synopsis, "Proteus—Mask Correction to Accommodate Systematic Distortions in Fabrication," *www.synposis.com.*
19. Jaccodine, R.J., "Use of Modified Free Energy Theorem to Predict Equilibrium Growing and Etching Shapes," *Journal of Applied Physics*, Vol. 33, No. 8, 1962, pp. 2643-2647.
20. DeLapierre, G., "Anisotropic Crystal Etching: A Simulation Program, *Sensors and Actautors A- Physical*," Vol. 31, 1992, pp. 267-274.
21. Buser, R.A. and de Rooij, N.F., "ASEP: A CAD Program for Silicon Anisotropic Etching," *Sensors and Actuators, A Physical*, Vol. 28, 1991, pp. 71-78.
22. Sequin, C.H., "Computer Simulation of Silicon Anisotropic Crystal Etching," *Sensors and Actuators, A Physical*, Vol. 34, No. 3, 1992, pp. 225-241.
23. Anttonsson, E. and Cagan J. (Ed.), *Formal Engineering Design Synthesis*, Cambridge University Press, 2001.
24. Lee, C-Y., and Antonsson, E. K., "Self-Adapting Vertices for Mask Layout Synthesis," *Proceedings of the Modeling and Simulation of Microsystems Conference*, M. Laudon and B. Romanowicz (eds.)., SanDiego, March 27-29, 2000, pp. 83-86.
25. Li, H., and Antonsson, E. K., "Evolutionary Techniques in MEMS Synthesis," *Proceedings of the ASME Design Engineering Technical Conferences*, 1998, Atlanta, Georgia.
26. Ma, L., and Antonsson, E. K., "Automated Mask-Layout and Process Synthesis for MEMS," *Proceedings of the Modeling and Simulation of Microsystems Conference*, M. Laudon and B. Romanowicz (eds.), SanDiego, March 27-29, 2000, pp. 20-23.
27. Yarberry, V. and Jorgensen, C., "A 2D Visualization Tool for SUMMiT V Designs," *Modeling and Simulation of Microsystems* Conference, 2001, www.cr.org.
28. Yarberry, V. R., "MEMS Design Rule Checking: A Batch Approach for Remote Operation," *SPIE 3328*, 1998, pp. 32.39.
29. Ananthakrishnan, V., Sarma, R., and Ananthasuresh, G.K., "Part to Art: Basis for a Systematic Geometric Design Tool for Surface Micromachined MEMS," *CD-ROM Proceedings* of the 2000 ASME Design Technical Conferences, Baltimore, MD, Sep. 10-13, paper # DETC2000/DAC-14251.
30. Perrin, A., Ananthakrishanan, V., Sarma, R., and Ananthasuresh, G.K., "Voxel-Based Heterogeneous Geometric Modeling for Surface-Micromachined MEMS,"

Proceedings of the Fourth International Conference on Modeling and Simulation of Microsystems held at the Hilton Oceanfront Resort, Hilton Head Island, South Carolina, U.S.A., March 19-21, 2001, pp. 124-127.

31. Ananthakrishnan, V., "Part to Art: The basis for a systematic geometric design tool for surface micromachined MEMS," *Masters Thesis*, University of Toledo, Toledo, OH, U.S.A., 2000.
32. Rappoport, A., "Geometric Modeling: A New Fundamental Framework and its Practical Implications," *Proceedings of the ACM Solid Modeling Conference*, Salt Lake City, Utah, 1995, pp. 31-41.
33. Mäntyla, M., *Introduction to Solid Modeling*, W. H. Freeman and Co., New York.
34. Osterberg, P. M., and Senturia, S. D., "MEMBUILDER: An Automated Solid Model Construction Program for Micromechanical Structures," *Proceedings of the International Conference on Solid State Sensors and Actuators and Eurosensors IX*, Vol. 2, 1995, pp. 21-24.
35. Kumar, V., and Dutta, D., "Approach to Modeling and Representation of Heterogenous Objects," *ASME Journal of Mechanical Design*, Vol. 120, No. 4, 1998, pp. 659-667.
36. Chandru, V., and Manohar, S., "Volume Modeling for Emerging Manufacturing Technologies," *Sadhana*, Vol. 22, Part 2, April 1997, pp. 199-216.
37. Press, W. H., Teukolsky, S. A., Vetterling, W. T., and Flannery, B. P., *Numerical Recipes in C: The Art of Scientific Computing*, Cambridge University Press, 1993.
38. Strang, G., *Linear Algebra and its Applications*, Harcourt Brace Jovanovich, 1988.

Chapter 10

SYSTEM-LEVEL SYNTHESIS

Tamal Mukherjee* and Gary K. Fedder*†
*Department of Electrical and Computer Engineering, and †The Robotics Institute, Carnegie Mellon University, Pittsburgh, PA, 15213, USA.

Synopsis: Microelectromechanical Systems such as resonators, accelerometers, gyroscopes, IR sensors, RF filters, electrothermal converters and force sensors can be composed out of beam springs, plate masses, dampers, and electromechanical comb sensors and actuators. MEMS design involves iteratively designing each of these submodules and the entire transducer including the electronics, to meet given design specifications. System-level synthesis helps automate much of this design problem for a fixed MEMS transducer topology. First, geometric layout design variables are identified to describe the topology. Next, functional constraints that map these variables to engineering performance specifications are obtained by static and dynamic mechanical as well as electrostatic analysis. Then, the variables and constraints are used to formulate a mixed-integer non-linear optimization problem, which is solved to synthesize the transducer layout from high-level engineering specifications. A variety of objective functions can be used to automate the exploration of the entire design space given specific user-specified engineering constraints, allowing the designer to understand the complex design trade-offs inherent to the design problem.

1. MEMS DESIGN REPRESENTATIONS

Design of complex integrated MEMS, like many other engineering design problems, relies on two fundamental principles: *divide and conquer* to simplify the design problem, and *iterative refinement* to optimize the design. To understand how these principles can be applied, we need to first consider the relevant MEMS design representations. Physical representations linking process sequences, material properties, and mask layout into a 3D structural view for continuum prediction of behavior [1], formalized the first

MEMS design representations (Figs. 1a and 1b). As MEMS design moved from the microstructure to microsystem levels, hand-crafted lumped-parameter models [2] such as the graphically represented behavioral schematic (Fig. 1c) became necessary. To get around the hand-crafted models, the MEMS "circuit" schematic of Fig. 1d was proposed [3]. Common MEMS circuit elements are identified, and parameterized with respect to geometry to model the element behavior. Circuit simulation frameworks integrate elements for lumped parameter circuit simulation.

These design representations may be linked into an integrated MEMS [4] design flow. The goal of the flow is to lead a design team from product definition to fabrication as quickly as possible (due to time-to-market pressures) with a minimal number of fabrication iterations (due to cost related constraints). Historically, the decade-long investment into the layout (Fig. 1a) and 3-D (Fig. 1b) representations has implied that the primary design flow available to MEMS is centered around those representations.

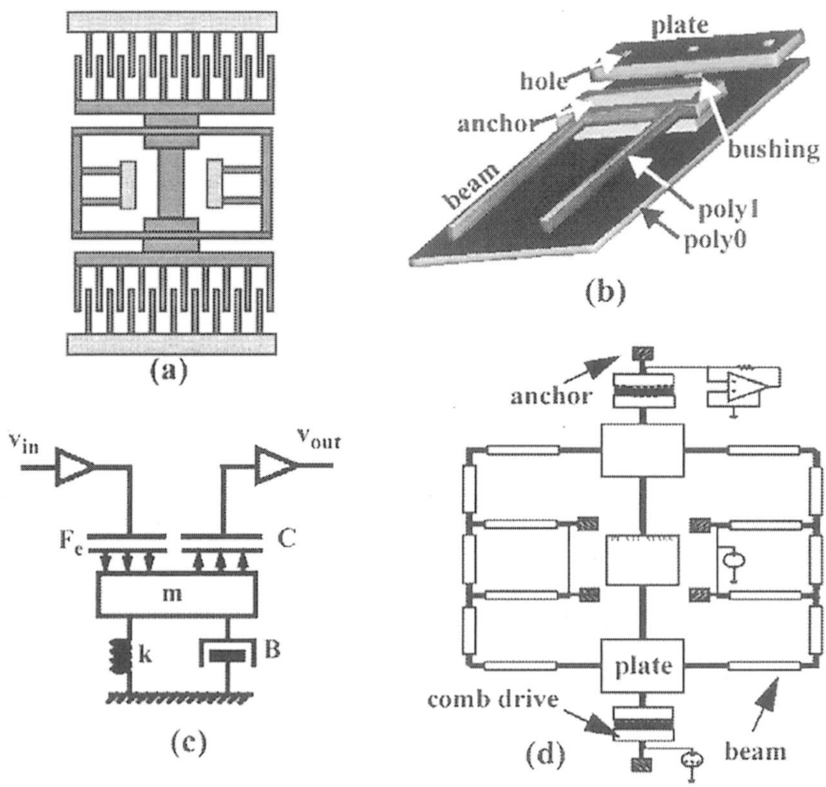

Figure 1. MEMS design representations of a micromechanical resonator filter with electromechanical transducer at input and output (a) layout, (b) 3D model (of a portion of the layout), (c) equation-based behavioral schematic, and (d) MEMS "circuit" schematic.

The 3D view has classically been the design entry mode of choice for the mechanical designer and was therefore co-opted for microstructure simulation. As most MEMS devices and systems intrinsically involve multiple energy domains, coupled field solvers [5,6] are needed. MEMS design environments enabling 3D model construction, coupled simulation using finite and boundary element analyses (FEA/BEA), and visualization of the results, are now widely available. 3D design optimization tools for various classes of MEMS components such as cantilever actuators [7], compliant mechanisms (flexible structures that generate a wide variety of mechanical motions through elastic deformations) [8], spring/active area trade-offs [9], comb-drives [10], and bulk micromachined structures [11] have also been developed.

The layout view is the preferred mode of design entry for microfabrication, as it efficiently maps design intent to manufacturing. However, as this view is ill-suited for simulation, MEMS design environments provide generation of 3D view directly from layout [12,13,14] to support simulation.

Design entry in the layout and 3D modes is cumbersome due to the level of detail needed to describe the design, especially for the rapid iterative design needs in the early stages of design exploration. For this stage of design, the use of analytical performance models (Fig. 1c) is the most widely used approach. Such analytical models are often entered into mathematical analysis/spreadsheets to enable design iteration for sizing optimization and will be the focus of this chapter. Emerging MEMS 'circuit simulation' approaches [3], in which a MEMS device is partitioned into *atomic* elements with elemental behavioral model is coupled to electronic simulation by using analog Hardware Description Languages (HDL), will lead to MEMS synthesis becoming increasingly similar to analog synthesis [15] to both design the schematic and generate layout. Layout to schematic translation will occur via extraction [16].

Here we focus on one part of the entire design flow: the *system-level synthesis*. The goal is to generate a design layout from high-level engineering specifications. We focus on the existing widely used equation-based behavioral schematic models and outline how to transform the models into a synthesis module. The well-known polysilicon resonator [17] is used to exemplify the steps involved in formulating a system-level synthesis module using analytical performance models [18]. A similar procedure can be applied for polysilicon accelerometers [19] and CMOS-MEMS accelerometers [20] as is outlined in [21, 22].

2. SYNTHESIS METHODOLOGY

The synthesis methodology involves the identification of all the degrees of freedom and the constraints in the design problem. The development of the synthesis tool is initiated by identifying the design variables that capture the degrees of freedom in the design. The design space is then defined by the maximum and minimum values of each design variable. Constraints, which the design variables should satisfy in order for the design to be acceptable, are then formulated. Different objective functions such as *minimize area*, are implemented in order to drive the synthesis towards preferred types of designs (compact designs, in the case of the *minimize area* objective function). The synthesis is achieved through an optimization algorithm which seeks to minimize the objective function while satisfying the constraints.

2.1 Design Variables

Fifteen design variables are identified for a microresonator [18]. The design variables are listed in Table 1 and depicted in Fig. 2.

Table 1. Design and style variables for the microresonator. Upper and lower bounds are in units of μm except N and V.

DESIGN VARIABLES							
Var.	Description	Min	Max	Var.	Description	Min	Max
L_b	Flexure beam length	2	400	w_{cy}	Comb yoke width	10	400
w_b	Flexure beam width	2	20	L_{cy}	Comb yoke length	2	700
L_t	Truss beam length	2	400	L_c	Comb finger length	8	400
w_t	Truss beam width	2	20	w_c	Comb finger width	2	20
L_{sy}	Shuttle yoke length	2	400	g	Comb finger gap	2	20
w_{sy}	Shuttle yoke width	10	400	x_o	Comb finger overlap	4	400
w_{sa}	Shuttle axle width	10	400				
N	# of rotor fingers	1	100	V	Voltage amplitude	1V	50V
STYLE VARIABLES							
w_{ba}	Beam anchor width	11	11	w_{ca}	Stator anchor width	14	14

These include 13 geometrical parameters (shown in Fig. 2), the number of fingers in the comb drive, *N*, and the effective voltage, *V*, applied to the comb drive. In practice, a dc voltage V_{dc} is applied to the shuttle, and a sinusoidal voltage source with amplitude V_{ac} is applied to one of the actuators. Under these conditions, we can simplify the applied voltage as an effective sinusoidal voltage with amplitude $V = \sqrt{2V_{ac}V_{dc}}$ for the electrostatic force calculation.

The minimum beam widths and minimum spaces between structures are set by technology-driven design rules. Maximum beam lengths are constrained to 400 μm to avoid problems with undesirable curling due to stress gradients in the structural film and possible sticking and breakage during the wet release etch. Beam widths are constrained to a maximum of 20 μm by the limited undercut of PSG to release the structures. The shuttle axle, the shuttle yoke and the comb yoke are at least 10 μm wide to ensure that they are relatively more rigid than folded-flexure beams (and can be ignored in the stiffness models). The comb yoke is allowed to extend up to 700 μm, to fill up the entire flexure length allowed for the resonator, even if the comb fingers occupy only a fraction of a length of the comb yoke. The number and type of design variables for other MEMS devices is similar to that for the resonator. For example, seventeen design variables are needed to define the geometrical layout topology for a polysilicon accelerometer [21] and fourteen for a CMOS-MEMS accelerometer [22].

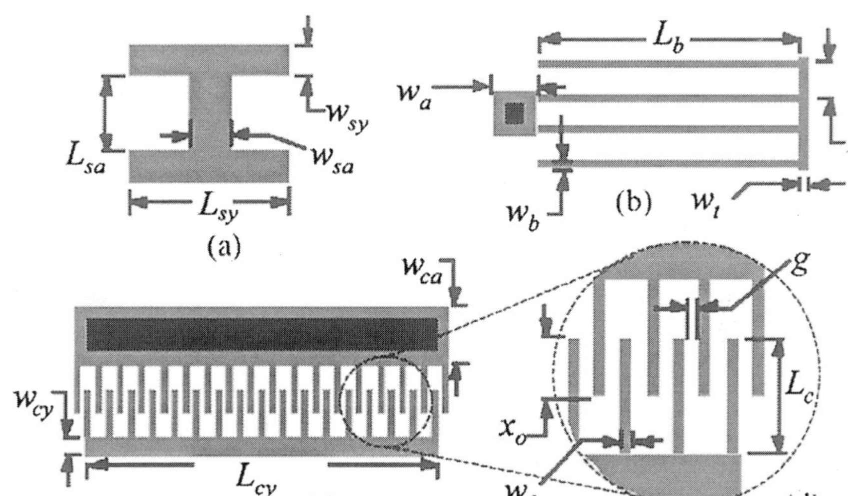

Figure 2. Dimensions of the microresonator elements. (a) shuttle mass, (b) folded-flexure, (c) comb drive with N movable 'rotor' fingers, (d) close-up view of comb fingers.

Geometric *style* variables are needed to complete the definition of the layout. Style variables do not affect the device behavior, and therefore, they are not considered to be *design* variables. They usually define the stationary parts of the device, such as the anchors for the stationary comb fingers and the folded-flexure beams in the resonator. Style variables are usually set to fixed values that exceed the process design rule minima. They can be considered to be device-specific design rules. Examples include the width of the anchor supports, w_{ba} and w_{ca}, the offset of attachment points of the flexure beams to the anchor edge, and the overlap around anchor cuts. The polysilicon accelerometer has 10 style variables [21] and the CMOS-MEMS accelerometer has nine style variables [22]. Some style variables are defined by the geometric topology of the device. Others are dimensions, whose impact on sensor behavior is yet to be quantified (*e.g.*, the size of the etch holes in a plate-mass in a CMOS-MEMS process). Therefore, they are commonly set to fixed values that have been shown to lead to high yield in similar devices in the past [19,20].

2.2 Constraints

The constraints can be classified into two types: *geometric* constraints, which are directly related to the physical dimensions of the microresonator and *functional* constraints, which are related to the behavior of the microresonator.

2.2.1 Geometric Constraints

Table 2. Geometric constraints

Constraint Description	Expression	Min [μm]	Max [μm]
Actuator length	$L_{cy}+2g+2w_c$	0	700
Comb-fill	$(2N+1)\ w_c + 2Ng - L_{cy}$	700	0
Flexure length	$L_{sy}+2L_b+2w_t$	0	700
Total resonator width	$3L_t+w_{sy}+4L_c-2x_o+2w_{cy}+2w_{ca}$	0	700
Comb clearance during motion	$L_c-(x_o+x_{disp})$	4	200
Minimum comb overlap	x_o-x_{disp}	4	200
Shuttle clearance during motion	$L_t-x_{disp}-(w_{sy}+w_b)/2$	4	200
Shuttle gap in y	$(L_{sy}-2w_{ba}-w_{sa})/2$	2	200

The geometric constraints illustrated in Fig. 3 and detailed in Table 2 ensure a functional resonator. The resonator width and length must not exceed an arbitrary fixed size, set at 700 μm in the example presented. Depending on the design, the overall resonator length is determined by either the flexure or the comb drive actuator. Therefore, both constraints need to be simultaneously satisfied. The actuator length constraint is linear (an alternative non-linear form of the constraint would have been $(2N+1)w_c+2Ng$). Linearized constraints aid in the efficiency of the optimization-based synthesis, and are preferred over non-linear constraints. The length of the comb finger yoke (L_{cy}) enables linearization and provides an extra degree of freedom, which allows the comb drive yoke to be longer than the minimum length required to accommodate all the comb fingers. The comb-fill constraint, $(2N+1)w_c + 2Ng \le L_{cy}$, prevents the comb drive length from being too small to hold all the fingers.

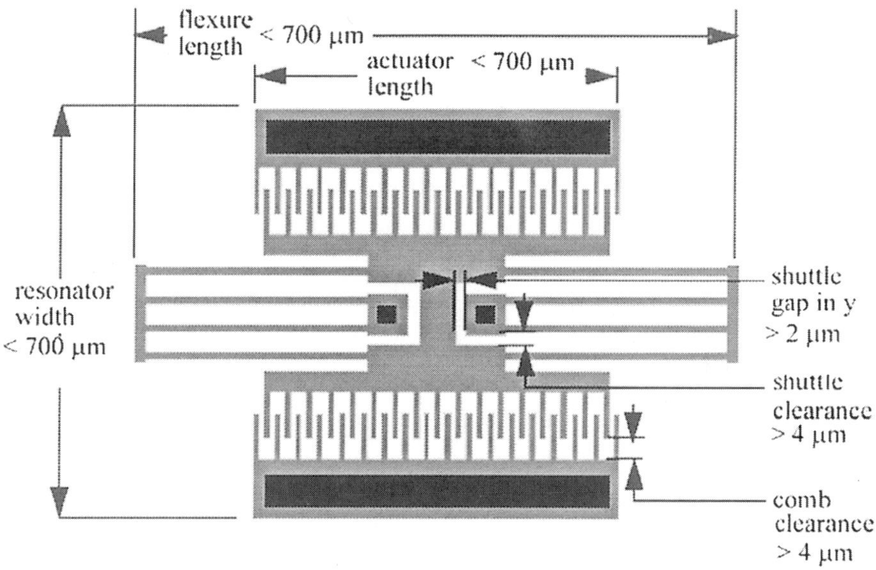

Figure 3. Geometric constraints. These constraints limit the overall size of the microresonator and also prevent the moving parts from colliding into the fixed parts of the microresonator.

Gaps between the comb fingers and between the shuttle and beam anchor must allow the shuttle to move freely and must accommodate the maximum possible stroke. The maximum expected displacement of the shuttle mass will be at resonance, and is encoded in the motion limit constraints using x_{disp}. First, we ensure that the comb fingers do not contact each other at the maximum displacement. Next, we constrain the minimum comb overlap at the maximum displacement, to maintain linearity of the comb drive actuation. We also constrain the resonator geometry to ensure an adequate shuttle clearance during movement of the shuttle. Finally, a shuttle gap constraint is defined to encode the technology-driven design rule for gaps between moving and anchored parts. Thus, we have four constraints due to the undisplaced device layout topology, two set by clearance limits in the comb topology and two by clearance limits in the shuttle topology.

Similar constraints exist in the polysilicon accelerometer (five arising from the undisplaced layout topology and six from clearance limits for the displaced proof mass) [21]. The CMOS-MEMS accelerometer has four geometric undisplaced layout constraints and three clearance limit constraints [22].

2.2.2 Functional Constraints

Constraints on the design specifications are assigned realistic values for synthesizing a valid resonator for use as a structure used for characterization. Alternative constraint values can be readily assigned depending on the design goals.

An essential specification is the resonant frequency of the lowest (preferred) normal mode. A valid layout must have a resonant frequency within 1% of the desired value (f_{spec}). Resonant frequencies of the other in-plane modes, f_{fl}, $f_{\theta z}$ and f_y, (collectively represented by f_i in Table 3 must be at least three times greater than f_x for adequate mode separation. For the out-of-plane modes of vibration, the quality factor is expected to be much lower than for the in-plane modes, since squeeze-film damping between the microresonator and the bottom layer will dominate over viscous air damping due to lateral motion. Therefore, even if these modes are closer to the x-mode than the in-plane modes, their oscillations will be more damped. Additionally, it is very difficult to achieve the *factor of three* in mode separation for the out-of-plane modes. Hence, the out-of-plane modes f_z, $f_{\theta x}$, and $f_{\theta y}$ (collectively represented by f_o in Table 3) are constrained to be at least twice f_x. For stability, the restoring force of the spring in the *y* direction must be three times greater than the destabilizing electrostatic force from the comb drive (i.e., $3k_{e,y} < k_y$). A similar stability constraint must hold for the rotational mode.

Table 3. Functional constraints

Constraint Description	Expression	Min	Max
Resonant frequency	f_x/f_{spec}	0.99	1.01
Stroke at resonance	x_{disp}	2 μm	100 μm
Quality factor in x	Q_x	5	10^5
y-axis stability	$k_{e,y}/k_y$	0	1/3
θ_z stability	$k_{e,\theta z}/k_{\theta z}$	0	1/3
In-plane mode separation	f_x/f_i	0	1/3
Out-of-plane mode separation	f_x/f_o	0	1/2
Ky accuracy	$k_y/k_{y,axle}$	0	1/10
Kx accuracy	x_{disp}/L_b	0	1/10
Buckling	L_b/L_{cr}	0	1/2

Assuming that the system is underdamped, the displacement amplitude at resonance is $x_{disp}=Q_xF_x/k_x$, where $F_x \propto NV^2$ is the comb drive force, $Q_x = \sqrt{m_x k_x / B_x^2}$ is the quality factor, and B_x is the damping coefficient. To enable easy visual confirmation of resonance, the displacement amplitude is constrained to be at least 2 μm. A quality factor constraint, $Q_x \geq 5$, is also implemented to ensure underdamped operation.

Assumptions used in deriving lumped-parameter macromodels need to be encoded as functional constraints. For example, in the y direction, the shuttle is forced to be stiffer than the flexure due to $k_{y,axle} > 10k_y$ (thus shuttle compliance is ignored in the model). Limiting the beam deflection to 10% of the beam length (k_x accuracy) ensures that a linear flexure stiffness model is accurate for synthesis. Without this constraint, a non-linear flexure stiffness model is needed for the synthesis result to accurately reflect device operation.

Finite residual stress in mechanical polysilicon films can cause released fixed-fixed suspensions to break in tension or buckle under compression. Polysilicon can be deposited either with compressive or tensile residual stress, depending on the processing conditions during deposition. For example, in the MUMPs process [23], residual stress is always compressive, having a nominal value of -10 MPa and worst-case value of -20 MPa.

Beams in the folded flexure are free to expand outward to relieve residual axial stress. However, as shown in Fig. 4, the central shuttle also expands an amount Δ due to the residual stress, creating additional axial stress in the outer beams and tension in the inner beams. A first-order value of the critical buckling length, L_{cr}, for the folded-flexure is given by the Euler column formula, $L_{cr} = \pi w\sqrt{2L_b/3\Delta}$, where $2L_b < L_{cr}$ to ensure no buckling, and w corresponds to the minimum of w_b and t.

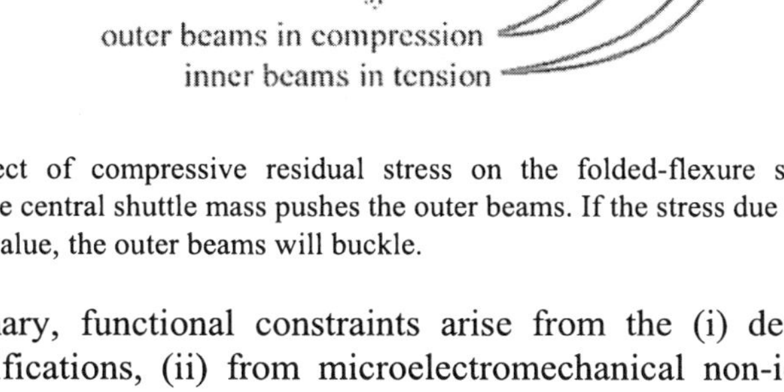

Figure 4. Effect of compressive residual stress on the folded-flexure suspension. The expansion of the central shuttle mass pushes the outer beams. If the stress due to this is greater than a critical value, the outer beams will buckle.

In summary, functional constraints arise from the (i) device specific design specifications, (ii) from microelectromechanical non-idealities that need to be alleviated, and (iii) model accuracy limitations. The polysilicon and CMOS-MEMS accelerometer both have the same five accelerometer specifications (sensitivity, minimum detectable acceleration, maximum range, cross-axis sensitivity and bandwidth) [21,22]. The polysilicon acclerometer has four constraints to alleviate the non-idealites similar to the stability, mode separation and buckling constraints for the resonator [21]. No model accuracy limitations are needed as the accelerometer displacements are much smaller than that of the resonator. The CMOS-MEMS accelerometer has five constraints to alleviate the non-idealites, and one model accuracy constraint [22]. The model accuracy constraint does not come from the device spring model as in the resonator, but rather from the circuit level model (parallel combination of springs).

2.3 Synthesis Formulation

Synthesis of the microresonator will result in one of two possible outcomes. Several designs may satisfy the above constraints, or no designs may meet the constraints (indicating a null design space). Our synthesis approach is to select the design that minimizes an objective function and therefore, may be considered optimal. The synthesized result depends very strongly on the choice of the objective function.

Generally, devices that occupy smaller area are preferred for cost reduction. Smaller operating voltages are also preferred for integrated devices. Therefore, three objective functions to be minimized are implemented: total active area, amplitude of the comb drive voltage, and the sum of normalized area and normalized voltage (normalized to maximum possible area and voltage respectively). The amplitude of oscillation is very crucial and large amplitudes are required for better sensing capabilities. To achieve this, a fourth objective function: maximize displacement at resonance, is also implemented. Engineering design is usually limited to finding a physical design. The choice of objective can therefore be arbitrarily chosen from the design parameter space (area, voltage) or design specification space (displacement). When the synthesized design is a component of a much larger system, there is often an incomplete model for how the component specifications interact to achieve the system specification. In such designs, the design specification objectives tend to be more useful. On the other hand, if the design question is to identify process limitations to motivate the development of a new micromechanical process, then a design parameter related to the process (such as thickness) is more useful [24].

Whatever the objective is, the overall synthesis problem is translated into a constrained optimization formulation that is solved using a non-linear constrained optimization technique. During the optimization, designs are evaluated by the values of the constraint functions and the objective functions for the current values of the design variables. Depending on the choice of the objective function, there can be more than one minimum point in the optimization, due to the complex non-linear characteristics of the individual equations in the lumped-element models. Furthermore, since our goal is synthesis, we need to be independent of any choice of starting point for the optimization.

In order to increase the probability of finding a better design (i.e., move closer to the global optimum) a gridded multi-start algorithm coupled with a gradient-based constrained optimization (NPSOL) [25] efficiently solves for the global minimum of the objective function. Alternate optimizers such as the built in solver in Microsoft Excel can also be used for many simple MEMS design problems using this synthesis methodology [26]. The use of a starting grid eliminates the need to provide good starting points to the gradient-based optimization. The starting grid is formed by assigning 3 values to each design variable (as described in Section 2.1, there are 15 design variables) leading to 3^{15} starting points. Each of these points in the design space is evaluated and 100 designs which best meet the constraints are selected. These 100 points are used as the starting points for the gradient-based optimization. A number of these 100 optimization runs may converge to the same design. From among the different designs resulting from these 100 optimization runs, the best design is chosen as the final synthesis result. This approach is feasible for both the resonator [18] and the accelerometer [21, 22] synthesis as they all have about a dozen variables. Alternative approaches are needed for reasonable synthesis execution times if the number of variables exceeds 50.

The non-linear constrained optimization formulation can be written as:

$$\begin{aligned} \min_u \quad & z = \sum_{i=1}^{k} w_i f_i(\overline{u}) \\ s.t. \quad & \overline{h}(\overline{u}) = 0 \\ & \overline{g}(\overline{u}) \leq 0 \\ & \overline{u} \in U_P \end{aligned}$$

where $\underline{u}$ is the vector of independent design variables given in Table 1. $\underline{f}(\underline{u})$ is a set of objective functions that codify performance specifications the designer wishes to optimize, *e.g.*, area. Scalar weights, w_i, balance competing objectives. $\underline{h}(\underline{u}) = 0$ and $\underline{g}(\underline{u}) \leq 0$ are each a set of functions that implement the geometric and functional constraints given in Table 2 and

Table 3. The decision variables can be described as a set $\underline{u} \in U_P$, where U_P is the set of allowable values for $\underline{u}$ (described by the bounds in Table 1).

The MEMS design problem cannot be completely modeled in the non-linear constrained optimization formulation. Some of the design variables in the design (such as the number of comb fingers) are integer in nature. The number of comb fingers is initially treated as a continuous variable. When the optimization (called the *relaxed* problem) terminates successfully, the number of comb fingers is truncated to the nearest integer and removed from the list of design variables. The optimization is run again (called the *non-relaxed* problem) with the result of the *relaxed* problem as the starting point, resulting in the final synthesized design. Furthermore, all the geometry parameters will directly affect the physical microresonator layout. Therefore, they should be represented as integers with centi-micron units rather than as real numbers, as is the case in the classical non-linear constrained optimization formulation. To implement this, the values of the design variables that result from the *non-relaxed* problem are rounded off to the nearest centi-micron units.

2.4 Layout Generation

Once the optimization results in a valid design, i.e., a set of values for the design variables, which satisfy all the requirements, these values are fed to a parameterized layout generation tool, CAMEL [27]. CAMEL produces a CIF file, which contains the mask information required for fabrication of the synthesized microresonator. CAMEL was modified to produce simplified layouts of the microresonator so that the number of style variables is reduced. An alternative is to use parameterized layout generation tools available in most electronic design automation frameworks [28].

3. PERFORMANCE MODELS

The functional constraints are based on a lumped parameter spring-mass-damper system model for the resonator that is constructed from models in the existing literature. Physics-based models for the effective spring stiffness of the folded-flexure suspensions, the effective masses of the shuttle mass, comb drives and the folded-flexure, and the viscous air damping are used in the synthesis tool. For completeness, the eight primary modes of vibration of the microresonator need to be modeled. Each oscillation mode is described by a lumped second-order equation of motion. For any generalized displacement ζ, we can write:

$$F_{e,\zeta} = m_\zeta \ddot{\zeta} + B_\zeta \dot{\zeta} + k_\zeta \zeta \tag{1}$$

where $F_{e,\zeta}$ is the external force (in the x-mode this force is generated by the comb drives), m_ζ is the effective mass, B_ζ is the damping coefficient, and k_ζ is the spring constant. Now, for example, the x-mode frequency is given by $\omega_x = 2\pi f_x = \sqrt{k_x/m_x}$. The other modes are modeled similarly.

Linear equations for the spring constants are derived using energy methods [29]. A force (or moment) is applied to the free end(s) of the spring in the direction of interest, and the displacement is calculated symbolically (as a function of the design variables and the applied force). In these calculations different boundary conditions are applied for the different modes of deformation of the spring. The spring constant of the folded-flexure in the x-direction is [29]

$$k_x = \frac{2Etw_b^3}{L_b^3} \frac{L_t^2 + 14\alpha L_t L_b + 36\alpha^2 L_b^2}{4L_t^2 + 41\alpha L_t L_b + 36\alpha^2 L_b^2} \tag{2}$$

where E is the Young's modulus of polysilicon, t is the polysilicon thickness, and $\alpha = (w_t/w_b)^3$. The geometrical layout parameters L_t, L_b, w_t and w_b are as shown in Fig. 2.

In order to calculate the effective mass accurately, it is not only necessary to take into account the velocities of the beams in the folded-flexure in the x-direction, but also the velocities of the truss beams in the y-direction. With this, the net effective mass of the microresonator in the x-direction (m_x) can be written as

$$m_x = m_{shuttle} + m_{t,eff} + m_{b,eff}$$

$$m_{b,eff} = \frac{m_{beams}}{140\beta}\left(832L_t^4 + 16121\alpha L_t^3 L_b + 92706\alpha^2 L_t^2 L_b^2 + 138348\alpha^3 L_t L_b^3 + 62208\alpha^4 L_b^4\right)$$

$$m_{t,eff} = \frac{m_{truss}}{280\beta}(57L_t^6 + 1020\alpha L_t^5 L_b + 4644\alpha^2 L_t^4 L_b^2 + 1120 L_t^4 L_b^2 + 17920\alpha L_t^3 L_b^3 +$$
$$91840\alpha^2 L_t^2 L_b^4 + 161280\alpha^3 L_t L_b^5 + 90720\alpha^4 L_b^6)$$

$$\beta = (4L_t^2 + 41\alpha L_t L_b + 36\alpha^2 L_b^2)^2$$

where m_{shuttle} is the shuttle mass, $m_{\mathrm{t,eff}}$ leads to the effective mass of all truss sections, $m_{\mathrm{b,eff}}$ leads to the effective mass of all the long beams, m_{truss} is the total mass of all truss sections, m_{beams} is the total mass of all the long beams.

Viscous air damping dominates the energy dissipation mechanisms in microresonators at atmospheric pressure. The total damping force in the x-direction is mainly composed of the forces due to Couette flow below the resonator, Stokes flow above the resonator, and air flow in the gap between comb fingers. The expression for the damping coefficient is [31]:

$$B_x = \mu\left[\left(A_s + 0.5A_t + 0.5A_b\right)\left(\frac{1}{d} + \frac{1}{\delta}\right) + \frac{A_c}{g}\right] \tag{3}$$

where μ is the viscosity of air, d is the fixed spacer gap between the ground plane and the bottom surface of the comb fingers, δ is the penetration depth of airflow above the structure, g is the gap between comb fingers, and A_s, A_t, A_b, and A_c are layout areas for the shuttle, truss beams, flexure beams, and comb finger sidewalls, respectively. It has been suggested [31] that, for calculating the viscous damping force, elements with small cross-section (like comb fingers) should be weighted thrice as much as large plate masses to take into account edge and finite-size effects. Instead of weighting the damping force on different elements differently, the same phenomena are modeled by extending each device dimension by 4 μm. The damping factors of the modes of vibrations other than the x-direction are not modeled, but are assumed to be larger due to the squeeze-film damping.

General analytic equations for the lateral comb drive force, F_x, as a function of comb finger width, w_c, air gap between comb fingers, g, structure thickness, t, and sacrificial spacer thickness, d, are derived in [32]. For the special case of equal comb finger width, gap, thickness, and spacing above the substrate ($w_c = g = t = d$), each comb drive generates a force that is proportional to the square of the voltage, V, applied across the comb fingers.

$$F_x \approx 1.12\varepsilon_0 N \frac{t}{g} V^2 \tag{4}$$

where ε_o is the permittivity of air, N is the number of fingers in the movable comb drive, and V is the instantaneous voltage applied across the comb drive.

If the comb fingers are not perfectly centered, a y-directed electrostatic force is also present. In the absence of restraining springs, this force will result in snapping of the movable comb fingers and the stationary comb fingers. Assuming a small perturbation δ_y in the y-direction, the destabilizing force, $F_{e,y}$, is proportional to displacement such that $F_{e,y} = k_{e,y}\,\delta_y$, where $k_{e,y}$ is an 'electrical negative spring constant'.

$$k_{e,y} \approx 2\varepsilon_0 N V^2 x_0 \frac{t}{g^3} \tag{5}$$

If there is a small rotation δ_θ about the z-axis, a destabilizing electrostatic torque, $\tau_{e,\theta} = k_{e,\theta}\,\delta_\theta$ is generated by the comb drive. The rotational spring constant is found by realizing that the destabilizing force acts through a moment arm, X_c, on the center of the resonator, giving:

$$k_{e,\theta} = k_{e,y} X_c^2 \approx 2\varepsilon_0 N V^2 x_0 \frac{t}{g^3} X_c^2 \tag{6}$$

where $X_c = 0.5\ L_{sa} + w_{cy} + L_c - 0.5\ x_o$ (see Fig. 2 for definitions of the geometric design variables).

4. SYNTHESIS RESULTS

The synthesis methodology (variables and constraints) and the performance models are integrated into synthesis module. This module was used to generate layouts for various frequencies and design objectives. For the first set of microresonators, only the in-plane mode separation constraints described in Section 2.2.2, were imposed. The trends in the synthesis results with changing frequencies and objective functions are discussed. Out-of-plane mode separation constraints (also described in Section 2.2.2) are then included in the synthesis and the generated results are discussed.

4.1 Synthesis with In-plane Mode Separation Constraints

Layouts were synthesized using various input frequency specifications ranging from 3 kHz to 300 kHz. These microresonators were synthesized for 4 different objective functions. The synthesis results for the 10 kHz and 100 kHz frequencies are shown in Fig. 5. As the frequency increases, the overall size of the microresonator decreases reducing mass, and the folded-flexure beams lengths reduce, increasing spring stiffness.

The minimize-area microresonators are smaller than the microresonators designed with other objectives. In some cases, the *minimize-area* microresonators have only one comb finger. The *minimize-voltage* microresonators have longer comb drives (because they have more comb fingers) than the other sets of microresonators. To produce adequate force with a small voltage, more comb fingers are required, since, the force produced is directly proportional to the number of comb fingers. The minimize-area and voltage microresonators are larger than the minimize-area microresonators and at the same time, have shorter comb drives than the minimize-voltage microresonators. The *maximize-displacement* microresonators have long comb fingers in order to accommodate the large motion amplitude of the shuttle during resonance.

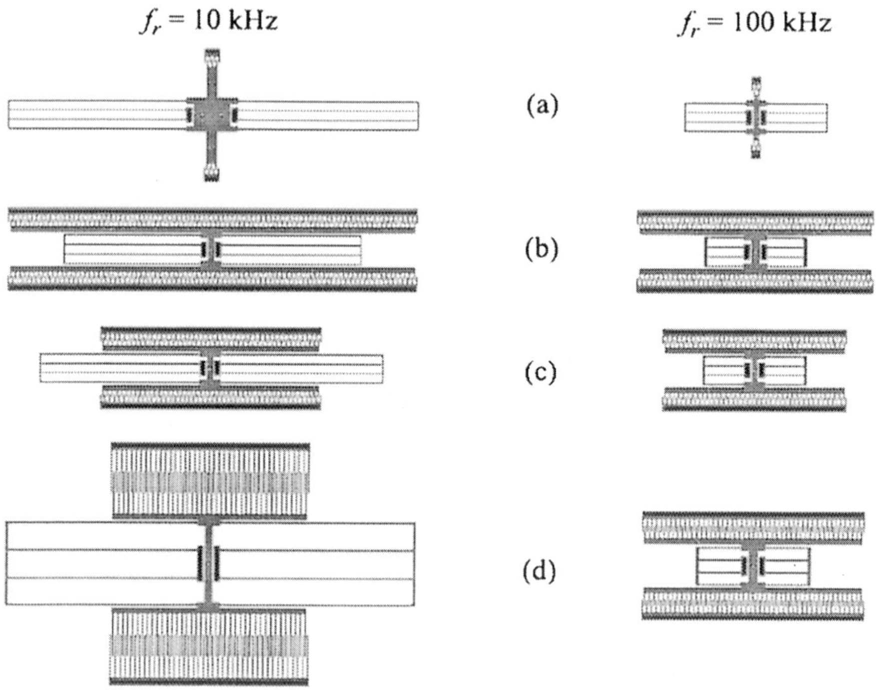

Figure 5. Layouts synthesized with in-plane mode separation constraints for 2 different frequencies (10 kHz on the left and 100 kHz on the right) and 4 different objective functions. (a) minimize area (b) minimize voltage (c) minimize area + voltage (d) maximize displacement at resonance.

4.2 Synthesis with Out-of-plane Mode Separation Constraints

To understand the effect of the out-of plane mode separation constraints, synthesized layouts for minimized area and voltage are compared. Figs. 6a and 6b show layouts for three frequencies with just a single in-plane (*x*-mode only) and all three in-plane (*x*-, *y*- translation and rotation about z) modes considered. These resonators were fabricated in the 2 μm thick MUMPS process [23]. After the incorporation of the out-of-plane mode separation constraints in the synthesis tool, an attempt was made to synthesize layouts as before. The 2 μm thickness limitation was not sufficient to meet these new constraints. Higher structural thickness are needed for stiffer out-of-plane modes and, therefore, higher resonant frequencies.

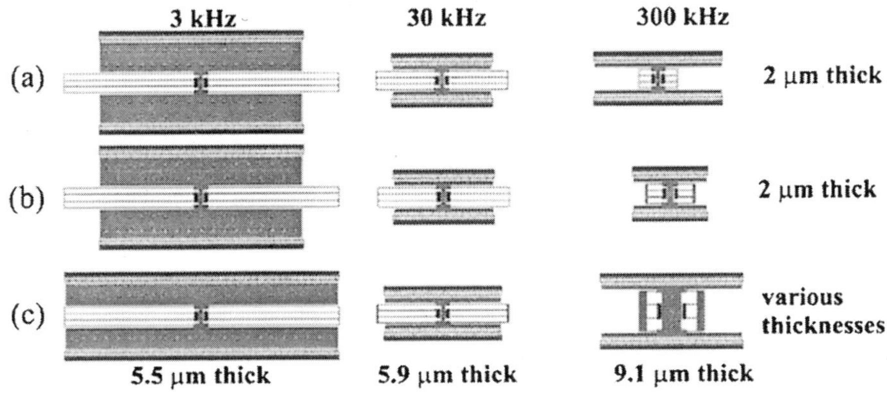

Figure 6. Comparison of layouts generated using increasing number of mode separation constraints. (a) 1 mode separation constraint (b) 3 in-plane mode separation constraints (c) 3 in-plane and 4 out-of-plane mode separation constraints. Layouts are optimized for area, voltage and thickness.

Structural thickness was introduced as a new design variable and a new objective function, viz., *minimize a normalized sum of microresonator area*, applied voltage and structural thickness, was implemented. The layouts generated are shown in Fig. 6c, and have varying thickness. The mode separation constraints are more significant near the design corners, *i.e.*, for the 300 kHz resonator. It is seen that the 300 kHz resonator in Fig. 6b has the least number of fingers and, therefore, has a smaller moment of inertia about the *z*-axis (pointing out of the page). This is necessitated by the rotation-about-*z* mode separation constraint. On the other hand, the 300 kHz resonator in Fig. 6c has more comb fingers. However, since the design has a much thicker and wider truss beam, the mode separation constraint can still be met even though the moment of inertia is relatively large.

5. SUMMARY

A system-level synthesis methodology based on identifying the design variables, geometric and functional constraints, with formulation as a mixed-integer non-linear optimization problem and layout generation is outlined. The functional constraints use a set of analytical equations that capture device performance as a function of device geometry. The resulting synthesis module can be used to explore the design space, and understand the complex trade-offs involved in MEMS design.

ACKNOWLEDGEMENTS

This research was sponsored by the Defence Advanced Research Projects Agency (DARPA) and U. S. Air Force Research Laboratory, Air Force Materiel Command, USAF, under agreements F30602-96-2-0304 and F30602-97-2-0323, and by the National Science Foundation through grants MIP-9625471and CCR-9901171.

REFERENCES

1. S. D. Senturia, "Microfabricated structures for the measurement of mechanical properties and adhesion of thin films," *Proc. 4th Int'l. Conf. Solid-State Sensors and Actuators (Transducers '87)*, pp. 11-16.
2. H. A. C. Tilmans, "Equivalent circuit representation of electromechanical transducers: I. Lumped-parameter systems," *J. Micromech. Microeng.*, vol. 6, no. 1, pp. 157-176, 1996.
3. T. Mukherjee, G. K. Fedder, D. Ramaswamy and J. White, "Emerging Simulation Approaches for Micromachined Devices," *IEEE Transactions on CAD*, "Special Issue: EDA at the Turn of the Century", vol. 19, no. 12, Dec. 2000, pp. 1572-1588.
4. S. T. Picraux and P. J. McWhorter, "The broad sweep of integrated microsystems," *IEEE Spectrum*, vol 35, no. 12, pp. 24-33, Dec. 1998.
5. D. Ramaswamy, N. Aluru, and J. White, "Fast Coupled-Domain, Mixed-Regime Electromechanical Simulation," *Proc. 10th Int'l. Conf. on Solid-State Sensors and Actuators (Transducers '99)*, pp. 314-317.
6. M. Bachtold, J. G. Korvink and H. Baltes, "The Adaptive, Multipole-Accelerated BEM for the Computation of Electrostatic Forces," *Proc. CAD for MEMS*, Zurich, 1997, pp. 14.
7. D. L. DeVoe and A. P. Pisano, "Modeling and optimal design of piezoelectric cantilever microactuators," *J. Microelectromech. Syst.*, vol. 6, no. 3, pp. 266-270, 1997.
8. L. Yin and G. K. Ananthasuresh, "A novel topology design scheme for the multi-physics problems of electro-thermally actuated compliant micromechanisms," *Sens. and Act. A*, 97-98, pp. 599-609.
9. D. Haronain, "Maximizing microelectromechanical sensor and actuator sensitivity by optimizing geometry," *Sens. and Act. A*, 50 (1995), pp. 223-6.
10. W. Ye, S. Mukherjee, and N.C. MacDonald, "Optimal Shape Design of an Electrostatic Comb Drive in Microelectromechanical Systems", *J. Microelectromech. Syst.*, March 1998, vol. 7, pp. 16-26.

11. H. Li and E. L. Antonsson, "Genetic algorithms in MEMS synthesis," *Proc. 1998 ASME Intl. Mech. Eng. Cong. and Exp. (IMECE '98)*, pp. 299-303.
12. T.J. Hubbard and E.K. Antonsson, "Emergent faces in crystal etching," *J. Microelectromech. Syst.*, vol. 3, no. 1, pp. 19-28, 1994.
13. P. M. Osterberg and S. D. Senturia, "Membuilder: An automated 3D solid-model constructionporogram for miecroelectromechanical structures," *Proc. 8th Int'l. Conf. Solid-State Sensors and Actuators (Transducers '95/ Eurosensors IX)*, Stockholm, Sweden, 25-29 Jun 1995, vol. 2, pp. 21-24.
14. Z. Zhu and C. Liu, "Anisotropic Crystalline Etching Simulation using a Continuous Cellular Automata Algorithm," *ASME Symposium on Computer Aided Simulation of MEMS*, Anaheim, CA, Nov 1998.
15. G. G. E. Gielen and R. A. Rutenbar, "Computer-aided design of analog and mixed-signal integrated circuits," *Proc. of the IEEE*, vol. 88, no. 12, Dec 2000, pp. 1825 - 1854.
16. B. Baidya, S. K. Gupta, and T. Mukherjee, "An extraction-based verification methodology for MEMS," *J. Microelectromech. Syst.*, vol. 11, no. 1, Feb 2002, pp. 2-11.
17. W. C. Tang, T.-C. H. Nguyen, M. W. Judy, and R. T. Howe, "Electrostatic Comb Drive of Lateral Polysilicon Resonators," *Sens. and Act. A*, 21 (1990) 328-31.
18. T. Mukherjee, S. Iyer, and G. K. Fedder, "Optimization-based synthesis of microresonators," *Sens. and Act. A*, 70 (1998), pp 118-127.
19. M. Lemkin and B. E. Boser, "A micromachined fully differential lateral accelerometer," *Proceedings of the IEEE 1996 Custom Integrated Circuits Conference*, pp. 315-318.
20. Hao Luo, Gang Zhang, L. R. Carley, and G. K. Fedder, "A post-CMOS micromachined lateral accelerometer," *J. Microelectromech. Syst.*, vol. 11, no. 3, June 2002, pp. 188-195.
21. T. Mukherjee, Y. Zhou, and G. K. Fedder, "Automated Optimal Synthesis of Microaccelerometers," *Tech. Dig. of Twelfth IEEE Intl. Conf. on Micro Electro Mechanical Systems (MEMS 99)*, Orlando FL, Jan. 17-21 1999, pp. 326-331.
22. V. Gupta and T. Mukherjee, "Layout Synthesis of CMOS-MEMS Accelerometers," *Tech. Proc. of Third Intl. Conf. on Modeling and Simulation of Microsystems (MSM 2000)*, San Diego, CA, March 27-29, 2000, pp. 150-153.
23. D. A. Koester, R. Mahadevan, K. W. Markus, *Multi-User MEMS Processes (MUMPs) Introduction and Design Rules*, Cronos MEMS Business Unit, 3026 Cornwallis Road, Research Triangle Park, NC 27709.
24. S. Iyer, T. Mukherjee and G.K. Fedder, "Multi-Mode Sensitive Layout Synthesis of Microresonators," *First Intl. Conf. on Modeling and Simulation of Microsystems, (MSM 98)*, Santa Clara CA, April 6-8, 1998.
25. P. E. Gill, W. Murray, M. A. Saunders and M. H. Wright, *User's Guide for NPSOL (Version 4.0): A Fortran Package for Nonlinear Programming*, Technical Report SOL 86-2, Stanford University, January 1986.
26. A. Ongkodjojo, and F. E. H. Tay, "Global optimization and design for microelectromechanical systems devices based on simulated annealing," *J. Micromech. Microeng.*, vol. 12, no. 6, Nov. 2002, pp. 878-897
27. *CaMEL Web Page*, http://www.memsrus.org/svcscml.html, Cronos MEMS Business Unit, 3026 Cornwallis Road, Research Triangle Park, NC 27709.
28. T. Mukherjee, "CAD for Integrated MEMS Design," *Proc. Design, Test Integration, and Packaging of MEMS/MOEMS (DTIP 2000)*, Paris, France, May 9-11, 2000, pp. 3-14, (invited).

29. G. K. Fedder, *Simulation of Microelectromechanical Systems*, Ph.D. thesis, University of California at Berkeley, September 1994.
30. J. M. Gere and S. P. Timoshenko, *Mechanics of Materials,* 4th ed., Boston: PWS Publishing Co., 1997.
31. X. Zhang and W. C. Tang, "Viscous Air Damping in Laterally Driven Microresonators," *Sensors and Materials*, v. 7, no. 6, 1995, pp.415-430.
32. W. A. Johnson and L. K. Warne, "Electrophysics of Micromechanical Comb Actuators," *J. Microelectromech. Syst.*, v.4, no.1, Mar 1995, pp. 49-59.

INDEX